# THE INTERNET:

## A Historical Encyclopedia

### VOLUME I

## BIOGRAPHIES

# THE INTERNET:

## A Historical Encyclopedia
## Biographies

Laura Lambert

Edited by
Hilary W. Poole

A B C • C L I O

Santa Barbara, California    Denver, Colorado    Oxford, England

© 2005 by MTM Publishing, Inc.

ABC-CLIO, Inc.
130 Cremona Drive, P.O. Box 1911
Santa Barbara, California 93116-1911

Text and editorial development by MTM Publishing, Inc.
445 West 23rd Street, Suite 1D
New York, New York 10011
www.mtmpublishing.com

| Publisher | Valerie Tomaselli |
|---|---|
| Executive Editor | Hilary W. Poole |
| Editorial Assistants | Leah Hoffmann, Nicole Cohen Solomon |
| Copyeditor | Carole Campbell |

Library of Congress Cataloging-in-Publication Data

The Internet : a historical encyclopedia / edited by Hilary W. Poole.
    p. cm.
  Includes bibliographical references and index.
  ISBN 1-85109-659-0 (set : hardback : alk. paper)—ISBN 1-85109-664-7 (eBook : alk. paper)  1. Internet—History. 2. Telecommunications engineers—Biography. 3. Cyberspace.  I. Poole, Hilary W. II. Lambert, Laura. III. Woodford, Chris, 1943- IV. Moschovitis, Christos J. P.

  TK5105.875.I57I5372 2005
  004.67'8—dc22

09 08 07 06 05    10 9 8 7 6 5 4 3 2 1

This book is also available on the World Wide Web as an eBook. Visit abc-clio.com for details.

ABC-CLIO, Inc.
130 Cremona Drive, P.O. Box 1911
Santa Barbara, California 93116-1911

This book is printed on acid-free paper ∞ .
Manufactured in the United States of America.

# Contents

# Introduction

Microchips and circuit boards, copper cables and lines of code—all loom large in any history of the Internet. Yet the story of the Net is, and has always been, primarily a story of people.

In the 1960s, two of the ARPANET's founding fathers, J.C.R. Licklider and Robert Taylor, shared a vision of how computer networks would one day link people together. Elsewhere, and quite independently, others were thinking along parallel lines. While the brilliant but erratic polymath Ted Nelson was grappling with something he called "hypertext" and his vision of a global information "docuverse," media theorist Marshall McLuhan struggled to communicate his concept of the "global village" years before any such thing was technically possible.

By the 1990s, these disparate visions had converged around and within Tim Berners-Lee's World Wide Web—and a new chapter in the Internet's history was beginning. Others were now picking up the torch. Amazon.com boss Jeff Bezos was helping to switch both shoppers and shops on to electronic commerce; Shawn Fanning, the founder of Napster, was showing millions how they could share music online; streaming media pioneer Rob Glaser was predicting television and radio would become obsolete as broadcasting migrated to the great "celestial jukebox"; and Finnish computer student Linus Torvalds was demonstrating how the Net could help people to collaborate—with results as impressive as the Linux operating system.

Today, as historians just begin to put the Internet's importance into perspective, what has become clear is that this remarkable invention is, in at least two ways, a story of people. There's the story of hundreds of millions of ordinary people whose lives have been changed—mostly for the better—by the extraordinary series of technical innovations collectively known as the Internet. This is the story most of us already know. But there's another story, the story of a relatively small number of digital pioneers and their struggle to change the world, a history of the Internet seen through the eyes of the people who made it possible. This book tells that story.

Fiction writers rarely assemble casts of characters as intriguing and accomplished as the Internet's own dramatis personae. From entrepreneurs like Steve Case and Ann Winblad, who have helped to inspire the commercial side of the Net, to such champions of Internet community as Richard Stallman and John Perry Barlow; from the Net's technical architects, luminaries such as Vinton Cerf and Charles Goldfarb, to programmers like Marc Andreessen and Shawn Fanning, whose everyday software drew millions online; from cryptographers like Whitfield Diffie, who helped make the Net secure, to hackers like John T. Draper (Cap'n Crunch) and Kevin Mitnick, who proved that it was anything but—the cast of digital pioneers could hardly be more diverse.

Yet, for all their variety, the people who have laid the foundations of our digital future have much in common. Many of them showed early promise. EarthLink founder Sky Dayton wrote a business plan for a candy store as a ten-year-old. As a boy, Tim Berners-Lee—son of two British computer pioneers—built make-believe computers out of cardboard boxes and tape. 3Com founder Robert Metcalfe went one better, making a computer with real flashing lights in eighth grade. By the age of 13, Bill Gates had embarked on the programming career that would rapidly ensure his fortune. Java inventor James Gosling cobbled together a tic-tac-toe game from a discarded TV set and some telephone switches when he was just twelve. Aged only 19, Shawn Fanning shook the entertainment industry to its foundations when he launched Napster.

Many were outsiders, geeks, or terminal nerds. Theirs were perhaps the last photos you'd have picked from the high school yearbook if you were trying to forecast fortune or fame. Ted Nelson was typical: as early as seventh grade, he'd developed the four maxims of contrariness that still guide his life: most people are fools, most authority is malignant, God does not exist, and everything is wrong. Richard Stallman similarly outgrew his childhood friends, finding he had more in common with his teachers and professors. The archetypal nerd, Bill Gates, was the class clown and problem student whom many thought would never amount to much.

Others were outsiders of a quite different kind. Lawrence Lessig was a promising lawyer whose programming experience enabled him to make particularly incisive contributions to the debate about copyright and the Internet. Another lawyer, Charles Goldfarb, turned his attention to the problem of publishing structured information and came up with the concept of mark-up language that underpins the World Wide Web. Esther Dyson was an analyst and journalist who used insights from those fields to understand and communicate to others how the drama of the Internet was unfolding. Jeff Bezos was a Wall Street stockbroker who realized that more money was to be made putting Main Street online. Outsiders like these have cross-fertilized the Internet's technical innovations and kept them firmly rooted in and relevant to the real world.

Most, if not all, of the digital pioneers are liberally imbued with hacker spirit—not the kind of criminal hacking that hits the headlines, but a gentler determination that drives technically gifted people to extraordinary feats of creative achievement. Richard Stallman famously lived, ate, and slept at MIT's artificial intelligence laboratory while he created (or "hacked") the GNU software that would play such a key part in the later success of Linux. The leading character in the GNU-Linux story, Finnish programmer Linus Torvalds, did much the same, developing an operating system that would eventually threaten the dominance of Microsoft Windows by hacking away for hours on end in his bedroom. Marc Andreessen's groundbreaking Web browser and Shawn Fanning's Napster music-sharing program were similarly great creations or "hacks."

The Net has been built not just by hacker programmers and engineers, but by hackers of all kinds. Whitfield Diffie's public-key cryptography, a way of securing Internet communications using nothing more than long strings of numbers, is a classic mathematical hack. Stewart Brand's pioneering online community, the WELL, might also be described as a hack—something that

combines the essential hacker values of creativity, caring, freedom, passion, and community. Even the more diffuse work of Anita Borg, a tireless champion of women in computing, could be seen as a hack, because hacking is, first and foremost, hard work in pursuit of a great idea—an idea whose time is bound to come.

Hard work and persistence are traits all digital pioneers have in common—Amazon founder Jeff Bezos stuck to his guns through years of financial losses and through the dot-com boom and bust to be rewarded, eventually, with one of the world's biggest retail empires. Often described as the "father of the Web," Tim Berners-Lee has discovered that parenthood is a lifelong commitment; still guiding the evolution of his invention today, he insists "the Web is not done." For some, the struggle to have their ideas accepted by the corporate mainstream has been rewarded only after years of effort—think of Bill Joy and James Gosling battling to convince Sun Microsystems that technically superb innovations such as Java were worth the risk. For others, the battle has taken place in the wider commercial marketplace. Marc Andreessen took on Microsoft with his Netscape browser—and lost; Rob Glaser and RealNetworks have been fighting a similar battle for the Internet broadcasting market—and are still in the game after more than a decade.

Persistence is one of the key qualities that the Internet's pioneers all share, vision is another. For the Net's original architects—Vinton Cerf and Robert Kahn among them—the vision was substantially a technical one. Ted Nelson and Tim Berners-Lee shared a dream of people, information, and ideas connected by a global information space. Some had visions of cyberspace as a fairer, more equitable community than the real world: John Perry Barlow famously promised a "civilization of the mind," while feminists such as Anita Borg and Aliza Sherman set their sights on gender equality. Others, following the lead of Bill Gates, have seen the electronic frontier as a land of profit, as well as promise.

Working in tandem, vision and persistence get things done: without exception, the Internet pioneers demonstrate that individuals really can change the world. Nonetheless, not everyone sees the digital future as benign. Fiction writer William Gibson painted a famously bleak picture in *Neuromancer,* the sci-fi novel that turned *cyberspace* into a household word. Bill Joy voiced the fears of many—technology could make humans obsolete—when he published a lengthy essay,

"Why the Future Doesn't Need Us," in 2000. Some digital pioneers—J.C.R. Licklider, Anita Borg, and Jonathan Postel, to name just three—will never know what the future holds; they have already slipped into cyber-immortality, their legends dissipated across a thousand Websites, their achievements part of the collective triumph that is today's Internet. Others are very much alive and charge into the future with optimism and bravado. Their innovations—software like Marc Andreessen's Web browsers, security innovations like Whitfield Diffie's cryptography, the striving toward more usable Websites championed by Jakob Nielsen, the vision of the Net as an open community, promoted by John Perry Barlow, Lawrence Lessig, and Richard Stallman, among others—almost without exception continue to put human interests, values, and concerns at the center of the Internet. Digital pioneers like these have ensured that the Internet will always remain primarily a story of people.

# Marc Andreessen (1971–)

## NETSCAPE CREATOR

**Marc Andreessen was an** icon of the mid-1990s Internet boom. In 1993, while still in college, he created one of the first and most popular browsers, Mosaic, which jump-started the popularity of the World Wide Web. After college, he went on to found the company that would become Netscape. On August 9, 1995, Netscape held one of the most successful initial public stock offerings (IPOs) for an Internet company ever, which, many believe, brought on the Internet boom by igniting investor interest in Net stocks. Less than five years later, though, after an intense battle against Microsoft known as the "browser wars," Netscape was subsumed by America Online, and many believed Andreessen's days as an Internet superstar were over. The Internet world has since waited and watched to see if the wunderkind of the 1990s can deliver an equally impressive second act.

*Everybody should be in a business once in their lives that competes with Microsoft, just for the experience.*

—*Andreessen to* The Economist, *March 9, 2002*

## MEET MR. MOSAIC

Andreessen grew up in Wisconsin, near New Lisbon, a small, semirural town. His mother was a customer service representative for Land's End, a Wisconsin-based clothing company; his father was a sales manager for a seed company. Andreessen, using a Radio Shack computer, taught himself to program at age 12, while recuperating from an operation. In high school, he was known as both bright and somewhat arrogant, and he outshone his fellow classmates (of which there were fewer than forty). In 1989, he left Wisconsin to major in computer science at the University of Illinois at Urbana-Champaign.

According to Andreessen, Urbana-Champaign had, at the time, one of the top three electrical engineering departments and one of the top ten computer science departments in the country. In a 1995 oral history interview for the Smithsonian, Andreessen explained that he was "considerably more interested in applied programming—in actually doing things that people found useful—as opposed to just studying theory." He gained some hands-on experience while interning in the early 1990s at IBM in Austin, Texas. In 1992, he began a part-time, $6.85-an-hour software-programming job at the National Center for Supercomputing Applications (NCSA), a government-funded research lab located at the college. While the pay was meager, the job introduced him to the worlds of high-end computing and networking.

Urbana-Champaign was one of the first dozen nodes of the original ARPANET, a U.S.-government sponsored computer network. Andreessen entered a world where networked computers, email, and distributed software were the norm. "I remember getting started with email in 1989," Andreessen recalled in a 2001 interview with *Fast Company.* "Originally, it was just a way of testing network connectivity. . . . And then pretty soon, we were communicating with people who were one office over: 'Want to have lunch?'" Most people on the Internet at that time had similar experiences.

When Tim Berners-Lee introduced his invention, the World Wide Web, in 1991, networking via the Internet began to change. Scientists at NCSA were some of the first to explore what the Web could do. Indeed, at the time, scientists were among the few who could even use it. Navigating the Web with early text-based browsers required knowledge of arcane commands. Often, to access a page, one had to type in the exact address; images had to be downloaded separately from

*Marc Andreessen. (James Leynse/Corbis)*

text; the browsers themselves often crashed. Berners-Lee's own browser, dubbed WorldWideWeb, featured a graphical user interface (GUI) and point-and-click technology but was only available on NeXT computers.

After experimenting with early browsers such as Midas and ViolaWWW, Andreessen and a fellow student, Eric Bina, began developing a simple to use GUI browser application for the popular UNIX computer. Over the course of three months in the fall of 1992, Bina developed the basic structure of the browser, while Andreessen designed the user interface and the networking capabilities. The NCSA provided the essentials—the resources, the machines, and the network; Andreessen, Bina, and a handful of other programmers provided the man-hours. They wrote computer code for days on end, slept, then started up again.

Andreessen and Bina finished the first draft of the browser in January 1993. They called it Mosaic. On March 14, 1993, Andreessen posted a free version of Mosaic for UNIX-based machines on NCSA's Internet site with the message, "NCSA Mosaic provides a consis-tent and easy-to-use hypermedia-based interface into a wide variety of information sources." Users flocked to download the software, and, by April 1993, Mosaic had an estimated 10,000 users.

Mosaic was popular because it made the Web easier to use. Mosaic's predecessors introduced various innova-tions—for example, the ViolaWWW browser from the University of California–Berkeley supported graphics and animation applets, and several other browsers fea-tured sound and full color. Mosaic pooled these innova-tions into a simple, attractive interface, and added other key features. Mosaic introduced the concepts of a book-mark and a window history, both of which allowed users to navigate through the Web pages they had already vis-ited. Perhaps the most important feature of Mosaic, however, was the "image" tag, which allowed images and text to appear on the page at the same time. While these features are commonplace today, to users of the Web circa 1993, they were extraordinary.

While Andreessen believed that Mosaic was a neces-sary next step in developing the Web, he did not realize the impact it would have. In a 2000 interview with *Wired,* Andreessen recalled, "I just thought, 'I may as well work on this now, and then when I get out of college I can go work for Silicon Graphics or Time Warner or TCI.'" Indeed, when one of NCSA's directors suggested to Andreessen that he start a company, he dismissed the idea because he did not know how to go about it. By December 1993, Mosaic had made it to the front page of the business section of the *New York Times.* Mosaic, now available in Mac- and PC-compatible versions, was being downloaded from NCSA's Website at a rate of 1,000 times per day. The number of Web servers (computers that host and serve up Web sites) had grown by a factor of 10 since the beginning of the year, from 50 to more than 500. Even so, when Andreessen graduated that month, his only plans were to start work at a small software company called Enterprise Integration Technologies (EIT) and abandon Mosaic altogether.

After a month or so at EIT, Andreessen received an email from Jim Clark, the founder of Silicon Graphics, Inc. (SGI). Clark, a seasoned businessman, wanted to go into business with the 22-year-old who had transformed how people experience the Web.

## THE DOG YEARS

The next several years of Andreessen's life, like those of Silicon Valley itself, could be measured in dog years—meaning he took just one year to do what others might only accomplish in seven. In February 1994, shortly after his email to Andreessen, Clark resigned from SGI and began brainstorming full-time with Andreessen. Their primary idea—an online gaming service for the Nintendo 64 game system—was abandoned when they realized that Nintendo would not ship the new system until late 1995, more than a year away. Other ideas—such as interactive television—were abandoned as well. Then, sometime after midnight on March 25, 1994, after a late dinner and several bottles of wine, Andreessen suggested that they remake Mosaic into a commercial product. Andreessen uttered the now-famous sentence, "We could always create a Mosaic killer—build a better product and build a business around it."

Days later, Andreessen and Clark were in Illinois recruiting the original creators of NCSA Mosaic. In a 1994 interview with the *Los Angeles Times,* one programmer recalled, "They said, 'Do you want to come work for us for something like five times what you're making now, be free of university bureaucracy and finish what you started?'" Indeed, associates of Andreessen have stated that he was eager to reclaim the browser from NCSA's control. Within twenty-four hours, they had drawn all but one of the old team into the fold. (The single holdout had already been hired by Microsoft.) That April, Andreessen officially quit EIT to form the Mosaic Communications Corporation (MCC) with Clark. By May, a half dozen of Mosaic's original programmers were hard at work in MCC's new offices in the City Center building in Mountain View, California. Andreessen was the unofficial product manager. The first version of the software became available as a free download on the Internet in October.

Meanwhile, NCSA was still licensing its version of Mosaic for more than $100,000. Companies such as Spyglass, Fujitsu, and SPRY Inc. had taken NCSA's Mosaic and created their own enhanced products. When MCC began developing its Mosaic browser without a license, the University of Illinois filed a lawsuit, claiming that MCC had, in effect, stolen the program and demanding that MCC not only change its name but also cease distributing its browser. MCC countered that, despite the company's name and employee roster, its browser was based on entirely new code. After a settling with the university in December 1994 for an undisclosed sum (estimated at well over $2 million), MCC changed its name to Netscape Communications Corporation and continued to develop the browser software under the name Netscape Navigator, while NCSA continued to license the original Mosaic. Thus Netscape's Navigator would be competing against products like Spyglass Mosaic—in other words, competing against itself. (In an ironic twist, Spyglass Mosaic would later provide the basis for one of Netscape's most vicious competitors: Microsoft's Internet Explorer).

By January 1995, Netscape Navigator had become the browser of choice for the majority of Web users. Netscape had adopted an "almost free" pricing schedule for noncommercial users, which helped Netscape secure market share. (Although Navigator supposedly cost $39, free versions were readily available for download.) Netscape earned revenue by charging commercial clients for the software. Programmers continued to develop new features for Navigator, including specialized tags that could only be viewed through the Navigator browser. This further encouraged the spread of the browser, since people had to download Netscape in order to view cutting-edge Web pages. On the other hand, it annoyed critics who argued that one of the Web's most valuable features was its universal accessibility.

Less than a year old, Netscape boasted more than 2 million users worldwide, and the company was just hitting its stride. Andreessen was given the official title of Chief Technical Officer. Jim Barksdale was hired away from AT&T to become Netscape's CEO. New software was being released every six months—an unheard-of rate in the industry at the time—which prompted Andreessen to boast that product development now adhered to a new clock: Netscape Time. It was clear that Netscape was becoming the fastest-growing software company in history.

## WHAT GOES UP, MUST COME DOWN

On August 9, 1995, Netscape held one of the most impressive IPOs (initial public offerings of stock) in Internet history. Even with its "almost free" pricing strategy, the company had achieved more than $16 million in sales for the first half of 1995. Nevertheless, Netscape had not yet posted any profits. Going public while still in the red seemed to many a revolutionary act—but one that paid off. In a frenzied day of trading, the stock shares soared from $14 to a high of $75, before closing at $58. Andreessen had woken up late in the day, checked the stock prices online, blanched, then went back to sleep. "We thought the IPO would be successful," he later recalled in an online chat with *Business Week* in 1998. "We didn't realize it would be that successful. . . . It became a calling card for the company, in terms of brand recognition." Netscape stock reached an all-time high of $85.50 just four months later. Six months after that, more than three-quarters of all Web users browsed with Netscape Navigator, and Netscape Communications Corporation was pulling in revenues of more than $100 million. The company's startling success flung the door open for the tide of Internet companies rushing into public markets.

> *One of the interesting things you see in the history of software over and over and over again is that products that one way or another redefine how people think of things tend to be something that a couple kids put together in three months.*
>
> —Andreessen to Newsweek, April 21, 2003

However, in the background of such blinding success lurked Microsoft. Just days after Netscape's stock reached a record high in early December 1995, Microsoft announced plans to refocus its Internet strategy. Netscape had been expecting a challenge by Microsoft. Even before Microsoft's announcement, Netscape's corporate counsel began quietly investigating Microsoft's interactions with its other competitors. In May 1995, Barksdale had presciently asserted that Microsoft would challenge Netscape's Navigator by giving away its browser for free—which is exactly what Microsoft did. What Netscape executives did not foresee was Microsoft's other strategy—its browser, Internet Explorer (IE), would be shipped with every Windows 95 operating system, creating an unprecedented guaranteed market share. Although early versions of IE were clearly inferior to Netscape Navigator, by early 1996, IE began making small but significant gains on Netscape Navigator's share of Web users, and by year's end, Netscape's ballooning revenue began to slow. Indeed, by the fourth quarter of 1996, the company posted quarter-to-quarter growth lower than 20 percent for the first time, and Navigator's share of the browser market had fallen to roughly 50 percent.

In a letter to the government dated August 8, 1996, Netscape asserted that Microsoft was violating a 1994 settlement agreement by bundling IE with the Windows 95 operating system. Netscape also asserted that Microsoft had embedded programming into its operating systems that allowed Explorer to run faster than other browsers, resulting in enthusiastic reviews about Explorer's capabilities. (Even so, many critics believe that IE did not actually surpass Navigator's technical capabilities until IE 5.0, released in 1998.) Many at Netscape believed Microsoft was also bribing Internet service providers and other businesses with money and services if they used or featured Explorer over Navigator and threatening those who refused with canceling vital software contracts. The letter spurred the government to launch another antitrust investigation of Microsoft.

As allegations flew back and forth between Netscape and Microsoft, Netscape continued to falter financially. In July 1997, in an attempt to turn the tide, Andreessen was promoted from Chief Technical Officer to the more hands-on position of executive vice president in charge of the product development group. In a 1998 *Business Week* interview, CEO Jim Barksdale said, "Marc is extremely important to the success of the company. . . . We're depending on him to get the products out." According to many analysts, however, the situation was already hopeless. With no end of the browser wars in sight, Microsoft released Internet Explorer 4.0 to rave reviews in September 1997, while Netscape's own browser development had stagnated. That quarter, Netscape experienced its first revenue decline, posting a loss of more than $85 million at a time when analysts had predicted a significant profit. "Netscape is dead" stories peppered both mainstream and industry publications.

Even as Netscape continued to sink, for Andreessen this era of Netscape was pivotal. The chubby boy wonder who had once posed on the cover of *Time* magazine wearing his traditional business attire of shorts and a T-shirt had transformed himself into a fit businessman in sleek Italian suits. No longer relegated to the relatively insulated post of CTO, as a vice president, Andreessen began to deal directly with the harsh realities of competition in Silicon Valley—first, firing more than 10 percent of Netscape's employees to cut costs, then, in a last-ditch effort to regain footing in the browser market, convincing Barksdale to offer standard Navigator packages for free. In the *Business Week* chat, Andreessen dismissed the huge fourth-quarter loss like a seasoned businessman. "I didn't cry myself to sleep. I just got to work."

By 1998, while shares of companies like America Online (AOL), Amazon.com, and Yahoo! posted gains of up to 60 percent in the stock market, Netscape's stock fell by nearly 25 percent. By the end of the year, AOL announced plans to purchase the battle-weary Netscape for $4.2 billion. Andreessen, surprising

*Fundamental change comes out of left field. It has to be an idea that's viewed as crazy at the time. If any idea looks like a good idea, there's lots of big companies out there like Microsoft that would already be doing it.*

*—Andreessen to Newsweek, April 21, 2003*

many, stayed with the company and, when the deal was made final in March 1999, became AOL's CTO. It was a short-lived tenure, as Andreessen was distracted by thoughts of the future. By September of that year, he had announced his resignation. The following month, the *U.S. v. Microsoft* antitrust trial began. Indeed, though Netscape lost the browser wars, it had succeeded in bringing the industry together to bear witness to Microsoft's illegal competitive practices. To some, this is Netscape's greatest legacy.

## LEAVING THE BROWSER WARS BEHIND

Shortly after leaving AOL, Andreessen, together with several colleagues from Netscape, announced their new Web services company called Loudcloud. The plans had been months in the making. "For a while we were four founders in search of a start-up idea," he told the BBC in 2001. After dozens of discarded ideas, one former Netscapee (as they sometimes called themselves) suggested they create a company that put *other* companies on the Net. By consolidating resources, the founders predicted that they could put a Website up in one-third the time it would normally take and maintain or upgrade the site at a fraction of the cost.

In short order, they began to develop technology—dubbed Opsware—that would automate the design, development, and maintenance of Websites. Rather than programming each element from scratch, Loudcloud customers could choose from a menu of preexisting "clouds"—a database cloud, an application server cloud, a mail cloud, a security cloud—to develop the initial site. Additions, such as a shopping basket, could be added with a handful of keystrokes. Maintenance, including software updates and any necessary adjustments, would be provided round-the-clock.

By October 1999, Loudcloud had moved from Andreessen's house to a meager office outfitted with store-bought folding tables and extension cords in Menlo Park,

California. Operations began in February 2000. By June 2000, Loudcloud, now housed in a plush office building Andreessen dubbed "the Taj," boasted a growing list of accounts, including Nike, and valued itself at over $1 billion (its actual sales hovered around $2 million).

In March 2001, at a time when many Internet companies were going under, Loudcloud went public at $6 per share. (According to *Forbes,* Loudcloud has the dubious distinction of being one of the last Internet IPOs.) Some venture capitalists predicted that Andreessen would, as he had once done with Netscape, reignite investor interest in Internet companies—but they were wrong. (Loudcloud earned roughly $450 million during its IPO, less than half of what it had predicted.) Three months later, Loudcloud was forced to lay off nearly 20 percent of its workforce.

By early 2002, the company had burned through nearly $200 million in financing. Predictions that Loudcloud would break even by 2003 were withdrawn. After selling the information technology (IT) services sector of the company to Electronic Data Systems for more than $60 million in August 2002, Andreessen scaled the company back, renaming it Opsware. Andreessen has since focused his energies on what he sees as the next wave in Internet business: selling computing power like a utility, e.g., electricity.

April 2003 marked the 10th anniversary of Mosaic, the simple browser that began the roller-coaster ride that has been Andreessen's Internet career. In interviews marking the event, his tone revealed a level of maturity and wisdom gained in the battle against Microsoft and the ensuing struggle to save Netscape. (More than once, Andreessen spoke almost kindly of Microsoft, with whom Loudcloud/Opsware was doing business.) Indeed, Andreessen had changed from a fresh-faced college grad to a seasoned Internet entrepreneur still in his early thirties, making predictions for what was to come in 2013. Even as Opsware struggles to find its footing in the new Internet economy, the industry still bends an ear when Andreessen had something to say.

## FURTHER READING

### In These Volumes

Related Entries in this Volume: Berners-Lee, Tim; Gates, Bill; Nelson, Ted; Yang, Jerry and David Filo

Related Entries in the Chronology Volume: 1990: The World Wide Web Is Invented; 1993: Mosaic Is Developed; 1994: Marc Andreessen and Jim Clark Found Mosaic Communications; 1996: The Browser War Heats Up; 1998: Department of Justice Files Suit Against Microsoft; 1998: America Online Announces Deal to Buy Netscape

Related Entries in the Issues Volume: Cookies; E-commerce; Usability

### Books

Clark, Jim. *Netscape Time: The Making of the Billion-Dollar Start-Up That Took On Microsoft.* New York: St. Martin's Press, 1999.

Cusumano, Michael A. *Competing on Internet Time: Lessons from Netscape and Its Battle with Microsoft.* New York: Free Press, 1998.

Quittner, Joshua, and Michelle Slatalla. *Speeding the Net: The Inside Story of Netscape, How It Challenged Microsoft and Changed the World.* New York: Atlantic Monthly Press, 1998.

### Articles

Alden, Christopher. "Bill Gates With a College Degree." *Red Herring,* January 1, 1996.

Anders, George. "Marc Andreessen, Act II." *Fast Company,* February 2001, http://www.fastcompany .com/magazine/43/andreessen.html (cited September 16, 2004).

Corcoran, Elizabeth. "Growing Up Is Hard to Do." *Forbes,* April 29, 2002.

Gimein, Mark. "Goodbye, Internet Poster Boy." *Salon.com,* September 10, 1999, http://www.salon.com/tech/log/1999/09/10/ andreessen (cited September 16, 2004).

Glasner, Joanna. "Conversation with Marc Andreessen." *Wired News,* February 14, 2003, http://www.wired.com/ (cited September 16, 2004).

Hamm, Steven. "The Education of Marc Andreessen." *Business Week,* April 13, 1998, http://www.businessweek.com/1998/15/topstory .htm (cited September 16, 2004).

Harmon, Amy. "Now You Don't Have to Be a Geek to Use the Net." *Los Angeles Times,* November 13, 1994.

Levy, Steven. "The Killer Browser." *Newsweek,* April 21, 2003.

———. "Out of Left Field." *Newsweek,* April 21, 2003.

Maney, Kevin. "10 Years Ago, Who Knew What His Code Would Do?" *USA Today,* March 9, 2003, http://www.usatoday.com/tech/news/2003-03 -09-internet_x.htm (cited September 16, 2004).

"On The Record: Marc Andreessen." *San Francisco Chronicle,* December 7, 2003.

Sheff, David. "Crank It Up." *Wired* 8.08, August 2000, http://www.wired.com/wired /archive/8.08/ loudcloud.html (cited September 16, 2004).

Wolfe, Gary. "The (Second Phase of the) Revolution Has Begun." *Wired* 2.10, October 1994, http:// www.wired.com/wired /archive/2.10/mosaic.html (cited September 16, 2004).

## Websites

NCSA Mosaic for X. Archived Website of the original NSCA Mosaic for X, the original UNIX-based browser, http://archive.ncsa.uiuc.edu/SDG/ Experimental/demoweb/old/mosaic-docs/ help-about.html (cited September 16, 2004).

Opsware, Inc. Home of Andreessen's most recent software venture, Opsware, http://www .opsware.com/ (cited September 16, 2004).

Smithsonian Institution Oral and Video Histories: Marc Andreessen. Site features a lengthy interview with Andreessen in June 1995, in the early days of Netscape's ascent, http://americanhistory.si.edu/csr /comphist/ma1.html (cited September 16, 2004).

# John Perry Barlow (1947–)

## COFOUNDER OF THE ELECTRONIC FRONTIER FOUNDATION

A former Wyoming rancher and lyricist for the rock band The Grateful Dead, John Perry Barlow lived many lives before becoming an Internet pioneer. Upon founding the Electronic Frontier Foundation (EFF), the first civil rights organization in cyberspace, in 1990, Barlow became a key figure in the struggle to civilize the digital frontier. Since then, he has helped the EFF evolve into one of the leading organizations at the forefront of legal battles and debates surrounding Internet issues such as encryption, privacy, copyright, and the First Amendment.

## HOME ON THE RANGE

Barlow was born near Cora, Wyoming, the third generation of Barlows to have grown up on the family's 7,000-acre ranch, the Bar Cross Land & Livestock Company. Though he was first educated in a one-room schoolhouse in Sublette County, Wyoming, at age 15 Barlow was sent to Fountain Valley School, a boarding school in Colorado Springs. (In 1994 his mother joked to the *Denver Post* that sending her troublemaker son away to school was the "best thing that ever happened" to their hometown.) At Fountain Valley Barlow met Bob Weir, who would later become the guitarist for the Grateful Dead. Their high school friendship would last for years.

After high school, Barlow attended Wesleyan College, where he studied comparative religion, organized for the Students for a Democratic Society, and explored what the 1960s countercultural movement had to offer. He graduated with high honors in 1969 and traveled throughout India, Europe, and the United States over the next couple of years. In 1971, with the family ranch in financial disarray and his father ill, he returned to Wyoming.

Though Barlow only intended to "help out" at the ranch for six months, he eventually took charge and stayed on for seventeen years. "I was still a hippie," Barlow told the *Denver Post,* "but I was a hippie that was running cattle and trying to keep the land."

While he learned to manage the ranch and raise cattle, Barlow began writing songs with Bob Weir on the side. Between 1971 and 1995, Barlow and Weir collaborated on more than thirty songs for the Grateful Dead, including such hits as "Cassidy," "Mexicali Blues," and "I Need A Miracle." (The royalties helped Barlow keep the ranch financially afloat for many years.) Under Barlow's guidance, Bar Cross Land & Livestock also became a place to send bright but somewhat troubled kids to work for the summer. John F. Kennedy, Jr., for example, spent time there in 1978. (A longtime friend, Barlow would later write for JFK, Jr.'s magazine, *George.*)

Although Barlow's hippie roots and leftist political leanings may have seemed an odd fit for rural Wyoming, the small-town feel and fiercely independent Wild West spirit of his hometown appealed to him. He married and began a family; in 1987, Barlow ran as a Republican for the seat in the Wyoming Senate that had been held by his grandfather and father before him. (He lost by one vote.) Then, John Perry Barlow got online.

## DRAWING FROM THE WELL

Barlow purchased his first computer in 1987. Although he originally used it to keep track of finances for the ranch, Barlow quickly discovered the Internet and, in particular, the WELL. The WELL (a.k.a. the Whole Earth 'Lectronic Link) was founded in the mid-1980s as a progressive online community comprising mostly writers, artists, activists, programmers, and, oddly enough, Deadheads (as the most fervent followers of the Grateful Dead called themselves). Indeed, the Deadhead community led Barlow online. (He claims that Grateful Dead lyrics were among the first nonmilitary materials to be posted on ARPANET, the precursor to the Internet.) As Barlow explained to the *Rocky Mountain News,* "[Deadheads] went [to the WELL] because they were looking for community at places where they could have random interaction—just like in a small town." Barlow shared that sensibility.

The Internet opened up an entirely new world for Barlow, even as his off-line world was falling apart. In 1988, Barlow was forced to sell the family ranch to pay $500,000 in debt. He found a new community, however, online. By the late 1980s, Barlow was posting regularly to the WELL, mostly essays and philosophical musings on the notion of the Internet as a digital frontier and the connections between cyberspace and open space. Indeed, his WELL postings earned him a reputation, in certain circles, as an Internet guru.

In 1989, *Harper's* magazine invited Barlow to take part in an online forum, hosted on the WELL and later covered in the magazine, on the subject of computer security and hacking. There, Barlow faced off with members of the new generation of hackers, including the well-known hackers Phiber Optik (né Mark Abene) and Acid Phreak. In a 1990 interview with journalist David Gans, Barlow remembered the forum as a clash "between the old techno-hippies and these new sort of digital skateboarders," which culminated in Phiber Optik downloading Barlow's credit history for everyone in the forum to see. In an essay posted to the WELL, Barlow later wrote, "To a middle-class American, one's credit rating has become nearly identical to his freedom. I've been in redneck bars wearing shoulder-length curls, police custody while on acid, and Harlem after midnight, but no one has ever put the spook in me quite as Phiber Optik did at that moment."

In early 1990, Barlow traveled to New York City to visit the teenage Phiber Optik. Barlow learned that, in January of that year, federal agents had raided Abene's apartment, which he shared with his mother and younger sister, along with the homes of several other New York–based hackers. Abene's story proved to be a harbinger of things to come—several months later, a federal agent appeared at Barlow's home in Wyoming, inquiring, among other things, whether he knew anything about a group of "info-terrorists."

## OPERATION SUN DEVIL

Barlow was just one of the dozens of people questioned as part of Operation Sun Devil, the government's na-

*Being an Internet guru isn't what it used to be.*

*—Barlow to the* San Francisco Chronicle, *June 11, 2002*

tionwide crackdown on computer hackers. Through the efforts of more than 150 Secret Service agents, over a period of two years, the government served twenty-eight search warrants in fourteen cities, seizing more than forty computer systems and 23,000 floppy disks—much of which was never returned. Operation Sun Devil was unlike anything the computer world had ever seen. During the New York raids, federal agents barged into homes with guns. Barlow's personal experience was not so dramatic, just surreal—and alarming.

On May 1, 1990, Special Agent Richard Baxter, from Rock Springs, Wyoming, phoned Barlow to see if he could come to the ranch to ask Barlow a few questions, though he remained vague about the topic of investigation. Baxter's specialty was cattle rustling, not computers—a fact that became abundantly clear to Barlow within minutes of Baxter's arrival. (In a 1991 article for *Byte* magazine, Barlow quipped, "the only chips he knew about were the kind cows make.") Indeed, Barlow spent the better part of an afternoon explaining to Baxter the difference between, among other things, codes and chips, and outlining the nature of the Internet.

Baxter was interested in what Barlow knew about the nuPrometheus League—which Baxter mistakenly called the "New Prosthesis League"—which had apparently stolen and distributed parts of the source code for Apple's Quick-Draw program. Though many industry figures had received a disk containing the source code in the mail, Barlow was not one of them. The San Francisco office of the FBI fingered Barlow because he had taken part in the Hacker's Conference, a yearly meeting of prominent computer programmers founded in the mid-1980s by Stewart Brand, who also founded the WELL. The FBI believed the Hacker's Conference to be a meeting of "outlaws," when, in reality, it was a meeting of computer luminaries, many with high profile, corporate addresses.

The agent's faulty intelligence and general misunderstanding of the workings of cyberspace alarmed Barlow. "I realized that what we were looking at . . . was a microcosm of a whole set of things that could now begin to happen with the government and with society and computers," Barlow recalled in an interview with journalist David Gans. That night he wrote an account of Baxter's visit and posted it to the WELL, under the title, "Crime and Puzzlement." Thousands of miles away, Mitch Kapor, founder of Lotus Development Corporation, read Barlow's posting with growing concern—he, too, had been visited by the FBI.

## The Founding Fathers

Like Barlow, Kapor was an eclectic renaissance man—a brilliant computer programmer in the 1970s, wildly successful entrepreneur and founder of Lotus Development Corporation in the 1980s, a former disc jockey, psychologist, and transcendental meditation instructor who became a cybermillionaire after leaving Lotus in 1987. He was also one of the industry figures to receive a floppy disk from the nuPrometheus League. Initially, he thought the unlabelled diskette might contain a virus, so he put it in a drawer. Then, when news of the stolen source code spread, Kapor put the disk in an unused computer and quickly realized that he, too, had the code. After conferring with his attorney, Kapor sent the disk to Apple Computer. Months later, the FBI knocked on his door.

Kapor dismissed the experience until he read Barlow's posting on the WELL. Days later, on a cross-country flight in his private jet, Kapor phoned Barlow and asked if he could "touch down" in Wyoming for a meeting.

The two men had met once before, when Barlow interviewed Kapor for an article in *MicroTimes*. Even in that one meeting, they found that, despite their wildly different backgrounds, they shared a sensibility. "Here we are, two very different guys—one from Long Island, who has been a leader in this field for years and me, a small town Wyoming rancher who just figured out how to turn [his computer] on," Barlow told the *Denver Post* in 1994. Nevertheless, he added, "I felt like I'd known Mitch all my life."

Each man talked about his own bizarre brush with the law and the New York raids earlier that year. As Kapor recalled, in the Gans interview, "I had a real concern that if the hackers got screwed over in some way, not only would they be losers—which was bad enough—but we all would be losers in some fashion, because one of them might be the next Steve Wozniak." [Wozniak is the former hacker who founded Apple Computer.] Then, the conversation turned to some of the more egregious Operation Sun Devil cases, includ-

ing cases against Craig Neidorf, the college-age publisher of *Phrack,* an online 'zine dedicated to hacking and Steve Jackson Games, a computer game publisher in Austin, Texas. According to Barlow, "We realized that there was a not so much planned and concerted effort to subvert the Constitution, but the natural process that takes place whenever there are people who are afraid and ignorant and issues that are ambiguous regarding constitutional rights."

The following day, Barlow posted the initial plans to form an organization to fight for constitutional rights—particularly those protected by the First and Fourth Amendments—in the digital realm. The men borrowed Barlow's metaphor of the digital frontier for the name. "Out on the frontier, there aren't established laws or practices," Kapor explained in the Gans interview. "We're making it up as we go along. But ultimately, we've got to civilize the frontier. We have to allow ordinary folks to come and settle. We need to build the equivalent of railroads, because if we don't take the lead in doing it and it kind of happens by itself, it's probably not going to come out in a way that any of us really like it." Several weeks later, on July 10, 1990, the Electronic Frontier Foundation (EFF) was born.

In the beginning, the fledgling EFF was propelled by a considerable amount of discourse and input from the WELL community, a few of whom joined the effort. John Gilmore, a millionaire Internet entrepreneur, and Steve Wozniak, founder of Apple Computer, together pledged several hundred thousand dollars to the foundation; they are considered to be cofounders. (EFF has remained a nonprofit organization, funded by its membership.) Stewart Brand, founder of the WELL, joined the first board of directors. One of the EFF's first actions was to award a $275,000 two-year grant to the Computer Professionals for Social Responsibility (CPSR), a like-minded public advocacy group formed in the early 1980s, to be used for their Privacy and Civil Liberties Project. Then, the EFF went to court.

## Opening Salvos

The EFF's first two legal cases both involved a stolen Bell South document, the E911, which described part of the emergency 911 telephone system. Federal authorities claimed that hackers could use E911 to penetrate the phone system and wreak havoc on the country's emergency services. Thus, Secret Service agents

*John Perry Barlow. (Ed Kashi/Corbis)*

were given warrants to raid anyone suspected of having the document—including two young men: Craig Neidorf and Steve Jackson.

Neidorf was a pre-law student at the University of Missouri, and the editor-publisher of *Phrack,* a hacking 'zine. In issue 24 of his 'zine, he published aspects of E911. In February 1990, he was indicted on charges that included wire fraud and interstate transportation of stolen property. Steve Jackson was the owner of Steve Jackson Games, a small computer-game publisher. The Secret Service raided his offices in Austin, Texas, in March 1990, confiscating several computers, numerous hard drives, hundreds of disks, and a large bag of screws in search of the E911 document, which may or may not have appeared on an electronic bulletin board maintained by Lloyd Blankenship, one of Jackson's employees, on one of the company's computers.

For the Neidorf trial, which began in late July 1990, the EFF hired security expert Dorothy Denning as an expert witness; she showed, first, that the published ma-

terial could not be used to compromise the phone system, and second, that the E911 document was currently available for legal purchase for roughly $13.50. The government dropped its case after four days.

The Jackson case was more complicated. The government, under a sealed search warrant, kept Jackson's computer equipment for more than four months. The computers contained all the materials for Jackson's primary book, *Gurps Cyberpunk: High-Tech Low-Life Roleplaying Sourcebook,* scheduled to be released that spring; Jackson, unable to publish, nearly lost his company. When the computers were returned, hundreds of email messages from electronic bulletin boards hosted on Jackson's computers had been accessed, copied, and erased. The EFF successfully sued to establish, under the Electronic Communications Privacy Act, that authorities would need a separate search warrant for each email message or email account it wanted to search for evidence of a crime.

To supporters, the First and Fourth Amendment issues brought up by the Jackson and Niedorf cases—

protecting freedom of speech, freedom of the press, and freedom from unreasonable search and seizure—highlighted the timeliness of the EFF. (Indeed, the American Civil Liberties Union [ACLU] did not handle its first Internet case until 1995.) However, these early victories earned the EFF a reputation as a "hackers defense fund," an image the EFF quickly sought to shed. To do so, EFF headed for the nation's capital.

In 1992, the EFF moved its offices from Boston, where Kapor was based, to Washington, D.C., in an effort to focus on lobbying. The foundation recruited Jerry Berman, an ACLU lawyer, as its new executive director. Barlow told *Newsbytes,* "With the movement of the offices to Washington, we were concerned with the natural gravitational pull of the Beltway mentality. The board felt that my day-to-day involvement would counter this tendency. The Bohemian credentials are pretty well established."

During the Washington years, the EFF lobbied for fair and reasonable computer laws at both the state and national levels, sponsored conferences and roundtable discussions on Internet issues with industry and Washington figures, and worked with the press to mitigate the growing fear of hackers brought on by the Operation Sun Devil raids. Surprisingly, the group was welcomed in Washington because, according to Kapor, it was one of very few groups knowledgeable about technology issues without being an industry trade association. The EFF enjoyed close relations with the Clinton administration—until the Clipper Chip debate emerged in 1993. The government's proposal, which stipulated a wiretapping loophole for federal agencies in the proposed standard for encryption technology, enraged Barlow. He saw the Clipper Chip as a gross incursion on personal privacy. Asked why the pro-Internet Clinton administration supported the Clipper Chip, Barlow told author Howard Rheingold, "They've drunk the Kool-Aid of national security."

In 1994, after bitter internal disputes over the EFF's handling of the Clipper Chip debate—Barlow and other core members felt that Berman was working too closely with the government—Berman left the foundation to form the Center for Democracy and Technology (CDT), which continued to focus on lobbying efforts, while the EFF poured its energies into legal battles. Barlow and Kapor both stepped down as chairman and vice chairman, respectively, though they remained on the board of directors. In 1995, Kapor resigned, and the EFF left Washington, D.C., for San Francisco, setting up shop in the basement of John Gilmore's Victorian house.

The organization regrouped and evolved into a fifteen-member office, with dozens of interns and volunteers, working with an annual budget of approximately $2 million. Barlow returned to his vice chairmanship and, eventually, Kapor also reengaged with the organization. Over the next several years, the EFF enjoyed some significant victories, including a late 1990s case, *Daniel Bernstein v. United States,* in which the EFF successfully challenged the government's encryption export controls and established that computer code is protected as speech by the First Amendment. Other victories for freedom of online speech were won against the Communications Decency Act of 1996 and the Child Online Protection Act of 1988 (both restricted speech online) by the EFF, which launched its Blue Ribbon campaign against censorship. That campaign made www.eff.org—the foundation's virtual headquarters—one of the most linked-to sites on the Internet.

> *The Internet is anarchy. It doesn't have a government, it doesn't have a head, it doesn't even have a map. It's information and connections.*
>
> —*Barlow to* Rocky Mountain News, December 6, 1994

## THE WRITING ON THE VIRTUAL WALL

Throughout, Barlow continued to write. Ever since he first applied the word *cyberspace*—coined by William Gibson in the sci-fi book *Neuromancer*—to the Internet back in the early 1990s, Barlow has been a highly regarded Internet scribe. In May 1994, he published a treatise on the future of copyright in *Wired* entitled "The Economy of Ideas," in which he predicted a brave new world of information-sharing and distribution. "In the absence of the old containers [books, CDs, etc.], almost everything we think we know about intellectual property is wrong . . . [M]ost human exchange will be virtual rather than physical."

Since its publication, "The Economy of Ideas" had become a staple of copyright law courses in universities throughout the country. Barlow later predicted that "The Powers That Were" (as he calls the government and major industrial-age corporations) would use copyright as a weapon against the Internet. Indeed, as evidenced by several major cases later brought by the EFF—including one on behalf of the file-sharing service Napster—Barlow was right.

In 1996, Barlow penned another seminal work, "A Declaration of Independence for Cyberspace." Of the new online frontier, he wrote:

> We are creating a world that all may enter without privilege or prejudice accorded by race, economic power, military force, or station of birth. Where anyone, anywhere may express his or her beliefs, no matter how singular, without fear of being coerced into silence or conformity. Your legal concepts of property, expression, identity, movement, and context do not apply to us. They are all based on matter, and there is no matter here.

In later years—especially once e-commerce became one of the primary driving forces of the Internet—Barlow's breathless predictions and proclamations became a point of ridicule. But the manifesto had enormous impact when it was originally released, earning Barlow the moniker, Thomas Jefferson of Cyberspace. Indeed, nearly 20,000 Websites have posted copies of the text.

Barlow has continued to write on various topics of import to the Internet community and to society at large. In January 1998, Barlow put himself at the forefront of the growing debate about the "digital divide" between developed and developing countries with his article, "Africa Rising," published in *Wired.* (In 2000, he helped found Bridges.org, a Washington-based organization dedicated to bridging the Digital Divide). In the wake of the terrorist attacks of September 11, 2001, Barlow wrote—this time in a widely distributed email—that the attacks would open the floodgates for government incursions on privacy and free speech. He likened the attacks to " . . . the Reichstag fire that pro-

vided the social opportunity for the Nazi takeover of Germany"—a statement that inflamed some of his readers and prompted a death threat. He noted later that the death threat was exactly the type of anonymous free speech the EFF might fight to defend.

## LET THE MUSIC PLAY

Since the late 1990s, Barlow's greatest interest at the EFF has been copyright infringement, especially in the arena of music. Indeed, the EFF has taken on several high-profile copyright infringement and fair use cases, including a case involving *2600 Magazine,* a hacker's journal that challenged the 1999 Digital Millennium Copyright Act. (The EFF lost.) More recently, the EFF has launched a campaign to fight against the impending cases brought by the Recording Industry Association of America (RIAA) against music consumers who have used file-sharing software—à la Napster—to share music over the Internet.

> *Speech is speech. It doesn't matter if it is words on paper or bits on bytes.*
>
> —*Barlow to the* San Francisco Examiner, *July 11, 1990*

Barlow in particular is adamantly opposed to litigation against file-sharing services. To his detractrors, he often points to his days as a lyricist for the Grateful Dead. At one time, the band contemplated cracking down on the bootleg tapes that Deadheads made of their shows, arguing, much like record companies do in the Napster-related debates, that they were being robbed of revenue. However, Barlow explains, the band decided to allow the bootleg tapes to flourish, and, instead of decreasing revenue, the bootleg tapes helped create a devoted community of listeners and actually increased record sales. It was a lesson only John Perry Barlow could teach.

The EFF, which had grown to a twenty-three-person staff at the turn of the 21st century, expanded once again in 2002, partly in response to the growing concern over the Bush administration's proposed Total Information Awareness Initiative, as well as consumer backlash against the music and entertainment industry crackdown on file-sharing services. To celebrate the growth in interest and membership, the EFF launched a campaign, "Let The Music Play," with an ad in *Rolling Stone* magazine—at a price tag equal to the yearly salaries of two EFF employees.

In addition to his duties with the EFF, Barlow keeps busy as a teaching fellow at Harvard Law School's Berkman Center for Internet and Society, where, at one point, he led a weekly study group, "Cyberspace vs. Meatspace." He also commits a significant amount of time and energy as an industry consultant, writer, and public speaker. In an effort to distill his various lives, interests, and activities into a single phrase, his business card simply reads, "Cognitive Dissident." Indeed, many believe Barlow's origins as an outsider to the computer world have helped him see the benefits, risks, and threats of the Internet more clearly than most.

## FURTHER READING

### In These Volumes

Related Entries in this Volume: Gilmore, John; Lessig, Lawrence

Related Entries in the Chronology Volume: 1990: Operation Sun Devil; 1990: The Electronic Frontier Foundation Is Established; 1997: *Bernstein v. U.S. Department of Commerce*

Related Entries in the Issues Volume: Digital Divide; Encryption; Hackers; Online Communities

### Works By John Perry Barlow

"Crime and Puzzlement," June 8, 1990, http://www.sjgames.com/SS/crimpuzz.html (cited September 16, 2004).

"The Economy of Ideas." *Wired,* March 1994, http://www.wired.com/wired/archive/2.03/economy.ideas_pr.html (cited September 16, 2004).

"Declaration of the Independence of Cyberspace," February 8, 1996, http://www.eff.org/~barlow/Declaration-Final.html (cited September 16, 2004).

"Africa Rising," *Wired,* January 1998, http://www.wired.com/wired/archive/6.01/barlow.html (cited September 16, 2004).

### Books

Sterling, Bruce. *The Hacker Crackdown: Law and Disorder on the Electronic Frontier.* New York: Bantam Books, 1992.

### Articles

Bottoms, David. "Cyber-Cowboy . . . Or Prophet?" *Industry Week,* December 4, 1995.

Cobb, Nathan. "Cowboy on the Cyber-Frontier." *The Boston Globe,* June 3, 1998.

Dickinson, Tim. "Cognitive Dissident." *MotherJones,* February 3, 2003, http://www.motherjones.com/news/qa/2003/02/we_268_01.html (cited September 16, 2004).

Dougherty, Steve. "John Perry Barlow: A Wyoming Rancher Turned Digital Visionary Fights to Keep Fences off the Electronic Frontier." *People,* December 4, 1995.

Gans, David, with John Barlow and Mitch Kapor. *Wired* interview, http://www.eff.org/Publications/John_Perry_Barlow/HTML/barlow_and_kapor_in_wired_interview.html (cited September 16, 2004).

Gerstner, John. "Cyber Cowboy: An Interview with John Perry Barlow." *Communication World,* November 1, 1995.

Harrington, Maureen. "Cyberspace Cowboy." *Denver Post,* April 3, 1994.

Harris, Scott. "Freedom Fighters of the Digital World." *Los Angeles Times,* January 13, 2002.

### Websites

Barlow Home(stead) Page. Barlow's personal Website. The "library" includes links to several profiles, interviews and some of his most celebrated works, http://www.eff.org/~barlow/ (cited September 16, 2004).

Bridges.org. An organization, cofounded by Barlow and other Internet luminaries, including Esther Dyson, committed to bridging the Digital Divide, http://www.bridges.org (cited September 16, 2004).

Center for Democracy and Technology. Online home of the CDT, which was formed in 1994 by EFF members who splintered from the organization, http://www.cdt.org (cited September 16, 2004).

Computer Professionals for Social Responsibility. Online home of the CPSR, an early sister organization to the EFF, http://www.cpsr.org/ (cited September 16, 2004).

Electronic Freedom Foundation. The online home of the Electronic Freedom Foundation offers exhaustive archives on previous court cases, as well as a thorough history of the organization, http://www.eff.org (cited September 16, 2004).

# Tim Berners-Lee (1955–)

## INVENTOR OF THE WORLD WIDE WEB

**In 1989, Tim Berners-Lee** invented a global hypertext program that allowed people to communicate, collaborate, and share information using the Internet. Today, his invention—the World Wide Web—is ubiquitous. Indeed, the names of the protocols and specifications he developed for the Web are embedded in every Web address—the "http://" for hypertext transfer protocol, how information is exchanged over the Web, and "www" for the World Wide Web itself. Not only did Berners-Lee lay the foundations of the Web and, eventually, revolutionize the way people use the Internet, his foresight allowed the Web to grow unfettered by commercial interests. In 1994, he founded the World Wide Web Consortium (W3C), a neutral organization designed to set standards and specifications for the Web's growth. Through the W3C, Berners-Lee remains at the forefront of the Web's development, steadfastly working to ensure that the Web continues to grow and evolve.

Although Berners-Lee is notoriously closemouthed about his personal life, the few public details about his childhood suggest it was no accident that he invented the World Wide Web. His parents, both mathematicians, met at Manchester University (U.K.) while developing the Mark I, the first commercially available computer (which arrived on the market in February 1951 and was sold by Ferranti Ltd.; it is also known as the Ferranti Mark I). As a child, Berners-Lee made make-believe computers out of cardboard boxes and computer tape. Later, while studying physics at Oxford University, Berners-Lee made his first real computer, using a soldering iron, spare computer parts, and an old television set.

Though its development was years away, the first notions of the World Wide Web came to Berners-Lee in high school, during a discussion with his father about the advantages of the human brain over computers. They spoke about the way a human brain can create meaning and relationships out of any two seemingly disparate objects or ideas. They also contemplated the possibilities of an intuitive computer. Berners-Lee then began to understand that a computer could become infinitely more powerful—or, more brain-like—once it could link any two pieces of previously unrelated information. Little did he know at the time, this is the core principle of hypertext.

In 1976, after Berners-Lee graduated from college, he worked for several years as a programmer in Great Britain's telecommunications industry; first, for Plessey Telecommunications Ltd., then, in 1978, for D.G. Nash. For a brief time in the early 1980s, Berners-Lee worked as a freelance consultant and spent six months at the Conseil Européen pour la Recherche Nucleaire (CERN), a prestigious physics lab located near Geneva, Switzerland. (CERN is also known as the European Particle Physics Laboratory.) After several years as a programmer for Image Computer Systems, Ltd., in 1984 he returned to CERN on a fellowship. There Berners-Lee proposed the global hypertext project that became the World Wide Web.

## ENQUIRE WITHIN

The concept of hypertext, on which the Web is based, was more than forty years old. An electrical engineer, Vannevar Bush—whom some call the Godfather of the Internet—wrote an article, "As We May Think," for the *Atlantic Monthly* in 1945. Bush describes a "Memex" machine that allowed a user to store, retrieve, and cross-reference documents linked by association. These "links" were made through microfilm—meaning they were physical links, not digital—but the concept was the same as modern hypertext.

Many scientists were deeply influenced by Bush's writing. In 1965, Ted Nelson, a somewhat radical computer scientist, coined the term *hypertext* to describe a way of linking ideas in electronic text based on Bush's

*Tim Berners-Lee. (Associated Press)*

Memex machine. According to legend, Nelson's grandfather read "As We May Think" aloud to him when he was a child. Douglas Englebart, another computing pioneer, read "As We May Think" while stationed in the Philippines during World War II. In the early 1960s, while doing research at the cutting-edge Stanford Research Institute (SRI), Englebart used the Memex model to create the first successful implementation of hypertext in digital form, using a program he called NLS (oNLine System).

In 1980 Berners-Lee began his own work with hypertext while at CERN. CERN maintains a string of research facilities nestled among the Jura Mountains between Switzerland and France. The institution works with a global network of individual scientists who are working on various projects. To manage such vast amounts of information, Berners-Lee developed what he called a "memory substitute" to help him recall who worked where and on what project. He called the program Enquire—a name drawn from a Victorian household encyclopedia, *Enquire Within Upon Everything*, that Berners-Lee had stumbled upon in his youth. (Some suggest that the Web is actually the culmination of young Berners-Lee's belief that *Enquire Within Upon Everything* could truly answer any question he might have had.) The program proved to be personally useful

for Berners-Lee, but it was never published. When Berners-Lee returned to CERN in 1984, however, he immediately began to seek approval for a more expansive hypertext program.

At the time, the Internet had been around for more than a decade. Email existed, as did protocols for transferring files and exchanging information. However, the Internet was still largely the provenance of academic and research institutions. Many programs were text-driven and difficult to use. No one had yet developed a way to make the Internet widely available.

## RIGHT PLACE, RIGHT TIME

While at CERN, Berners-Lee promoted a global hypertext system as a way for physicists all over the world to collaborate and pool their information in one single information space, instead of downloading copious amounts of information to each of their individual computers. The system would also address the increasingly frustrating problem of incompatibility between the many different computer systems in use at CERN. The initial response to his ideas was tepid. Only in 1989, five years later, did Berners-Lee's proposals begin to be taken seriously. His boss, Mike Sendall, put Berners-Lee in contact with Robert Cailliau, a CERN scientist who had independently proposed a similar hypertext project. Cailliau had been experimenting with Hypercard, an early, highly flexible database application developed by Apple Computer in 1987. Cailliau believed hypertext was the next step, though he had been thinking strictly in terms of the CERN computer network. It was Berners-Lee who envisioned hypertext on the Internet, and the global information space it could create. In a 1997 interview with *Internet Computing Online,* Cailliau said, "It was obvious that there was no use trying to do anything else but push his proposal through."

Though Berners-Lee had already developed much of the prototype, in the fall of 1990 the two men began to hash out the essential aspects of the project—including the name. Among the ideas Berners-Lee considered were Mine of Information (MOI) and The Information Mine (TIM)—both were rejected for having excessively egotistical acronyms. Ultimately he settled on World Wide Web, in order to reflect his goal that the system be global in nature. By December 1990, after only a month's work, Berners-Lee had created the first World Wide Web point-and-click hypertext editor-browser, using a NeXTStep computer. On December 25, 1990, Berners-Lee connected his computer to Cailliau's; the two computers communicated through info.cern.ch—the first World Wide Web server.

This was a key technological achievement, but for the World Wide Web to be successful, Berners-Lee had to make it not only practical and useful, but also widespread. The first popular application of the World Wide Web was the CERN directory. Berners-Lee, in his 1996 paper, "The World Wide Web: Past, Present and Future," recalled "many people at that point saw the Web as a phone book program with a strange user interface." Still, it caught on. (Indeed, some scientists took to keeping a window open to the Web at all times, just to have access to the phone book.) At that time, the Web was decidedly simple—an internal network of documents at CERN connected through links accessible to anyone with the right specifications.

## NUTS AND BOLTS

Over the next year, Berners-Lee strove to make the Web as accessible as possible. He built his specifications upon existing technologies and practices in the computing world. For example, HTML (hypertext markup language) was based on SGML (standardized general markup language), which had been the international standard for coding computer languages since 1986. HTTP (hypertext transfer protocol) was based on the widely used FTP (file transfer protocol). Both HTML and HTTP were designed to run like their predecessors, only more quickly, simply, and efficiently in order to function in the new hypertext environment.

URIs (universal resource identifiers), however, were the most important and unique element of the Web's architecture. As Berners-Lee told *Wired News* in 1999, "The idea that any piece of information anywhere should have an identifier, which will not only identify it, but allow you to get hold of it, that idea was the basic clue to the universality of the Web." Today, URIs are most commonly referred to as URLs (universal resource locators) or Web addresses. (Originally, Berners-Lee re-

*For me the fundamental Web is the Web of people. It's not the Web of machines talking to each other. It's not the network of machines talking to each other. It's not the Web of documents.*

*—Tim Berners-Lee in a speech at the MIT Laboratory for Computer Science, April 1999*

ferred to URIs as UDIs—universal document identifiers. By changing "document" to "resource," however, he expanded the definition of what elements could be represented on the Web.)

In each of his endeavors, Berners-Lee's primary hope was interoperability. In a 1999 speech at the Massachusetts Institute of Technology (MIT), he explained, "The raison d'être . . . for getting the Web protocols out, was to be independent of hardware platform: to be able to see the stuff on the mainframe from your PC and to be able to see the stuff on the PC from the Mac. To get across those boundaries was at the time so huge and strange and unbelievable." His efforts were aided by Nicola Pellow, a recent college graduate working at CERN, who, in 1990, developed a browser that was simple enough to run on almost any computer.

In August 1991, Berners-Lee posted a notice to the Internet on Usenet's alt.hypertext newsgroup, announcing that the software for the World Wide Web server, the browser, and all the required specifications were available for download by FTP. New mailing lists related to Web programming—such as comp .infosystems.www—cropped up as more people were drawn to the Web. Between 1991 and 1993, Berners-Lee continued to refine his specifications for URIs, HTTP, and HTML in response to feedback from these forums.

In June 1992, Berners-Lee embarked on a three-month tour of the United States, promoting the Web at such venerable research and academic institutions as Xerox Palo Alto Research Center and the MIT. At the time, the Web's success was not yet guaranteed, and Berners-Lee found himself explaining the underlying concepts of the Web to many puzzled audiences. The Web also faced some competition. Internet Gopher, an information retrieval system developed at the University of Minnesota in 1991, vied for popularity among Internet users. Though some avoided the Web because of the perceived complexity of HTML, soon many were drawn to it because of the power and flexibility inherent to the Web model, as opposed to the typical "tree" model employed by hierachical systems like Internet Gopher. (Berners-Lee also attributes the downfall of Internet Gopher to the University of Minnesota's decision to license the technology, believing that, had he done the same, the Web would have never taken off.)

Overall, Berners-Lee's goal was to get a critical mass of users on the Web. He believed that the more people who were online, the more incentive there would be to put up servers; the more servers, the more information there would be online; the more information online, the more powerful and useful the World Wide Web could be.

## A CALL TO ORDER

As news of the Web spread, first through the academic and research communities and then through commercial industries, programmers—most of whom were still students—began to develop software. Berners-Lee's original browser—which was called World-WideWeb– was based on a NeXT computer, a not entirely popular computing platform. For the Web to be truly universal, browsers had to be developed for UNIX, Macintosh, and PC machines. In quick succession, students from the Helsinki (Finland) University of Technology wrote Erwise, a browser for UNIX-based computers; a Macintosh-based browser, called Samba, was developed at CERN; a student at the University of California–Berkeley created a browser named Viola.

Then, Marc Andreessen and Eric Bina led a team of software engineers at the National Center for Supercomputer Applications (NCSA) in developing a browser called Mosaic. By August 1993, Mosaic had been released for free for both Macintosh and PC computers. (A UNIX version had been released earlier.) The popularity of Mosaic caused a massive upswell in Web traffic.

While the new browsers drummed up increasing support for the Web, they also took the Web in a direction somewhat different from what Berners-Lee had originally imagined. (For one, Berners-Lee balked at the addition of the "image" tags to HTML by Marc Andreessen at Mosaic.) Indeed, the buzz surrounding Mosaic nearly overshadowed the Web. In his 1999 book, *Weaving the Web,* Berners-Lee writes, "The people at NCSA were attempting to portray themselves as the center of Web development, and to basically rename the Web as Mosaic. At NCSA, something wasn't 'on the Web,' it was 'on Mosaic.'" The proprietary atmosphere created by the new browsers became a growing concern to Berners-Lee, who believed the Web should be allowed to develop free from such pressures and constraints.

On April 30, 1993, Berners-Lee convinced CERN to officially release the code for the Web into the public domain—but not without some trepidation. Since the beginning of the browser wars, Berners-Lee had feared that fragmentation and the splintering of Web standards would threaten the Web's evolution. "It was never clear that it wouldn't just stop," Berners-Lee told *Wired News* in 1999. "Any time during that exponential growth, it could have stalled." But even as he worried, the Web was catching fire. Indeed, the load on info.cern.ch had grown by a factor of 10 each year since the site's debut in 1991.

Bolstered by the increasing success of the Web, Berners-Lee called together nearly 400 scientists and Web enthusiasts from around the globe to gather at CERN in May 1994 for the first World Wide Web Conference. Berners-Lee suggested to this newfound community that a neutral organization was needed in order for the Web to evolve in a free, nonproprietary, open environment. Later that year, on a bus traveling to an industry conference in northern England, Berners-Lee met a scientist from the Laboratory for Computer Science (LCS) at MIT, who suggested that another man at LCS, Michael Dertouzos, could help. Indeed, he could. In July 1994, Dertouzos convinced the LCS to house what would become the World Wide Web Consortium (W3C), and he helped Berners-Lee nurture the fledgling organization. Months later, in October 1994, Berners-Lee moved to Cambridge, Massachusetts, to become its director and the W3C was officially born.

> *If I'd started "Web Inc." it would have been just another proprietary system. You wouldn't have had this universality. For something like the Web to exist, it has to be based on public, nonproprietary standards.*
>
> —Tim Berners-Lee to Wired, *March 1997*

Initially, CERN served as the European headquarters for the W3C, with MIT as the American base. In late 1994, the Institute Nationale pour la Récherche en Informatique et Automatique (INRIA), also known as France's National Institute for Research in Computer Science and Control, joined the ranks and served as the European base until 2003. Today, the W3C has headquarters on three continents: at the European Research Consortium for Informatics and Mathematics (ERCIM), in France; at Keio University, in Japan; and at MIT. In addition, the W3C has grown to include more than a dozen international offices (located in Australia, Finland, Germany, Greece, Hong Kong, Hungary, Israel, Italy, Korea, Morocco, the Netherlands, Spain, Sweden, and the United Kingdom) and more than 450 member organizations that collectively and democratically guide the fate of the Web.

In accordance with Berners-Lee's vision, the W3C has served as a vendor-neutral forum where individuals from the technical, commercial, and creative sectors of the computing world meet to hash out Web standards and specifications as well as tackle long-term issues such as universal access and security. Its official mission is "to realize the full potential of the Web." By the turn of the 21st century, the W3C had developed more than fifty technical specifications for the Web, including new versions of HTML, developing XML (extensible markup language), which has been hailed as the new lingua franca of the Web, creating P3P (Platform for Privacy Preferences), a privacy standard, and exploring the possibilities of one of Berners-Lee's pet projects, the Semantic Web.

## A SEMANTIC ARGUMENT

As director of the W3C, Berners-Lee found himself at the forefront of many debates on the future of the Web. He is probably best known, however, as a fervent champion of the Semantic Web. The Semantic Web can be thought of as an extension of the World Wide Web—a Web imbued with another layer of meaning that can be understood and manipulated by computers. In this context, *semantic* means computer-processable.

*People keep asking me what I think of it now that it's done. Hence my protest: "The Web is not done!"*

—*Berners-Lee to* Wired News, *October 23, 1999*

The Semantic Web was an early—though long-term—goal of Berners-Lee's. He had hoped that not only would the Web provide a global space where information could be exchanged, but that, in time, computers could analyze *how* that information was exchanged. If a computer could understand, in a sense, the pattern of interaction on the Web, certain activities could be automated or enhanced. In that way, the Semantic Web would open the door for technological tools called software agents to be useful online. (Berners-Lee is quick to point out that the Semantic Web is not a form of artificial intelligence, even though the development of software agents has grown out of the artificial intelligence sector.)

Some crucial elements of the Semantic Web have already been developed. With the advent of XML, programmers can describe what data *are,* not just what that data should look like. Another W3C project, RDF (resource description framework) is a metadata specification that allows elements in a document to be described in three ways—akin to the subject, verb, object in a simple sentence. The explicit nature of this descriptive framework allows computers to mine the elements for meaning. Indeed, supporters of the Semantic Web believe that by making the Web more meaningful to computers, computers will make the Web more useful to humans.

For example, in a 2001 article about the Semantic Web published in *Scientific American,* Berners-Lee explained how two adult siblings, both using software agents on the Semantic Web, could easily negotiate taking their mother to the doctor. The software agents could seek out the best doctors within a certain neighborhood who were covered by the proper insurance plan and then collaborate with the software agent on the doctor's office Website to set up appointments. Using the highly descriptive tags in the Semantic Web, the software agents could process terms necessary to carry out these actions—terms referring to location, dates, and concepts like "ratings" and "insurance plans" and "appointments."

Berners-Lee has been criticized for his dogged focus on the Semantic Web, especially while other more pressing—and often commercial—interests lie before the W3C. Indeed, some critics have dubbed the Semantic Web the

"Pedantic Web." Regardless, it remains a long-term goal for both the W3C and Berners-Lee.

It is difficult to overstate Berners-Lee's contribution to the Internet—especially in light of that fact that, unlike many of his peers, he chose not to earn millions by making the Web a proprietary venture. Though hypertext and the Internet had both existed for a decade or longer before Berners-Lee developed the Web, Berners-Lee was able to combine the two into a revolutionary information space. The development of the Web helped alleviate the frustrations of incompatibility between computer systems and created the first real "open architecture" available on the Internet. By creating HTML—which can be learned in an afternoon—Berners-Lee, in essence, wrested control of the Internet from the hands of elite academic and research institutions and put it in the hands of the masses. More recently, as commercial interests have come to dominate much of the Web, Berners-Lee's continued vigilance against corporate or commercial control of the Web has kept Web standards patent- and royalty-free. And perhaps most important, he has not lost sight of his vision for the Web as a means for human communication and collaboration. In an interview with *National Public Radio*, Berners-Lee said, "I'd like . . . computers to completely disappear. I want people to be able to forget that they've got a keyboard and a mouse and a screen and just be immersed in the abstract communication, the abstract ideas that they're talking about, the bridge they're designing and, of course, be very aware of the other person at the other end."

## FURTHER READING

### In These Volumes

Related Entries in this Volume: Andreessen, Marc; Nelson, Ted; Nielsen, Jakob

Related Entries in the Chronology Volume: 1990: The World Wide Web Is Invented; 1991: The World Wide Web Is Developed at CERN; 1991: Members of the Internet Community Are Encouraged to Write Applications for the Web; 1994: World Wide Web Consortium Is Formed to Promote the Evolution of the Web; 1994: The First International World Wide Web Conference

Related Entries in the Issues Volume: Anonymity; Content Filtering; Privacy; Usability

### Works By Tim Berners-Lee

with Mark Fischetti. *Weaving the Web: The Original Design and Ultimate Destiny of the World Wide Web by Its Inventor.* San Francisco: HarperSanFrancisco, 1999.

with James Hendler and Ora Lassila. "The Semantic Web." *Scientific American,* May 2001. Also available online at: http://www.sciam.com/issue.cfm?issuedate=May-01 (September 16, 2004).

### Books

Gillies, James, and Robert Cailliau. *How the Web Was Born: The Story of the World Wide Web.* Oxford: Oxford University Press, 2000.

### Articles

Bush, Vannevar. "As We May Think." *The Atlantic Monthly,* July 1945, http://www.theatlantic.com /unbound/flashbks/computer/bushf.htm (cited September 16, 2004).

Kirsner, Scott. "The Modest Inventor." *Salon.com,* September 15, 1999, http://www.salon.com /tech/books/1999/09/15/berners_lee (cited September 16, 2004).

Logan, Tracey. "Net Guru Peers Into Web's Future." *BBC News,* September 25, 2003, http://news .bbc.co.uk/1/hi/technology/3131562.stm (cited September 16, 2004).

Oakes, Chris. "Interview with the Web's Creator." *Wired News,* October 23, 1999, http://www.wired .com/ (cited September 16, 2004).

Petrie, Charles. "Robert Cailliau on the WWW Proposal: 'How It Really Happened.'" *Internet Computing Online,* November 1997, http://www .computer.org/internet/v2n1/cailliau.htm (cited September 16, 2004).

Quittner, Joshua. "Time 100: Tim Berners-Lee." *Time,* March 29, 1999, http://www.time.com/time /time100/scientist/profile/bernerslee.html (cited September 16, 2004).

Schwartz, Evan. "The Father of the Web." *Wired* 5.03, March 1997, http://hotwired.wired.com/collections /connectivity/5.03_berners_lee1.html (cited September 16, 2004).

Wright, Robert. "The Man Who Invented the Web." *Time,* May 19, 1997.

### Websites

Biographical and historical information are also available, including: Berners-Lee's professional home page at W3C, with links to several of his papers and talks, http://www.w3.org/People /Berners-Lee/ (cited September 16, 2004).

History of HTML, taken from Chapters 1 and 2 of *Raggett on HTML 4.* http://www.w3.org/People/Raggett/book4/ch02.html (cited September 16, 2004). World Wide Web and CERN. CERN resources about the history of the Web, http://cern.web.cern.ch/CERN /WorldWideWeb/WWWandCERN.html (cited September 16, 2004).

Timeline of the Web: 1945–1995, http://www .w3.org/History.html (cited September 16, 2004).

World Wide Web Consortium (W3C). Home base of the W3C, with extensive links to information on current and past W3C projects, including HTML, XML, P3P, and the Semantic Web, http://www .w3.org (cited September 16, 2004).

# Jeff Bezos (1964–)

## FOUNDER, AMAZON.COM

When the editors of *Time* magazine chose Jeff Bezos as Man of the Year for 1999, they hailed him as the "king of cybercommerce." Indeed, five years earlier when he founded the online bookseller Amazon.com, Bezos set a high standard for customer service and put in place a "get big fast" mentality that pushed Amazon.com to grow ambitiously, even as business losses accrued. Throughout the late 1990s, detractors had predicted the demise of Amazon. Bezos kept his company focused on its goals throughout this criticism as well as during the dark days of the dot-com bust, when the price of all Internet stocks, not just Amazon's, plummeted. He and Amazon emerged triumphant, with the company posting its first quarterly profit in late 2001. Bezos's singular vision and seemingly endless optimism have brought Amazon.com to the pinnacle of e-commerce enterprises.

*Strip malls are history.*

*—Jeff Bezos to* Wired, *March 1999*

## SON OF PEDRO PAN

Jeff Bezos was born in Albuquerque, New Mexico, to Jackie Gise, a seventeen-year-old high school student. She divorced his biological father when Jeff was eighteen months old; two years later, she married Miguel Bezos, a Cuban immigrant and engineering student at the University of Albuquerque. Bezos adopted four-year-old Jeff in 1968 and has been the only father Jeff Bezos has ever known.

Miguel Bezos had fled the Castro regime at age 15, under a Catholic church–sponsored program in Florida known as Operation Pedro Pan. Without family of his own, he lived in a Catholic mission in Delaware, where he earned a high school diploma. After graduating from the University of Albuquerque with an engineering degree, he began working as a petroleum engineer for Exxon—a job that required frequent relocations.

The Bezos family moved, first, to Houston, Texas, where Jeff enrolled in kindergarten. Bezos was an extraordinarily bright student, and his parents soon enrolled him in a magnet program for gifted children at Houston's River Oaks Elementary School. Featured as a young prodigy in the book *Turning on Bright Minds: A Parent Looks at Gifted Education in Texas* (1977), he was described as a "friendly but serious" boy of "general intellectual excellence," though "not particularly gifted in leadership."

At River Oaks, Bezos got his first taste of computing. He impressed his fourth-grade teachers by figuring out how to use a teletype monitor to connect to a time-shared computer mainframe via modem. Bezos then taught friends how to use the computer so they could play a Star Trek video game after school. (To this day, Bezos remains an avid *Star Trek* fan.)

When Jeff was in sixth grade, job transfers brought the Bezos family to Pensacola, Florida, and then, eighteen months later, to Miami. The Exxon job meant a semi-rootless existence for the family, but Jeff visited his maternal grandparents every year, staying on their sprawling cattle ranch in Cotulla, Texas (pop. 3,000). From age four to sixteen, Bezos spent almost every summer on the ranch, helping lay pipes, brand and vaccinate cattle, and rebuild tractors as his grandfather, Lawrence "Pop" Gise, regaled him with tales of working on space technology for the Pentagon and the Atomic Energy Commission in the 1950s and 1960s. These reminiscences instilled in Bezos a lifelong fascination with space travel as well as a deeply held belief in hard work. "One of the things that you learn in a rural area like that is self-reliance," he told *Inc.* magazine in 2004. "People do everything themselves."

Back in Florida, Bezos attended Miami Palmetto Senior High School. Bezos was an excellent student,

*Jeff Bezos. (Natalie Forbes/Corbis)*

and he won the Florida State Science Fair for a project on the effect of zero gravity on houseflies. (The prize was a trip to NASA's Marshall Space Flight Center—a treat for the aspiring astronaut.) At home, he spent hours in the garage working on other science experiments. "I think single-handedly we kept many Radio Shacks in business," his mom joked in a 1999 *Wired* interview.

Bezos also worked throughout high school. After spending one summer scrambling eggs and making fries at a McDonald's on Florida's Route 1, Bezos and his high school girlfriend founded a business they called the Dream (Directed REAsoning Methods) Institute, for which they earned $600 per student for teaching middle school kids new ways to think about science and literature. In 1982, Bezos graduated as valedictorian and National Merit Scholar finalist.

By the time Bezos entered college at Princeton, his dream of becoming an astronaut had given way to becoming a theoretical physicist, along the lines of Albert Einstein. Once Bezos concluded that he might be, at best, a mediocre scientist, he turned to computers, graduating summa cum laude in 1986 with a degree in electrical engineering and computer science.

## GIVING 2300 PERCENT

When Bezos graduated, job offers rolled in from Intel Corp., Andersen Consulting, and Bell Labs. But Bezos decided to go with Fitel, a start-up firm in New York City that had recruited "the best computer-science graduates" through Princeton's student paper. Founded by Columbia University professors, Fitel was attempting to build a special type of telecommunications network—a mini-Internet—geared to international financial trading firms. Bezos was employee number eleven, head of customer service and software development. After two years, the network had still failed to take off. In April 1988, Bezos left for a job at the financial services firm Bankers Trust.

In his new position, Bezos developed software tools for the company's pension fund clients. Rising quickly through the ranks, Bezos became, in less than a year, one of the company's youngest vice presidents. Nevertheless, Bezos met with resistance from the "old guard," who routinely dismissed his idea of bringing personal computers into the world of banking. In 1990, he quit.

Although Bezos wanted to leave the financial-services sector, a corporate recruiter convinced him to interview with a new technology-oriented hedge fund, D.E. Shaw. The founder of the company, David Shaw, had a doctorate in computer science from Stanford, and the two men saw eye-to-eye on the potential of investing in high-technology stocks. Bezos worked endless hours at Shaw, sometimes curling up for the night in a sleeping bag he kept in his office. After four years, he had become indispensable, working as a general-purpose business engineer on a variety of projects. After two years, Bezos, only twenty-eight, was named senior vice president of new business opportunities.

By early 1994, Bezos and Shaw began to research business opportunities online. Bezos had first used the Internet almost a decade earlier, while still at Princeton, but the Internet's commercial potential did not become apparent to him until the World Wide Web brought an ever-increasing population of users online. In the spring of 1994, Bezos was bowled over by a simple statistic—Web usage was growing at a rate of 2300 percent per year. "That's what sort of set me off," Bezos said in a 1998 *USA Today* interview.

Bezos then researched the top twenty mail order businesses to figure out which goods might sell well in an Internet marketplace. Software, music, videos, and clothing were all possibilities, but Bezos methodically whittled the list down to one choice—books. An $82 billion market, books offered consumers a lot of choice (more than 3 million titles in print) and they were not costly. To boot, no one bookseller dominated the market—not even Barnes & Noble, which held just 12 percent. In addition, a virtual bookstore could offer many more titles than even the largest bricks-and-mortar Barnes & Noble Superstore. There was, it seemed, room for something new.

Bezos brought his findings to Shaw, who, though intrigued, was not ready to take his company in such a direction. Then talk turned to Bezos's own future at the company, which, Shaw asserted, looked incredibly bright. The online world's attraction was stronger, though, and Bezos quit in June 1994. Over the Fourth of July weekend, he and his wife headed west to break new ground in e-commerce.

## PEDDLING BOOKS

> *Success would be other companies in completely different industries pointing to Amazon.com and saying, "There's a new bar for customer experience, for customer-centricity, and if they can do it, we can do it." That's a never-ending mission.*
>
> —Bezos to Business Week, August 2004

Bezos's first stop was his family's home in Fort Worth, Texas, where his parents gave him a used Chevy Blazer for the cross-country drive, as well as $300,000 seed money, despite his warnings of 70 percent chance of failure. ("I wanted to make sure I could still come home to Thanksgiving dinner," he joked to the *Dallas Morning News* in 1999.) Leaving Fort Worth, with his wife driving, Bezos sat in the Chevy's passenger seat, typing up a business plan on his laptop and calling around the country on his cell phone in search of a lawyer and his first vice president.

Bezos had already decided that his destination would be Seattle, which was near the country's largest book distribution center, owned by the book wholesaler Ingram, and was bubbling over with engineers and Web talent. On the way, Bezos stopped in Santa Cruz, California, to interview Shel Kaphan, an engineer, who became employee number one. Five days after leaving Texas, the couple arrived at their rented house in the Seattle suburb of Bellevue, where they set up shop in the garage, using desks fashioned from doors bought at Home Depot and three used Sun Microsystems workstations.

Over the next year, Bezos and Kaphan worked on the Website. The company would be called Amazon, after the vast, branching South American river. The name appealed to Bezos because it was striking and powerful, and yet did not limit his business to one specific product or endeavor. The name Amazon beat out the Dutch word Aard, deemed too obscure, and Cadabra, which Bezos's lawyer vetoed for sounding too much like "cadaver." In the spring of 1995, Bezos invited 300 friends and acquaintances to test the site. On July 16, 1995, Amazon.com opened its virtual doors to the public at large.

The first book to sell was Douglas Hofstadter's *Fluid Concepts,* and others quickly followed. Sales were generated entirely by word of mouth as Amazon was not yet advertising and had not drawn any early press. In its first week, Amazon.com processed nearly $12,500 of orders. Within its first month, books were sold to customers in all fifty United States and forty-five other countries. Order volume was still small enough that Bezos himself drove the packaged books to the post office. But the company was growing fast. By September 1995, Amazon.com boasted weekly sales of $20,000.

In November 1995, Amazon.com moved to an inauspicious office located across the street from a Seattle pawnshop and a wig store. By then, Bezos had hired a dozen people for customer service and technical support; oftentimes, he would still man the phones and pack orders.

## BIG, BIGGER, BIGGEST

From the start, Bezos's goal was for Amazon to "get big fast"—a phrase splashed over T-shirts handed out at a 1996 company picnic. By the end of that year, Amazon boasted $15.7 million in sales—a 3000 percent gain over 1995. But losses, at $6.2 million, were also significant; the company had spent huge amounts of money developing software and expanding its business, and Amazon had miles to go before it would make a profit, which prompted many questions about Bezos's business model.

Speculation about Amazon.com's survival intensified the following year, when, in May 1997, Barnes & Noble entered the world of e-commerce by launching barnesandnoble.com. Despite Amazon's well-received initial public offering of stock, which had raised $54 million that very month, George Colony, head of the industry research firm Forrester Research, was quick to

*We may be the most customer-obsessed company to ever occupy planet Earth.*

—Bezos to the New York Times, March 1999

call Bezos's company "Amazon.toast" once Barnes & Noble went online. As Amazon's stock price slipped, *Barron's,* a financial newspaper, chimed in with a June 1997 headline that read "Amazon.bomb."

Four months later, however, Amazon was still getting bigger, faster. Bezos hand-delivered Amazon's millionth order—a Windows NT manual and *The Royals,* by Kitty Kelly—to a customer in Japan. Engineers continued to upgrade the site, introducing features like one-click shopping that were soon picked up by other e-commerce sites. By the end of 1997, the value of Amazon stock had risen 233 percent. The company expanded, moving in August 1998 to a stately twelve-story converted hospital building southeast of downtown Seattle.

The following year, as Bezos had planned, Amazon began branching out, selling music, video, electronics, toys, software, and other consumer goods. In October, the company went international with Amazon.co.uk in the United Kingdom and Amazon.de in Germany. Amazon.com also challenged successful online auction site eBay.com with its new Amazon.com Auctions service, as well as the zShops, an online marketplace for buyers and sellers, and a $45-million deal with the auction house Sotheby's. (Bezos's first desk, fashioned from a door, was one of the first things auctioned on Amazon.com, selling for $30,000 to the highest bidder—his mom.)

Amazon also let it be known that the company was willing to compete ferociously against all comers. To steal the thunder from barnesandnoble.com's announcement of an initial stock offering in May 1999, Amazon cut prices on best-sellers by 50 percent. Such growth and fierce competitive instincts kept Amazon in the headlines, and, in December of that year, earned Bezos *Time* magazine's Man of the Year award—the fourth-youngest person to have been awarded the title. He was hailed as the king of cybercommerce.

Indeed, with its dedication to customer service and ease of use, Amazon had come to define e-commerce at the end of the 1990s. But despite all the publicity, the company continued to spend far more than it earned, and, by the summer of 2000, Amazon's stock price had dropped by more than two-thirds. Grumblings about the firm's performance had begun as early as mid-1999, when, even as sales jumped 171 percent, net losses reached $82.8 million. Wall Street analysts suspected that Amazon had grown too big, too fast, and had spread itself too thin by branching out into so many new product categories. A particularly scathing report released by Lehman Brothers in June 2000 predicted that Amazon would run out of cash by 2001: "The party is over," it read. Other reports suggested that bankruptcy or a buyout were on the horizon.

In January 2001, a defiant Bezos told analysts his company would post a profit by the holiday season. That month, he laid off 15 per cent of the workforce, closed two warehouses, and shut down a customer service center. As Bezos had promised, in January 2002, after seven years of losses, Amazon.com posted its first profit ever for fourth quarter 2001: the net profit of $5.1 million translated into one cent per share. The following year, the company also posted fourth-quarter profits; in 2003, its first full-year profit.

## PATENTLY OFFENSIVE

While leading the pack of e-commerce sites, Amazon.com also made waves in the realm of software patents in 1999 and 2000. The U.S. Patent Office awarded Patent No. 5,960,411 to Amazon.com on September 28, 1999. The patent protected Amazon's "one-click" online purchasing technology, which it had begun using in September 1997. A month later, Amazon sued barnesandnoble.com, claiming that its "express lane" ordering system constituted patent infringement. In a written statement, Bezos explained, "We spent thousands of hours to develop our 1-Click process, and the reason we have a patent system in this country is to encourage people to take these kinds of risks and make these kinds of investments for customers." Amazon won a preliminary injunction against barnesandnoble.com that December—barnesandnoble.com had to add an extra "click" to its ordering software.

Critics—most notably Richard Stallman of the Free Software Foundation and computer book publisher Tim O'Reilly—claimed that an idea as obvious as "one-click" purchasing (which is based on the common use of cookies, which Websites use to store information about specific visitors) should be unpatentable. Stallman went so far as to call for boycotts against Amazon. Amazon again angered its critics in February 2002, when it announced another patent for its affiliate program, which offers revenue-sharing opportunities to sites that link to Amazon.com. The passions surrounding these software patents became so fierce, so quickly, that, on March 9, 2000, Bezos responded with an open letter calling for major patent reform, including reducing the length of a software patent from seventeen to three or five years. Later, Bezos and O'Reilly joined forces to lobby Washington for changes to the system.

## INSIDE THE BOOK AND OUT

While Amazon may have tempered its "biggest, faster" mentality following the dark days of early 2001, the company's more recent ventures prove that Bezos is as ambitious as ever. On October 23, 2003, Amazon.com launched the Search Inside the Book project. By digitizing the contents of more than 120,000 books, Amazon allows users to search the entire contents of its collection, creating one of the most powerful indexing and research tools for nonfiction and literature. As many have pointed out, Search Inside the Book has set Amazon on a course to develop a universal library.

In addition, a month earlier, the company founded A9.com, an online search engine subsidiary. The A9 engine, which uses technology similar to Google's Froogle, can be used to search for products and compare prices on e-commerce sites from all over the Web.

Bezos has also busied himself with a venture completely unrelated to Amazon—Blue Origins, a secretive space technology company. Located in a warehouse south of downtown Seattle, engineers at Blue Origins are working on a $30-million spaceship meant for private use. Bezos's interest in this work dates to high school, when as valedictorian, he told the *Miami Herald* that one day he hoped to put hotels and amusement parks in orbit.

Since its founding in 1994, Amazon has grown from a struggling online bookseller into a vast online shopping mall and marketplace, and, in so doing, has helped define and shape e-commerce. Much of Amazon.com's success can be attributed to Bezos, whose "heads down" approach to adversity and steadfast dedication to customer service has kept Amazon afloat while competitors sank. Amazon.com still tops the list for customer satisfaction among e-commerce sites, and Bezos continues to lead his company with his trademark barking laugh and aggressive approach to innovation online.

# FURTHER READING

## In These Volumes

Related Entries in this Volume: Stallman, Richard

Related Entries in the Chronology Volume: 1995: Amazon.com and eBay Are Founded; 2000: The Dot-Com Crash

Related Entries in the Issues Volume: Digital Libraries; E-books; E-commerce; Patents

## Books

Brackett, Virginia. *Jeff Bezos*. Philadelphia: Chelsea House, 2001.

Leibovich, Mark. *The New Imperialists*. Upper Saddle River, NJ: Prentice Hall, 2002.

Marcus, James. *Amazonia*. New York: New Press, 2004.

Spector, Robert. *Amazon.com: Get Big Fast*. New York: HarperBusiness, 2000.

## Articles

"Amazon.com and Beyond," *Wired* 8.07, July 2000, http://www.wired.com/wired/archive/8.07/bezos.html (cited September 20, 2004).

Bayers, Chip. "The Inner Bezos," *Wired* 7.03, March 1999, http://www.wired.com/wired/archive/7.03/bezos.html (cited September 20, 2004).

De Jonge, Peter. "Riding the Wild, Perilous Water of Amazon.com." *New York Times,* March 14, 1999.

Deutschman, Alan. "Inside the Mind of Jeff Bezos." *Fast Company,* August 2004, http://www.fastcompany.com/magazine/85/bezos_1.html (cited September 20, 2004).

Frey, Christine, and John Cook. "How Amazon.com Survived, Thrived and Turned a Profit." *Seattle Post Intelligencer,* January 28, 2004, http://seattlepi.nwsource.com/business/158315_amazon28.html?source=techdirt (cited September 20, 2004).

GeBalle, Bob. "Amazon.c(abo)om." *Seattle Weekly,* October 21, 1997, http://www.seattleweekly.com/features/9942/features-geballe.shtml (cited September 20, 2004).

Gleick, James. "Patently Absurd." *New York Times,* March 12, 2000.

Granberry, Michael. "At Amazon.com, He's the Mouth That Roared." *Dallas Morning News,* August 8, 1999.

Hansell, Saul. "A Surprise from Amazon: Its First Profit." *New York Times,* January 23, 2002.

Harbrecht, Doug, Frank Comes, and Kathy Rebello. "Chewing the Sashimi with Jeff Bezos." *Business Week,* July 15, 2002, http://www.businessweek.com/bwdaily/dnflash/jul2002/nf20020715_5066.htm (cited September 20, 2004).

Hazleton, Lesley. "Jeff Bezos." *Success,* July 1998.

Hof, Robert D. "Jeff Bezos: 'Blind Alley Explorer'." *Business Week,* August 19, 2004, http://www.businessweek.com/bwdaily/dnflash/aug2004/nf20040819_7348_db_81.htm (cited September 20, 2004).

Kaufman, Leslie. "Amazon II: Will This Smile Last?" *New York Times,* May 19, 2002.

Krantz, Michael. "Inside Amazon's Culture." *Time,* December 27, 1999.

Levy, Doug. "Amazon.com Amazes." *USA Today,* December 24, 1998.

Perez, Elizabeth. "Store on the Internet Is Open Book," *Seattle Times,* September 19, 1995.

Quittner, Joshua. "The Background and Influences That Made Bezos the Multi-billion-dollar Champion of E-tailing." *Time,* December 27, 1999.

Ramo, Joshua Cooper. "*Time* Person of the Year: Jeffrey P. Bezos." *Time,* December 27, 1999, http://www.time.com/time/poy2000/archive/1999.html (cited September 20, 2004).

Sandoval, Greg. "Bezos: Back on Top." *CNet News.com,* January 23, 2002, http://news.com.com/2008-1082-821168.html (cited September 20, 2004).

Streitfeld, David. "Booking the Future." *Washington Post,* July 10, 1998.

Walker, Leslie. "Looking Beyond Books." *Washington Post,* November 8, 1998.

Walker, Rob. "America's 25 Most Fascinating Entrepreneurs: Jeff Bezos." *Inc.com.* April 2004, http://www.inc.com/magazine/20040401/25bezos.html (cited September 20, 2004).

Wolf, Gary. "The Great Library of Amazonia." *Wired,* December 2003.

## Websites

Amazon.com. Amazon's e-commerce site, http://www.amazon.com (cited September 20, 2004).

A9.com. Amazon.com's beta search engine site, http://a9.com/ (cited September 20, 2004).

Blue Origin. Online home of Bezos's side project in space technologies, http://www.blueorigin.com/ (cited September 20, 2004).

# Anita Borg (1949–2003)

## SCIENTIST; FOUNDER OF SYSTERS

**Although Anita Borg (Naffz)** enjoyed great professional success as a computer scientist, her true legacy is the accomplishments of the women whose lives she touched, whose careers she encouraged, and whose burden she lightened by creating communities for women where none had existed, including the Systers mailing list and the Institute of Women and Technology. During a 1999 online chat with www.girlgeeks.org, an online community for girls interested in computers and technology, Borg explained her beginnings simply and succinctly. "I am very proud to call myself a feminist—a person who believes that all people should have equal opportunities to contribute fully in the world," she wrote. "It was a natural combination of my interest in computing and my feminism to think about my female colleagues."

## NO ORDINARY SECRETARY

Borg got her start in computing in the late 1960s as secretary in the data processing department of a small insurance agency in Manhattan. Bored by such menial work, she taught herself basic programming in COBOL, a common computer language at the time, using the IBM books she found in the office. She had entered college in 1967 but quit after two years, going to work to put her husband through graduate school. When the couple divorced in the early 1970s, Borg pulled together enough money to finish her undergraduate degree at New York University. Borg did not initially pursue a degree in computer science (she had considered a degree in Russian), but she excelled at programming. "I thought I'd see if I could get a master's degree," she told the *Boston Globe*. "NYU said, 'How about a Ph.D.?'"

NYU's Courant Institute of Mathematical Sciences awarded Borg a doctorate in computer science in 1981. Her research focused on synchronization efficiency in operating systems, a highly technical aspect of computer science. Her thesis adviser, Gerald Belpaire, would later re-

member how, even as a young graduate student, Borg strove to bring women together, lobbying for the few women in her department to have offices near each other.

After graduating, Borg spent several years building a fault-tolerant UNIX-based operating system for Auragen Systems Corporation, a small start-up in New Jersey. In 1985, she moved to northern Germany to work for Nixdorf Computer. There, the operating system she had begun developing at Auragen was successfully integrated into a Nixdorf product called Targon. Borg, however, would not stay at Nixdorf long. (In her polite way, she called living in small-town Germany after fifteen years in Manhattan a "character-building" experience.)

In March 1986, Borg attended the IBM Workshop on Fault Tolerant Distributed Computing at Asilomar, near Monterey, California. "It was March, and I'd never lived anywhere that you could walk on the beach barefoot in March," Borg reminisced later, speaking at a Women in Engineering Conference at the University of California–Berkeley, in 1993. "I decided that I *had* to live in California. After the workshop, I practically hung a sign around my neck that I was looking for work." Later that year, Borg was hired as a research engineer by the Western Research Lab of Digital Equipment Corporation (DEC), based in Palo Alto, where she stayed for the next twelve years.

## A MEETING IN THE LADIES' ROOM

In 1987, not long after Borg began at DEC, she attended the Symposium on Operating System Principles conference in Austin, Texas. Only a handful of the 400 participants were women. "We barely ran into each other except in the bathroom," Borg told the *Philadelphia Inquirer*. Lingering in the two-stall restroom, the women began talking about the issues they faced, which mainly centered on being the sole women in all-male offices and the dearth of women at the conference. "Every time someone came in, they joined the conversation." The half-dozen women from the ladies' room gathered the other women at the conference together for dinner that night, and the conversation continued. Borg returned to Palo Alto bearing twenty email addresses, and from those, the first mailing list solely for women in computing—dubbed Systers (a play on the words *sisters* and operating *systems*)—was born. Over the next decade, the initial 20 members grew to more than 2,500 women from more than twenty-five countries.

Since its inception, Systers has been a private list with no official moderator, though Borg herself "strongly guided" the list for many years. (She was known, on occasion, to send private, polite emails to members who strayed off-topic.) Systers was a forum where professional and academic women in computer science and related fields could ask questions, offer information, advice and support, and find community.

Systers was not a forum to talk about family, personal relationships, or any other topic that was unrelated to computer science as a profession. It was not a place to vent or to gossip, nor was it a place to ask overly technical questions. (Writes the current list manager, "Mail sent to the entire list should relate to women AND computing. Women on the list have access to other resources for technical information and for information about women in general.")

For hundreds of women, the list was a godsend. Isolated among the men in their labs, offices, or classes, women who came to the list were finally able to connect (virtually) with their female peers. On Systers, they found what they had once lacked: mentors, the insights and experiences of other professional women, a sympathetic audience. Borg herself decried the fact that, in the more than twenty years she had worked in computer science, she had never once worked closely with another female colleague.

Similarly, Systers offered a haven from the culture of "flaming" that existed on many other mailing lists. To flame someone is to insult or criticize them via email, and it happened all the time on mailing lists, especially to women. On open computing-related lists, women often found that the answer to a technical question would be, "Newbie, go home." Women who identified themselves as female on mailing lists often faced sexual harassment and gendered epithets, or they were ignored completely. In response, Systers was a list "organized

*Anita Borg. (Associated Press)*

around *conversations*"—with all the give-and-take that the word *conversation* connotes. Obnoxious behavior was rare and, when it happened, was not tolerated. The organizing principles of the list helped ensure that few, if any, arguments occurred. By 1995, only 2 of the nearly 2,000 members had been removed from the list. Systers was proof, Borg suggested, that email could be conducted ethically.

The early success of Systers left the group open to criticism. After a 1990 profile in an industry publication, Systers was accused of reverse discrimination. In her 1993 article, "Why Systers?" Borg defended this list's existence. "Until Systers came into existence," she writes, "the notion of a global 'community of women in computer science' did not exist." Indeed, at the time, women working in the field of computer science found themselves cut off from other women in their field, in labs dominated by men and by male ways of thinking, communicating, and working. Borg argued that women in technology were such a minority that Systers could not function in the way that all-male or all-white or all-Christian clubs had functioned historically: women did not have the advantage of privilege. She preferred to speak of the list as a "place of respite in our journey to equality."

Systers opened the door for other places of respite. Another like-minded mailing list, systers_students was founded at the University of Massachusetts, Amherst, to foster an international community of female undergraduate and graduate students in computer science. Other Systers members have started similar lists for high school students and women of color.

As Systers grew into an international community of more than 2,500 members, the sheer volume of messages generated by the list inspired Borg to shift the focus of her research at Digital Equipment to collaborative databases. She developed MECCA, a program that provided automatic administration of membership-based electronic mail communities like Systers, along with state-of-the-art features such as increased security, privacy filtering, and information filtering. (Systers used MECCA until the list moved to Web-based technology.) For her dedication to both the technological and social success of the list, Borg came to be known as "her Systers keeper."

## 50–50 BY 2020

In "Why Systers?" Borg wondered whether a virtual community like Systers would be necessary in a world where women were 50 percent of the computer science ranks. "When we get there," she wrote, "we can make that decision." Indeed, Systers was just the beginning of Borg's part of the struggle to "get there." In 1994, she and Dr. Telle Whitney, then director of software at Actel Corporation, founded the Grace Hopper Celebration of Women in Computing Conference (GHC), named after Admiral Grace Murray Hopper, a World War II computing pioneer. (Hopper is sometimes called the "mother of the computer.") From the beginning, the GHC has been the largest technical conference for women in computing in the world.

*What if only 30-year-old women developed technology—all of it—and that technology was geared mainly for 13-year-old girls? Technology would be out of whack, out of balance. But that's the world we live in: Men hold the power, and boys drive the market.*

*— Anita Borg, to* Fast Company, *in 1999*

The first conference, held in Washington, D.C., in June 1994, featured speakers on such topics as natural language trees, robots, parallel computing and distributed programs, and workshops that ranged from the "Lesbian and Bisexual Women's Exchange" to "Human-Computer Interaction." Subsequent conferences, held in 1997, 2000, and every two years since, have achieved the same mix of technical and professional aspects of "women in computing." By 2002, the GHC had grown to include more than 600 women from across the globe, a vibrant mix of industry leaders, academics, government researchers, professionals, and students. "We had tapped a need in the computing community," wrote Telle Whitney.

Borg brought her infectious brand of feminism to other industry conferences as well. At a National Science Foundation conference in December 1995, she threw down the gauntlet—"50–50 by 2020," a challenge to have half of all computer science degrees go to women by the year 2020. (The concept of 50–50 by 2020 has since become ubiquitous in women's engineering, math, and other science groups.) It was a bold vision: according to the National Center for Education Statistics, the number of women pursuing bachelor's degrees in computer science had dropped from 37.1 percent in 1984 to 28.4 percent in 1995. The number has continued to drop every year since, to less than 20 percent in 1999.

## BREAKING THROUGH THE SILICON CEILING

In 1997, Borg founded the Institute of Women and Technology (IWT), a nonprofit organization housed at Xerox's Palo Alto Research Center. The IWT began with just three staff members, including Borg, but drew millions of dollars in equipment and support from companies like Hewlett-Packard, Sun Microsystems, Xerox, and Lotus.

The aim of IWT was simple: to increase the impact of women on technology and to increase the positive impact of technology on the lives of women. In addition to housing Systers and the GHC, supporting policies and research regarding women in computing, and encouraging women to pursue computing degrees and professions, the IWT strives to increase the social relevance of technology by bringing nontechnical women into the equation at the level of research and design. "There are many ways that women can impact technology without being *in* technology," said Borg. "We need both kinds of women."

By 1999, Borg had collected both kinds of women for a handful of workshops where participants brainstormed about new, more practical technology, dubbed "Technology in Support of Families." During one workshop, Borg told a group of thirty women ranging from high school juniors to prominent researchers from Silicon Valley tech firms, "You are all experts. This is not a focus group. Each of us is here as an expert in our own experience."

Unlike male engineers who, when asked by a professor at Texas A&M to design a car with women in mind, simply increased the number of vanity mirrors, the women in these workshops came up with practical technology to solve any number of problems: electronic Tupperware "sniffers" that could tell when food went bad, multilayered electronic calendars that kept track of an entire family's schedule (unlike traditional personal digital assistants, which are made for one person), "smart" pipes that alert homeowners if they are clogged or frozen. Some of the ideas generated in Borg's workshops have seeped into the minds of those in power. Greg Papadopoulos, chief technology officer at Sun Microsystems, once joked that Borg had "ruined him for life" by making him obsessed with the idea of smart pipes.

Borg hoped that, by bringing "real women" into the design process at one end, she could affect the number of women choosing careers in technology at the other. As she told the *San Jose Mercury News,* "Part of the image of working on computers is working to create gadgets—techie stuff having nothing to do with people's lives. Young women want to have a positive impact on people. If we can get across that there are powerful ways to have a hugely positive impact on people, then maybe we can turn that image around."

> *[The workplace] is not a gender-free environment. It's nice to be able to yell out to a bunch of women ... "Is it me or what?"*
>
> —Systers member to the Los Angeles Times

Borg continued the brainstorming work begun at the "Technology in Support of Families" workshops through the Institute's university-based Virtual Development Centers (VDC), which brought college students together with local industry women. The first VDC opened at Purdue University in 1999. Eight more centers have since opened, including centers at Texas A&M, MIT, and UC–Berkeley, and more than fifty VDC-generated projects have been implemented in the centers' respective communities.

In 2002, Borg received the $250,000 Heinz Award for Technology, Economy and Employment. It was a cap to the long list of awards and honors Borg had garnered in her twenty-five years in the field of computer science, including the 1995 Augusta Ada Lovelace Award from the Association of Women in Computing, the 1995 Pioneer Award from Electronic Frontier Foundation, induction into the Women in Technology International Hall of Fame in 1998 and, in 1999, appointment to President Bill Clinton's Commission on the Advancement of Women and Minorities in Science, Engineering and Technology. Her greatest achievement, however, was connecting women with other women through Systers, IWT, and GHC. These institutions continue although Borg died of brain cancer on April 6, 2003. As her colleague Ellen Spertus stated at Borg's memorial service, "It's impossible to overstate her role in creating a community of female computer scientists that has helped students and professionals stay in the field, organized to bring more women in, and helped people achieve their full potential."

## FURTHER READING

### In These Volumes

## Works by Anita Borg

"Why Systers?" *Institute for Women in Technology.*
www.iwt.org, 1993. Available online at http://www
.systers.org/about.html (cited September 16, 2004).

National Council for Research of Women. *Balancing
the Equation: Where Are Women and Girls in Science,
Engineering and Technology?* New York: National
Council for Women, 2001.

## Books

Camp, L. Jean. "We Are Geeks, and We Are Not Guys:
The Systers Mailing List." In *Wired Women: Gender
and New Realities in Cyberspace,* edited by Lynn
Cherny and Elizabeth Reba Weise. Seattle, WA:
Seal Press, 1996.

Margolis, Jane, and Allan Fisher. *Unlocking the
Clubhouse: Women in Computing.* Boston: MIT
Press, 2001.

## Articles

Chaudry, Lakshmi. "Building the Digital Systerhood."
*Wired News,* February 15, 2000,
http://www.wired.com/news/culture/0,1284,34175,
00.html (cited September 16, 2004).

Corcoran, Cate T. "Profile: Anita Borg." *Red Herring,*
March 1999. http://www.redherring.com (cited
September 16, 2004).

Mieszkowski, Katharine. "Sisterhood Is Digital." *Fast
Company,* September 1999.

———. "Well-behaved Women Rarely Make History."
*Salon.com,* April 9, 2003,
http://archive.salon.com/tech/feature
/2003/04/09/anita_borg/ (cited September 16,
2004).

———. "Not Just Another 'Nerdy White Guy'."
*Salon.com,* September 9, 2003,
http://archive.salon.com/tech/feature/2003/09/10
/borg_memorial/ (cited September 16, 2004).

## Websites

The ADA Project. Clearinghouse for information and
resources related to women in computing,
http://tap.mills.edu/ (cited September 16, 2004).

Grace Hopper Celebration of Women in Computing.
Home of the largest women in computing
conference, http://www.gracehopper.org (cited
September 16, 2004).

Institute for Women and Technology (IWT). The IWT
is a nonprofit educational and research organization
dedicated to increasing the impact of women on
technology and the positive impact of technology
on women's lives. Original site, http://www.iwt.org
(cited September 16, 2004), newer site, http://www
.anitaborg.org (cited (September 16, 2004).

Systers: Online Community. The oldest virtual
community and mailing list for women in
computer science, founded by Anita Borg in 1987,
http://www.systers.org (cited September 16, 2004).

# Len Bosack (1952–) and Sandy Lerner (1955–)

## FOUNDERS OF CISCO SYSTEMS

**In 1984, Len Bosack** and Sandy Lerner, two college sweethearts from Stanford University, founded Cisco Systems. The company was built around a multiprotocol router, new type of networking technology. For several years, Bosack and Lerner built and sold routers out of their Silicon Valley living room. But as networking caught on, so did Cisco. In 1987, when venture capitalists and new management finally came on board, the company really began to blossom, and, in 1990, Cisco held a successful initial public offering (IPO) of stock. About six months later, Lerner was forced out, and Bosack soon followed. Now one of Silicon Valley's most successful companies, Cisco Systems has gone on to control between 80 and 90 percent of the router market, expanding its reach to include "anything that builds a network."

## THE DYNAMIC DUO

Len Bosack and Sandy Lerner met in 1977 in a Stanford University computer lab. Bosack, an alumnus of the University of Pennsylvania's electrical engineering program, was working on a master's degree in computer science. Lerner had transferred from Claremont College to study graduate-level statistics and computer science and was one of the only women to hang out in the lab. She quickly became one of the boys, and, to her, Bosack stood out from the rest. "Nerd culture at Stanford was pretty extreme," Lerner said in a 1997 *Forbes* interview. "There was no way I could have taken one of these people home to meet my family. But Len's clothes were clean, he bathed, and he knew how to use silverware. That was enough. I was enchanted."

The two married in 1980. Bosack accepted the position of director of computer facilities for Stanford's computer science department. Lerner earned her degrees in 1981, and then took the job of director of the computer facilities for Stanford's graduate school of business. Though they held similar positions and shared a passion

for computers and technology, Bosack and Lerner were wildly dissimilar: Bosack, shy and reserved, Lerner, aggressive and outspoken—but the relationship worked.

## A FAILURE TO COMMUNICATE

Lerner's department at the graduate school lay just 500 yards across campus from the computer science lab where Bosack worked. Each department had its own computer network through which emails and documents flowed. But the networks were not connected to each other. In fact, in the early 1980s, Stanford University had nearly two dozen incompatible email systems, and many of its 5,000 computers had no way of connecting to one another, aside from accessing the ARPANET (the predecessor of the Internet). That, however, required the use of a special computer—an Interface Message Processor (IMP)—which cost nearly $100,000 and was often in need of repair.

Lerner once described the computer situation at Stanford as "complete bedlam." Unable to communicate even 500 yards away, she, Bosack, and a handful of colleagues resorted to building a "guerilla" network of their own, weaving networking cables through manholes and sewer pipes to put one lab's computers in touch with another's. This gave way to a larger project in which, according to legend, Lerner and Bosack, intent on sending each other love notes via email, invented technology that could network the whole campus. In truth, the desire to connect Stanford's computer networks had been sparked years before.

In late 1979, Xerox Corporation gave Stanford (as well as the Massachusetts Institute of Technology and Carnegie Mellon University) a very important Christmas gift: Alto computers and Ethernet equipment. Both the Alto, considered the first personal computer, and Ethernet, a local area network (LAN) system for connecting computers, were cutting-edge equipment developed at Xerox's Palo Alto Research Center (PARC).

The Alto and Ethernet equipment were given to computer labs in Stanford's medical school and the computer science department. Ralph Gorin, then the head of the computer facilities at Stanford, pulled together a group of computer whizzes—mostly graduate students and research staff—to connect the medical school to the computer science lab, and then to the other computer labs on campus. Because nothing of the sort had been done before, Gorin asked for what he called a "network extension cord." What he got was something quite different—the multiprotocol router.

## ROSES ARE RED, ROUTERS ARE BLUE

Routers are an integral part of any Ethernet; even in the early 1980s, many computer companies had built proprietary routers to network their own brand of computers. What Gorin's request prompted was the development of new *universal* routing technology that could connect computers regardless of brand or computing platform. These devices came to be known as gateways or, more commonly, blue boxes. (Stanford's "blue box" routers should not be confused with another kind of blue box that emerged in the early 1970s. These other—perhaps more famous—blue boxes were tone devices that early hackers, known as "phone phreaks," used to gain access to the long-distance telephone system. Steve Jobs and Steve Wozniak, founders of Apple Computer, are famous for building and selling blue boxes when they were college students in the mid-1970s.)

Bosack helped build the boxes, while the internal computer board was built by Stanford student Andy Bechtolsheim. (Bechtolsheim later founded, with others, Sun Microsystems.) The software, perhaps the most important aspect of the technology, was developed by Bill Yeager, a research engineer based in the medical school lab. Yeager had already used Ethernet technology to connect computers at the medical school to those in the computer science department. By June 1980, he also had developed software that could connect the computer science department's Alto to the DEC minicomputers housed in the computer lab at the Graduate School of Business. By the summer of 1982, Yeager had

*Our myopia made us stronger.*

—Lerner to the Electronic Engineering Times, June 1989

tweaked the software to also route the emerging Internet protocol (IP), which is part of the TCP/IP protocol on which the Internet still runs. Once Yeager had accomplished that, not only could the multiprotocol router connect different LANs to each other, it could connect those LANs to the ARPANET, and, thus, to the rest of the research world.

Soon, Yeager, Bosack, and others had built more than twenty blue boxes, which were scattered across Stanford's sixteen-square-mile campus. It was one of the first successful wide-area network (or "WAN") experiments. When other departments—and then other universities—began to express interest in the blue boxes, Stanford's computer facilities department, where Bosack and his colleague Ken Lougheed were on staff, struggled to keep up with demand.

Around this time Bosack and Lerner realized that blue boxes could form the basis of a business. "We just walked up to the secretary of state's office with our $25, or whatever it was, and filled out papers," Lerner told *Electronic Engineering Times* in 1989. "I think we did not have a very good idea of what we were doing." In December 1984, they set up shop in the living room of their suburban home in Atherton, California, calling their company Cisco—short for San Francisco.

## PARTING WAYS

Both Bosack and Lerner worked at Stanford during the day and built their fledgling company at night, financing much of it on their credit cards. The first router took them nearly two years to devise and assemble.

During those two years, they struggled at Stanford. In early 1986, someone told one of Bosack's superiors that Bosack had used university time and resources to finance Cisco. When the university investigated, it appeared that Bosack had sold Xerox networking boards allegedly made at Stanford's expense and had not yet repaid the university. By May 1986, the issue had reached the dean of the school. Two months later, Bosack and Lougheed resigned from their positions at Stanford, joining Lerner and two other colleagues, Greg Satz and Richard Troiano, at Cisco's living room offices.

Lerner and Bosack had already proposed that Stanford should sell the routers, but as a nonprofit the university could not enter the commercial market. Lerner then began to feel as though Stanford was holding the technology hostage. Indeed, as Cisco continued to grow, the university clung to the research that had gone into developing the routing technology, at one point demanding $11 million to license the blue box technology to Cisco. Some university officials even urged a lawsuit. Bill Yundt, then the head of Stanford's computer facilities, was wary of legal action. In his view, it was still debatable whether the university could prove that Bosack and others had done anything wrong. So, in April 1987, Stanford licensed the software and hardware to Cisco for $19,300, with the additional promise from Cisco of $150,000 in royalties and considerable product discounts.

When Cisco began selling routers, the market was quite small—primarily other educational institutions and fellow engineers who were on the ARPANET—all of whom knew the source of the networking technology. "Everybody in the world who knew anything, knew this stuff had been done at Stanford," said Yundt in a 2001 interview with the *San Jose Mercury News.* Shortly after, however, that knowledge faded from memory. Once Cisco had cornered the market on routing technology, the true origins of the router were often obscured. Routers, according to the legend, were developed by two lovebirds who wanted to send each other email; names like Bill Yeager, for example, did not come up. In their defense, while Bosack and Lerner may have been guilty of the sin of omission—not always setting the record straight—neither openly claimed to have invented the router on their own. Indeed, both had often acknowledged the huge collaborative effort that went into creating Cisco's routers. In an email to the *San Jose Mercury News,* Lerner quipped, "The only person I'm certain had nothing to do with it is Al Gore."

## THE LEAN YEARS

In the midst of the threat of a lawsuit with Stanford, Bosack and Lerner faced struggles of their own at home. In a 1992 interview with *Fortune,* Lerner recalled that they had started Cisco "without a particular business vision." Bosack largely handled the technology side, while

> *I certainly hope to give a billion dollars to charity before I'm all done.*
>
> —*Bosack to* Nerds 2.0.1, *1998*

Lerner focused on customer services and the business aspect of the fledgling company; they kept different hours in order to keep Cisco running almost around the clock. The routers were developed in the bedroom and built in the living room. The couple kept a phone by the bed to answer 2:00 A.M. customer service calls. Meetings of Cisco's nine employees were held at the dining room table. The two-car garage was filled with a used mainframe computer, purchased on a credit card for $5,000. In fact, for two years, while the first routers were developed, the entire operation had been funded on credit.

After Bosack and Lougheed quit Stanford, Bosack lived without a car, and, during a particularly lean period in 1986, Lerner went back to work as a manager at a computer data processing lab. It was a difficult time, not just for Cisco, but for many companies in high technology. Venture capital to help finance new businesses was nowhere to be found. Cisco faced additional pressure when other companies, such as 3Com, moved into the realm of networking; "Clearly," Lerner informed the *Electronic Engineering Times,* "we were the ninth bridesmaid at the wedding." Meanwhile, the market was still relatively unclear. "In the academic community, no one believed that it was possible to build a good-performance, reasonably priced, multi-protocol router, and in the commercial world, no one knew they wanted one," Bosack told *InternetWeek* in 1991.

Nevertheless, by 1987, having sold their first router only a year earlier, Cisco was selling $250,000 worth of equipment each month. (The company was still cash-poor, however.) The primary customers were universities and government agencies, which were all connected via the ARPANET. But with the mainstreaming of the Internet (as early as 1987) and the subsequent explosion in corporate computer networking, the lean years would soon be over, and Cisco would be well on its way to becoming a multimillion-dollar company.

## ON THE OUTS

In 1987, Cisco found its first venture capitalist—Don Valentine, the founder of Sequoia Capital, a well-established Menlo Park firm that had invested early in Apple Computer. Bosack and Lerner had shopped Cisco around to various potential investors for a while, but more than seventy-five had already turned them down. The three-year-old company's $250,000 to $350,000 per month revenues, however, were enough to convince at least Valentine to take a chance. "The venture community are very practical people," Bosack told the *Electronic Engineering Times.* "If you're still in business, same concepts, year after year, you earn, if not their respect, their polite attention."

Valentine helped secure $2.5 million in first-round financing for Cisco and, in exchange, took over management of the company. Indeed, Valentine told them, "It's time to get real. Here's how other companies do it." By May 1998, Valentine had pushed out longtime executives in favor of a new vice president of sales and, most important, a new president and CEO, John Morgridge. Bosack and Lerner became, respectively, the chief technologist and vice president of customer services—having relinquished considerable control over their company.

Under Valentine's management, Cisco experienced skyrocketing growth, from $500,000 in sales in May 1998 to more than $1 million in August. But trouble was brewing inside the company, particularly between Lerner and the new regime. As the entrepreneurial half of the founding couple, Lerner had long defined the way Cisco was run. She and Morgridge were at odds from the beginning. In the documentary *Nerds 2.0.1,* Lerner said that Morgridge's first words to her were, "I hear that you're everything that's wrong with Cisco." (Morgridge has denied ever making that statement.) But under the agreement Lerner and Bosack had hastily signed with Sequoia Capital, in which Valentine retained the right to force the founders out at will, there was little room to challenge Cisco's new authority.

Although interpersonal dramas continued to rage at the office, in 1988, Cisco released the highest-speed network interface in the industry; by the end of 1989, it reported $4.2 million in profits. Toward the end of 1989, Morgridge, Lerner, Bosack, and Valentine could finally agree on one thing—Cisco should go public. Cisco held its initial public offering (IPO) of stock in February 1990. Shares opened at $18.00 and closed at

$22.50—then went on to become one of Wall Street's best-performing stocks.

Within six months of the IPO, however, seven of the company's vice presidents had approached Morgridge, and then Valentine himself, with an ultimatum—Lerner goes, or we go. The animosity between the old guard—Lerner and Bosack—and the new had increased exponentially. Finally, on August 28, 1990, Lerner was asked to leave the company she had founded six years earlier. Shortly thereafter, Bosack resigned from the board of directors in support of his wife and then left Cisco entirely.

## NOT THE RETIRING TYPES

Upon leaving Cisco, Bosack and Lerner sold their two-thirds stake in the company—about $170 million dollars worth of stock. They then moved to Redmond, Washington, in part because it reminded them of Silicon Valley in the late 1970s and early 1980s. In 1991, they founded an umbrella organization, Ampersand Capital, to fund their future business and charitable activities.

That year, Bosack founded a company, XKL Systems, and brought several original members of Cisco on board. The engineers set to work on building devices to improve the input-output ability of computer memory.

Together, they also founded the Leonard X. Bosack and Bette M. Kruger Charitable Foundation, named for his father and her mother. Endowed with $32 million, the foundation funded projects that reflected Bosack and Lerner's respective interests—"weird science," including a Harvard-based project on the search for extraterrestrial intelligence, and animal welfare, including an antipoaching air patrol program in West Africa. Bosack and Lerner handpicked the projects—often, the more obscure or underfunded, the better.

Financially secure, Bosack and Lerner might seem to have had the easy life after Cisco. However, giving birth to and then losing Cisco had taken its toll on the marriage. "I don't think people understand what 'give up everything' means," Lerner told the *Washington Post* in 1990, talking about the sacrifices the couple had made—personal and otherwise—to get Cisco up and

> *We had to ask ourselves the question, "Do we want to run a company or do we want to make money?" For us, it was the latter.*
>
> —Bosack to Forbes, March 1992

running. After an amicable divorce, Lerner left the West Coast to live on a 1,000-acre farm in Virginia, where she could raise rare breeds of animals, farm organically, and mix music in her private sound recording studio.

In October 1992, at a meeting of the Jane Austen Society of North America in Santa Monica, Lerner found out that Chawton House, an English manor once owned by Jane Austen's family, was available for purchase. (Lerner had discovered Jane Austen while at Stanford; she joined a Jane Austen fan club on campus, and devoured all of Austen's novels, rereading her favorite, *Persuasion,* more than seventy times.) The current owner of the dilapidated 17th-century estate, Richard Knight, a descendant of the Austen family, could barely afford to keep the land. To him, Lerner was, as he told the *Los Angeles Times* in 1998, "a knight in shining armor, galloping to the rescue." Indeed, by 2003, after eleven years of work, she had renovated the estate and founded the Center for the Study of Early English Women's Writing. In that time, Lerner also amassed one of the world's largest libraries of women's writing from the period between 1600 and 1830.

Soon after her fortieth birthday, Lerner added a cosmetics company to her idiosyncratic list of endeavors. One of Lerner's aunts suggested it was time for the still-tomboyish Lerner to start wearing makeup. But Lerner was frustrated with the pinks and reds of traditional nail polish and lipstick. "Fundamentally, I was just pissed off that [cosmetics firms] were telling women they had to look like Barbie," she said in a 1997 *Forbes* interview.

Lerner's answer to Barbie was a cosmetics line, Urban Decay, that featured lipstick and nail polish in edgy colors with names like Bruise (a purple hue), Road Stripe (a white), and Acid Rain (a green). She founded the company in 1995, with the help of Wende Zomnir, and launched an ad campaign that asked, "Does pink make you puke?" In its first year of business, Urban Decay grossed $5 million in sales.

Bosack has continued to work in the technology field and to support charitable endeavors. He remains especially interested in the search for intelligent life in space, telling *Nerds 2.0.1:* "It's one of the most important questions that a sentient being can ever formulate, and that is: Are we alone? Either answer, if you could obtain it, is of tremendous import."

Although both Bosack and Lerner founded other companies and embraced new lives as eccentric philanthropists, they remain bitter about leaving their company behind. "Len and I underestimated our skills," Lerner told *Forbes* in 1992. "I certainly could have run that business. I had my hands on the reins."

## CISCO TODAY

Soon after Lerner and Bosack left Cisco, Valentine brought aboard a new executive, John Chambers, whose brash and sometimes unpredictable management style has come to define Cisco's history ever since.

By the late 1990s, Cisco was among the most valuable companies in the United States, and, for a brief moment in March 2000, it surpassed Microsoft as the most valuable company *in the world.* (Microsoft's stock had dropped because of uncertainties surrounding its antitrust case.) Cisco, which had taken over various smaller companies and reinvented itself as an "anything that builds a network" company, was easily either the number one or number two competitor in almost every segment of the networking market—from traditional routers to advanced gigabit Ethernet technology. In most cases, Cisco cornered 80 to 90 percent of the market.

Chambers is most famous for steering Cisco successfully through the rough waters of the Internet bust, when sales and stock prices both plummeted. (Indeed, the drop in stock value was so brutal—from $80 to $14 per share—that Chambers began to keep bodyguards.) After intensive cost-cutting measures, Cisco bounced back and remains one of the top Internet technology companies today.

Two decades after Len Bosack and Sandy Lerner founded Cisco in their living room, Cisco has provided an ironic twist. Under the "Corporate Overview" section of its Website, the company has stated, "Cisco was founded in 1984 by a small group of computer scientists from Stanford University," obscuring Bosack and Lerner's role in founding the company, much as Yeager and others' roles had once been obscured. Still, Lerner and Bosack should be credited, if not for developing the router itself, then for having the foresight and courage to bring the router to market and help usher in the age of the commercial Internet.

## FURTHER READING

### *In These Volumes*

Related Entries in this Volume: Metcalfe, Robert

Related Entries in the Chronology Volume: 1990 (sidebar): Battle of the Networked Stars

## Books

Bunnell, David. *Making the Cisco Connection: The Story Behind the Real Internet Superpower.* New York: John Wiley & Sons, 2000.

Segaller, Stephen. *Nerds 2.0.1.* New York: TV Books, 1998.

Slater, Robert. *The Eye of the Storm: How John Chambers Steered Cisco Through the Technology Collapse.* New York: HarperBusiness, 2003.

Stauffer, Dave. *Nothing but Net: Business the Cisco Way.* Milford, CT: Capstone, 2000.

Waters, John K. *John Chambers and the Cisco Way: Navigating Through Volatility.* New York: Wiley, 2002.

## Articles

Asirvatham, Sandy. "Fast Lerner—Sandy Lerner." *Success,* October 1997.

Bellinger, Robert. "Sandy Lerner and Leonard Bosack of Cisco Systems Inc.: No Venture Capital, No Offices, Just Customers and Sales." *Electronic Engineering Times,* June 19, 1989.

Carey, Peter. "A Start-Up's True Tale." *San Jose Mercury News,* December 2, 2001, http://pdp10.nocrew .org/docs/cisco.html (cited September 22, 2004).

Clark, Don. "Cisco Systems Soars to Top Spot." *San Francisco Chronicle,* April 20, 1992.

Cringley, Robert. "Valley of the Nerds: Who Really Invented the Multiprotocol Router, and Why Should We Care?" *PBS.org,* December 10, 1998, http://www.pbs.org/cringely/pulpit/pulpit19981210 .html (cited September 22, 2004).

Flower, Bill. "The Cisco Mantra." *Wired 5.03,* March 1997.

Grice, Elizabeth. "Money's My Little Defining Thing." *Telegraph* (London), June 27, 2003, http://www .telegraph.co.uk/arts/main.jhtml?xml=/arts/2003/06 /27/baft27.xml (cited September 22, 2004).

Kirsner, Scott. "Nonprofit Motive." *Wired 7.09,* September 1999, http://www.wired.com /wired/archive/7.09/philanthropy.html (cited September 22, 2004).

Linden, Dana Wechsler. "Does Pink Make You Puke?" *Fortune,* August 25, 1997, http://www.forbes .com/forbes/1997/0825/6004058a.html (cited September 22, 2004).

Long, Katherine. "Fund Mission." *Seattle Times,* December 28, 1993.

Pitta, Julie. "Long Distance Relationship." *Forbes,* March 16, 1992.

Schofield, Jack. "Cisco Kids Ride High." *The Guardian,* April 20, 2000.

Sterngold, James. "Setting Her Sights on a Pinkless Palette." *New York Times,* November 12, 1995.

Tawa, Renee. "Lady Godiva." *Los Angeles Times,* March 1, 1998.

Weeks, Linton. "Network of One." *Washington Post,* March 25, 1998.

Workman, Bill. "A Woman of Means." *San Francisco Chronicle,* September 20, 1996.

## Websites

Chawton House Library and Study Center. Online home of Chawton House, one of Lerner's organizations, which celebrates women's writing before 1830, http://www.chawton.org/ (cited September 22, 2004).

Cisco Systems, Inc. Online home of Cisco Systems, includes some corporate information, http:// newsroom.cisco.com/dlls/corporate_timeline.pdf (cited September 22, 2004).

Nerds 2.0.1—Cisco. Online version of the segment of the documentary *Nerds 2.0.1* on the founding of Cisco Systems, http://www.pbs.org/opb /nerds2.0.1/serving_suits/cisco.html (cited September 22, 2004).

XKL, LLC. Website for XKL Systems, founded by Bosack, http://www.xkl.com/ (cited September 22, 2004).

# Stewart Brand (1938–)

## FUTURIST; AUTHOR; COFOUNDER OF THE WELL ONLINE COMMUNITY

**Before founding the** influential online community called the WELL (Whole Earth 'Lectronic Link) in 1985, Stewart Brand was best known as the founder, publisher, and editor of the *Whole Earth Catalog,* an irregularly published, wholly unique compendium of tools, reviews, and writing known to many as the "hippie Bible." An accomplished journalist and author since the early 1970s, Brand has displayed a knack for being at the forefront of new ideas—particularly in the early days of computing. Indeed, in 1972, Brand was arguably the first person to use the term *personal computer* in print, several years before Apple Computer developed the first Macintosh. He was also one of the first journalists to explore the world of early computer gamers and hackers, a decade before any other author addressed the subject. Still, to many in the Internet world, the WELL is, perhaps, Brand's most important legacy, as it became the blueprint for a successful, vital, and virtual community in the 1980s and 1990s.

## IF YOU'RE GOING TO SAN FRANCISCO . . .

Brand was born and raised in Rockford, Illinois, the youngest of four children born to an MIT-educated father and a Vassar-educated mother. After attending Phillips Exeter Academy, an elite East Coast prep school, Brand enrolled at Stanford University in Palo Alto, California, where he first got a taste for ecology. He graduated with a degree in biology in 1960, then spent two years in the army, mostly as a photographer for the Pentagon. Though he was drawn to the burgeoning counterculture movements of the early 1960s, Brand did not disparage his military experience. In a 2001 interview with the British newspaper, *The Guardian,* Brand explained, "At government expense, I was trained in leadership and small-unit management."

Those were vital tools for a man who would come to lead many organizations throughout his career.

In 1962, Brand moved to San Francisco to study design and photography. He took various jobs, including a summer stint as a photographer in central Oregon, where he documented a Native American reservation. His work there and on other reservations became a multimedia project entitled, "America Needs Indians"—the first of several multimedia events that Brand would stage throughout the 1960s. Brand also delved into San Francisco's psychedelic underground, participating in the then-legal LSD studies at the International Foundation for Advanced Study and the "acid tests" hosted by Ken Kesey, a Bay area counterculture guru (and widely known author of *One Flew Over the Cuckoo's Nest*) who was known for giving LSD parties as a way of freeing one's mind.

By the late 1960s, Brand was a vital figure in the San Francisco youth culture. Indeed, at age 27, he organized the Trips Festival, one of the first large-scale rock concerts and light shows, which helped launch the career of the influential hippie rock band, the Grateful Dead. Journalist Katherine Fulton, writing for the *Los Angeles Times* in 1994, linked the Trips Festival to projects in Brand's later life—including the *Whole Earth Catalog,* the WELL online community—claiming that the structure of each project was signature Brand. His recipe, she claims, was to "toss interesting people in the pot and turn up the flame, then dive in."

## FROM LSD TO ASCII

The idea that would later make Brand a household name in certain circles began during an acid trip in 1966. Having taken a small amount of LSD, Brand went up to his roof, in the North Beach neighborhood of San Francisco, and while looking at the skyline noticed that the

*Stewart Brand. (Ted Streshinsky/Corbis)*

astronomer Fred Hoyle was right in 1947 when he forecast, 'Once a photograph of the Earth, taken from outside, is available . . . a new idea as powerful as any in history will be let loose.'"

A small part of the new idea unleashed by the photograph of the Earth was Brand's *Whole Earth Catalog.* By the late 1960s, many of Brand's peers were interested in starting communes. They were college-educated hippies who lacked the farming and building abilities necessary to create their utopian communities. As Brand would often say, "The spirit was willing, but the skills were weak."

In 1968, to address this knowledge gap, Brand published the *Whole Earth Catalog*—part how-to book, part back-to-basics Sears catalog, part counterculture primer. Brand started by writing about the tools and skills he knew himself, then opened the book up to suggestions from anyone with anything to add. The edited version was an eclectic and exhaustive tome on everything from composting to building a yurt, a Mongolian tipi-like structure.

While the catalog's primary influence was the hippie spirit it promulgated, Brand, and others, believes its influence spread to publishing itself—both in print and online. Brand has suggested that the *Whole Earth Catalog* was the first instance of desktop publishing; the editorial team used an IBM Selectric Composer, which allowed for different fonts, and a 1960s Polaroid camera that functioned like a modern-day photocopier, which made copying and pasting images a snap. Brand later likened the collaborative, customer-generated endeavors of the Internet age, such as the Google search engine, to the process of publishing the catalog. Kevin Kelly, a former *Whole Earth* editor, says he borrowed heavily from the catalog when helping to launch *Wired* magazine in 1993. Alan Kay, an early scientist at Xerox's Palo Alto Research Center (PARC), said the catalog's layout and design—in particular, the way Brand once published an entire novel in sidebars that were embedded on each page of the catalog —helped inspire the development of the first PARC browser, an intellectual predecessor of Netscape Navigator.

The first edition of the *Whole Earth Catalog*, published in 1968, cost $1 and had only twenty-one pages. By 1971, the catalog had grown to 447 pages and sold more than 1.5 million copies. It garnered a National Book Award in 1972. By then, however, Brand had burned out. "I went into pretty severe depression and

downtown office buildings were not parallel. He began to think of how Earth's surface curved and wondered what the world would look like from higher than three stories above ground—from space, even. It had been nearly a decade since *Sputnik,* the first satellite, had been launched into orbit, so a view of Earth from space was not an impossibility. Soon after, Brand was wearing and selling buttons that read, "Why Haven't We Seen a Photograph of the Whole Earth Yet?"

Some people believe that Brand's "Whole Earth" project helped accelerate NASA's efforts to photograph the Earth. By 1969, American astronauts on *Apollo 8* had returned with the first color photos of the Earth from outer space. A year later, the first Earth Day was celebrated and the ecology movement in America was, in effect, born. Brand believes it was no coincidence. In 2000 he told the Denver *Rocky Mountain News,* "The just kept working," he told the British newspaper *The Guardian* in 2001. "Then I had the idea that we could just quit." Brand then founded the Point Foundation, a nonprofit organization that, in addition to its philanthropic activities, was the home base for all Whole Earth activities for the next thirty years, including the magazine *CoEvolution Quarterly* (later known as the *Whole Earth Review*) and subsequent *Whole Earth* catalogs.

At the same time as the Whole Earth enterprise captured the imaginations of hippies, Brand was exploring the nascent computer culture in and around the Bay area. In the early 1960s, a version of the first computer game, Space War, had made its way from the computer hackers at MIT and landed at Stanford. When Brand witnessed the gamers, he was taken aback. He later told *The Guardian,* "What I saw was an interaction around computers that was as intense as anything I saw around drugs or anything else that I knew. People were absolutely out of their bodies playing. It seemed that computers were doing everything that drugs had promised. Drugs were much more self-limiting than computers: the hackers had found something better than drugs, but theirs was the same bohemian frame of reference." From that moment, Brand was hooked.

In 1969, Brand collaborated with Doug Engelbart, a visionary computer scientist known for developing the first mouse, in a demonstration of Engelbart's pioneering computer project, called Augmented Human Intellect. In the early 1970s, Brand went on to write two prescient articles about computers: a 1971 *Rolling Stone* article entitled "The Fanatic Life & Symbolic Death Among the Computer Bums," in which he wrote "Ready or not, computers are coming to the people"; and a 1973 *Harper's* article entitled "Unbinding Conversations with Meta-naturalist Gregory Bateson," about cybernetics. In 1974, Random House published Brand's book, *II Cybernetic Frontiers,* which expanded on both the *Rolling Stone* and *Harper's* articles. The book is believed to be the first instance of the words *personal computer* appearing in print—a year before the first microcomputer, the Altair 8800, debuted on the market.

Brand spent the late 1970s and early 1980s wearing a variety of hats—in the mid-70s, he was editor of *Space Colonies,* a book about building communities in outer space, and *Soft Tech,* another book about energy conservation. He also served as an adviser to Jerry Brown, then governor of California. Brand revived the *Whole Earth Catalog* in 1980 and, three years later, accepted an astounding $1.3 million advance—the largest advance in paperback history to that date—from Random House for the *Whole Earth Software Catalog* (1984), which combined Brand's initial Whole Earth spirit with the personal computer movement. The book "failed conspicuously," as Brand puts it, nearly bankrupting the Point Foundation by the mid-1980s. Still, in 1995 Brand told *Fortune,* "[T]he disaster was tremendously freeing for me . . . because it meant that it was okay for me to leave."

## DIGGING THE **WELL**

In 1984, with the *Whole Earth Software Catalog* debacle well under way, Brand met Larry Brilliant, a physician and entrepreneur, at a conference in La Jolla, California. Brilliant wanted to start an online teleconferencing system and thought Brand was the man to help him do it. (Indeed, Brand had established himself in business circles as a risk taker.) Brilliant had the technology—from Network Technologies International (NETI), his computer conferencing company—but needed a community. The people who contributed to, published, and read the Whole Earth publications seemed, to Brilliant, to be the right fit. Indeed, Brand and others at the Whole Earth offices already had some experience with teleconferencing through the Electronic Information Exchange System—known colloquially as EIES (pronounced "eyes")—and some of the electronic bulletin board systems (known as BBSs) that had cropped up since the late 1970s.

Brand accepted Brilliant's offer; NETI and the Point Foundation would each own 50 percent of the business. NETI provided the VAX computer, modems, and software (an investment of $250,000) and Brand provided the people and the name—the WELL, for Whole Earth 'Lectronic Link. Together, Brilliant and Brand hashed out the structure of the WELL.

Brilliant initially wanted the WELL to be a real-time *Whole Earth Catalog,* but Brand had other things in mind. In order to set the WELL apart from the dozens of

> *Hackers are hippies who got it right.*
>
> —*Brand to* Fortune, *October 1995*

BBSs already online in the Bay area, Brand wanted it to be more like a French literary salon. There would be conferences on various discussion-inspiring subjects, hosted by a mediator, with smaller topic-based conversations branching off each conference. Unlike BBSs, anonymity would not be permitted—each member of the WELL would post an online biography, which was linked to a person's actual login name. Brand reinforced this concept with the credo, "You own your own words," which raised the ante on personal responsibility.

The technology employed by the WELL, a conferencing software called PicoSpan, further enforced Brand's credo. Postings could not be erased. Instead, entire conversations were archived for all to read. PicoSpan did not support filters, either. As journalist Katie Hafner wrote in *Wired* in 1997, "PicoSpan didn't simply foster openness but forced it on users." The software was difficult to use, favoring those who were familiar with UNIX-based software, and, as it was text driven, who were fast typists. Critics called it cryptic and, later, arcane. Steve Jobs, a founder of Apple Computer, called it the ugliest interface he had ever seen. For those who could master the system, however, PicoSpan was a great way to follow group thought. Because PicoSpan was based on the clothesline model of conferencing, each message was attached directly below the main topic, so that "conferences" could be read like conversations.

While Brand designed the "environment" of the virtual community that would become the WELL, Brilliant tended to the business aspects. Membership would cost $8 per month, with a $2 hourly rate. To accommodate heavy users, Brand developed a plan for the hourly rate to decrease the longer the user stayed online. Other commercial teleconferencing systems of that day cost as much as $25 per hour, which meant, in some ways, that the WELL was priced for the masses. (Once virtual communities became the norm, however, the WELL would come to be seen as expensive.) The idea was to make it accessible.

The WELL first went online, in March 1985, with six phone lines, six modems, and a VAX computer

> *The Internet is a runaway plague.... It's a self-accelerating system that makes itself go faster and faster and faster and faster. And if it's not framed properly, it will "overleap itself," as Shakespeare would say.*
>
> —Brand to Fast Company, May 2000

housed in the Point Foundation's dilapidated offices in Sausalito, California, a gentrifying fishing village just across the Golden Gate bridge from San Francisco. Initially, the only users were Whole Earth and NETI employees. It opened to the public on April 1, 1985, with little fanfare beyond a notice in the *Whole Earth Review.*

In the mid-1980s not many had the equipment or abilities required to participate in the WELL—a PC, a minimum 2400-baud modem, and a rudimentary knowledge of UNIX-based software. Soon, however, more people came online. The members were predominantly white, predominantly male, left-leaning baby boomers—a well-educated mix of activists, artists, writers, journalists, and teachers. In just a few short months, a handful of different conferences had emerged—from the sexuality conference, which, predictably, garnered quite a bit of attention, to more esoteric conferences that covered UNIX or the Mind (the Mind was Bay area writer Howard Rheingold's conference on how the mind works). Soon, in the words of journalist Andrew Brown, the WELL had "attracted almost everyone who had survived the sixties with anything interesting to say."

As the WELL found its footing, Brand started The Hacker's Conference, a yearly meeting of some of the most important and innovative minds in computing—early MIT-based hackers, many of whom were also WELL members. (Brand used *hacker* in the original spirit of the word, which was someone who explored computers, not as a synonym for computer vandal.) By 1985, Brand had stepped down as the editor of Whole Earth publications, and, in 1986, he headed east to Cambridge, Massachusetts, as a visiting professor at MIT. In 1987, he wrote a book about MIT's advancements in artificial intelligence and technology, titled *The Media Lab: Inventing the Future at MIT.*

Beginning in early 1986, with Brand across the country, the WELL began to show signs of financial instability. The existing VAX computer was not powerful enough to handle the influx of users—if more than eight or ten people logged on simultaneously, traffic was unbearably slow. The business itself was operating at a loss, until an unexpected group of WELLbeings—as many members called themselves—joined the fray. Starting as early as November 1985, three devoted fans of the Grateful Dead—David Gans, Mary Eisenhart, and Bennett Falk—began encouraging other fans to visit the "virtual village green" for Grateful Dead fans on the WELL. Every week, Gans made a similar announcement on his radio show, the *Deadhead Hour,* on KFOG. "You don't have to be a computer person," he announced, "just a person with a computer."

By the end of 1987, the WELL had grown to more than 2,000 users—nearly one-half of whom were Deadheads—and had turned a small profit. The WELL also posted a small profit the following year, but the aging technology continually weighed down the community. The WELL's managers asked users to prepay some of their fees in order to upgrade the system, and the community responded by pledging $28,000 toward buying a new Sequent computer. By the end of 1989, with membership near 3,000, the WELL was able to expand its staff to nearly six people.

By the WELL's fifth anniversary, it had firmly established its identity as a raucous, tightly knit, highly opinionated, and relatively diverse virtual community. Members would spar in conferences, often hurling cutting remarks at one another through the ether, then pull together to support each other through bad times. WELLbeings helped pay a teenager's tuition, replaced books lost in a house fire, and were virtual witnesses to the play-by-play of one member's lifelong search for his birth mother. WELL members discussed and responded to births, marriages, love triangles, bitter break-ups, and even suicide (details of which sometimes made it into mainstream press). As Katie Hafner wrote in her 1997 profile of the WELL in *Wired,* "It began to dawn on people that this was the sort of thing that happened in a small town."

The WELL continued to grow throughout the 1990s; however, the intimacy once characteristic of the group began to wane. In 1991, with the WELL in fi-

*Although people usually remember my most famous remark, "Information wants to be free"—if not always that I made it, they always forget that I went on to say, "Businesses want information to be expensive."*

—Brand to Red Herring, June 1999

nancial straits once again, Bruce Katz, the Rockport Shoe millionaire, bought NETI's half of the business for $175,000. Three years later, he acquired the rest of the company from the Point Foundation. The shift in ownership echoed a shift in the nature of the community—in 1992, the WELL connected to the Internet, which increased membership and put further pressure on the existing technology. The community grumbled. Many distrusted Katz and saw him as an outsider—a fate shared by many who tried to crack the inner circle of WELLbeings. Some members who disdained Katz's profit-driven business plans left to form the RIVER, another virtual community that was to be run as a co-op. (The RIVER never actually gained momentum as a community because its members did not completely break from the WELL.)

By its tenth anniversary, WELL membership had blossomed to 11,000. In comparison to the bustling virtual communities supported by companies like America Online, the WELL was barely a dot on the map; however, what it lacked in numbers, it made up for in influence. The WELL was where some of the earliest discussions and arguments about key issues in cyberspace—free speech, privacy, and anonymity—took place. (Many of these concerns were exported to the world of print by the many journalists on the WELL, who—thanks to Brand's early insight—were given free accounts.) Such issues were debated by a virtual (pun intended) "Who's Who" of Internet visionaries and entrepreneurs on the WELL, including John Perry Barlow, the first person to describe the Internet as "cyberspace"; John Gilmore, one of the earliest employees of Sun Microsystems; Mitch Kapor, founder of Lotus Development Corp.; Kevin Kelly, executive editor of *Wired;* and Howard Rheingold, virtual community guru, journalist, and author. Indeed, history was made on the WELL. It was the site of the founding of the Electronic Frontier Foundation, the first civil rights organization for cyberspace, by Barlow, Kapor, and Gilmore in 1990. It was where *Harper's* magazine hosted its landmark forum on computing, privacy, and hacking. In 1995, the WELL even played a role in the capture of infamous computer hacker Kevin Mitnick. By the late 1990s, however, the prestige had begun to fade.

In 1999, membership had fallen to fewer than 7,000. That year, Katz sold the WELL to *Salon.com,* one of the leading Internet-based media companies, also headquartered in the Bay area. In a 1999 article about the sale, a writer for the *Washington Post* quipped, "The move is a little like a newspaper buying the bar where its reporters hang out and drink after work—many Salon staffers spend a great deal of time in discussions on the WELL." For the long-suffering WELLbeings who had wanted to oust Katz from the start, the changeover to the like-minded *Salon.com* was welcomed.

For the WELL's 15th anniversary, in 2000, *Salon .com* published a handful of essays, written by WELL members, on the meaning of being a WELLbeing. One talked about the eerie disconnect between online personas and real-life people, realized at one of the many parties thrown by the WELL for its members; another, about lifesaving advice that had been shared on the WELL. Nearly all testified against the prevailing notion that the WELL was not the community it once was. In his essay, "Bleedin' Big Mouths," Farai Chideya wrote, "There is an incredible loyalty—an unthinkable loyalty—that leads members of the WELL, like me, to do things their momma told them not to do. Putting strangers up at my house is not among the stranger things. We love each other, we members of the WELL do, sometimes physically, sometimes metaphorically, sometimes while spitting verbal venom like cobras across the ether." Brand, the father of what was a mediocre business investment but a wildly successful sociological experiment, said, in a 2000 interview with *Mother Earth News,* "Looking back now, it seems pioneering, but at the time it was just kind of taking the tools that everybody knew about and doing one other thing with them."

## IT's ABOUT TIME

Brand, ever restless for new projects and new ideas, stepped away from day-to-day management of the WELL in 1986, just as it began to take shape on its own. After his stint at MIT, he became involved in the long-range planning group at Royal Dutch/Shell, the oil company. The concept of the long-range—with respect

to time—became the driving intellectual force behind much of Brand's later career. In 1988, he cofounded the Global Business Network (GBN), one of the first think tanks-cum-consultant firms concerned with globalization and innovative planning for the future. Brand believes that thinking in terms of decades, instead of years, helps instill a sense of responsibility in an organization.

In 1995, Brand and Danny Hillis, a famed supercomputer designer, cofounded the Long Now Foundation, which is home to several long-range projects, including a 10,000-year clock and the building blocks for a 10,000-year library, built to withstand nuclear catastrophes and natural disasters and home to a modern-day Rosetta Stone project, based on cataloging more than 1,000 languages for future generations. The projects sponsored by Long Now Foundation are meant to instill a sense of responsibility toward future generations and a spirit of patience and ecology toward the present. In an interview with John Brockman on Edge.org, Brand explained, "In a sense what we're doing with the clock is to do even more for time what the photograph of the Earth did for space." Brand's treatise on these issues, *The Clock of the Long Now: Time and Responsibility,* was released in 1999. In 2001, Brand cofounded the Long Bets Foundation, a forum where people wager at least $1,000 on, for instance, whether in the year 2030 airplanes will be pilotless, in an effort to further foster accountability in long-term thinking (most of the money wagered goes to charity).

Although Brand has gone on to embrace projects beyond the Internet, his presence can be felt in the thousands of virtual communities alive on the Internet today, which strive to be as cohesive, intellectually stimulating, and important as the WELL was in the 1980s and 1990s. Many point to Brand's gifts as a designer, evidenced as early as his multimedia events of the 1960s, to the success of the WELL. Matthew McClure, the WELL's first director, explained in a 1997 *Wired* article, "The WELL didn't just evolve, it evolved because we designed it to evolve." Brand's greatest gift to the Internet world was giving the grand social experiment of virtual community a chance to grow, founder, and define itself.

## FURTHER READING

### In These Volumes

Related Entries in the Chronology Volume: 1985: The
Whole Earth 'Lectronic Link Goes Online
Related Entries in the Issues Volume: Hackers; Online
Communities

## Works By Stewart Brand

*Whole Earth Catalog.* Menlo Park, CA: Portola
Institute, 1969.

*The Last Whole Earth Catalog.* New York: Random
House, 1971.

*II Cybernetic Frontiers.* New York: Random House,
1974.

*Space Colonies.* Ed. Sausalito, CA: Whole Earth
Catalog, 1977.

*Soft Tech.* Ed. New York: Penguin Books, 1978.

*The Next Whole Earth Catalog.* New York: Random
House, 1980.

*Whole Earth Software Catalog.* Garden City, NY:
Doubleday, 1984.

*Essential Whole Earth Catalog.* Garden City, NY:
Doubleday, 1986.

*News that Stayed News, 1974—1984: 10 Years of
CoEvolution Quarterly.* Ed. San Francisco: North
Point Press, 1986.

*The Media Lab: Inventing the Future at MIT.* New York:
Viking, 1987.

*How Buildings Learn: What Happens After They're Built.*
New York: Viking, 1994.

*The Clock of the Long Now: Time and Responsibility.*
New York: Basic Books, 1999.

## Books

Hafner, Katie. *The WELL: A Story of Love, Death and
Real Life in the Seminal Online Community.* New
York: Carroll & Graf, 2001.

## Articles

Brown, Andrew. "Whole Earth Visionary—Stewart
Brand." *The Guardian,* August 4, 2001.

Carroll, Jon. "Slouching Towards the Post-Millennial
Era." *San Francisco Chronicle,* December 31, 2000.

Fulton, Katherine. "How Stewart Brand Learns." *Los
Angeles Times,* October 30, 1994.

Hafner, Katie. "The Epic Saga of the WELL." *Wired*
5.05, May 1997, http://www.wired.com/wired
/archive/5.05/ff_well.html (cited September 29,
2004).

Krassner, Paul. "A Question of Lifestyle." *Los Angeles
Times,* December 18, 1994.

Schwartz, John. "Online Lothario's Antics Prompt
Debate on Cyber-Age Ethics." *The Washington Post,*
July 11, 1993.

Stipp, David. "Stewart Brand: The Electric Kool-Aid
Management Consultant." *Fortune,* October 16,
1995.

Weil, Elizabeth. "Time to Slow Down?" *Fast Company,*
May 1, 2000.

## Websites

ALL Species Foundation. Cofounded by Brand in 2000,
the ALL Species Foundation is attempting to catalog
all of Earth's surviving animal species, http://
www.all-species.org/ (cited September 29, 2004).

Global Business Networks. Cofounded in 1987 by
Brand, GBN helps multinational companies with
long-range planning, http://www.gbn.org (cited
September 29, 2004).

The Long Now Foundation. Founded in 1996 by
Stewart Brand and computer scientist Brian Hillis,
the Long Now Foundation supports several projects
designed to promote "slower/better" thinking,
http://www.longnow.org (cited September 29, 2004).

The WELL. Online home of the WELL, founded in
1985 by Stewart Brand and Larry Brilliant,
http://www.well.com (cited September 29, 2004).

*Whole Earth Magazine.* Online home of *Whole Earth
Magazine,* the successor to the *Whole Earth Catalog,*
founded, edited, and published by Stewart Brand,
http://www.wholeearthmag.com/ (cited September
29, 2004).

# Sergey Brin (1973–) and Larry Page (1972–)

## FOUNDERS, GOOGLE

**Once the verb** *to Google* entered the lexicon, around the year 2000, Sergey Brin and Larry Page had made their mark—not just on technology circles but the world. By then, Google, the deceptively simple search engine Brin and Page developed as Stanford doctoral students, had transformed online life. Since Brin and Page founded Google, Inc., in 1998, their search engine has been one of the primary means for Internet users—from beginners to experts—to make sense of the vastly expanding Web. So bright was Google's future that, in 2004, when the company announced its initial public offering of stock, financial analysts and technology pundits alike hailed the coming of a new economic age for the Internet. While those predictions were not immediately fulfilled, Google remains one of the most-watched Internet companies and boasts one of the most-visited sites online.

## TWO PATHS TO STANFORD

Sergey Brin and Larry Page may have shared similar childhoods—both were children of academics, born in the 1970s—but they began life worlds apart. Brin spent his earliest years in Moscow. When he was six years old, his family fled Russia, spurred by rampant anti-Semitism. Brin's father, holder of a doctorate in mathematics, took a teaching position at University of Maryland.

Although Brin dropped out of high school, his father helped him enroll at the University of Maryland, College Park, where he earned undergraduate degrees in both math and science. Brin was an exceptional student, winning a National Science Foundation Fellowship and earning high honors. One of a long line of academics—in addition to his father, Brin's great-grandmother had studied microbiology at the University of Chicago and his grandfather had been a math professor in Moscow—Brin went on to pursue his doctorate in computer science at Stanford in 1993.

Page was born in East Lansing, Michigan. His father, Carl Page, had been a pioneer in the fields of computer science and artificial intelligence before accepting a position as professor of computer science at Michigan State University. His mother taught computer programming at the university. Page began using computers at six: "I never got pushed into it," he said in a 2004 interview with the *Mirror,* a British newspaper. "I just really liked computers. I was probably the first student in my elementary school to turn in a word-processed homework assignment."

Following in his father's footsteps, Page pursued an undergraduate degree in computer science at the University of Michigan–Ann Arbor. He excelled in classes and even built a programmable ink-jet printer out of Legos. After graduating with honors, Page took a job at American Management Systems in Washington, D.C. In 1995, he, too, headed to Stanford to study for a doctorate.

Brin and Page met in March 1995, during a spring orientation of new computer science Ph.D. candidates. Brin, who had already been in the program for two years, was assigned to show some students, including Page, around campus. According to legend, the two men argued the entire weekend, but soon they were close friends.

## LOG ON, DROP OUT

By January 1996, Brin and Page were collaborating on a project they called BackRub, for a computer science paper on building a better search engine. At the time, most search engines worked by locating keywords and ranking results based on how many times the keyword appeared on the site. BackRub was to be a new kind of Internet technology, a search engine that analyzed the back links that connected one Web page to another. The more related links a Web page contained, they reasoned, the more useful and relevant it would be to a user. The Google search engine continues to be based on this very idea.

Over the next two years, Brin and Page toiled away in their dorm rooms, running experiments and analyzing the science, math, and even psychology involved in searching the Web. Brin's academic background was in data mining, which involves analyzing huge amounts of data and finding meaningful patterns and relationships that can then be expressed as mathematical algorithms. Page worked with the Web itself, drawing out massive amounts of data. As Brin explained in a 2001 interview with *Internet World,* "There is no more interesting source of data than the World Wide Web, which roughly represents human knowledge."

What began as a class project quickly grew into a legitimate search engine that Stanford staff and students used to make their way through the maze of the 25 million Web pages that were available by the late 1990s. To handle the growing traffic, Brin and Page set up a bare-bones business operation. Brin's office became the main office. Page, the more mechanical of the two, built BackRub's initial data center in his dorm room. He and Brin used whatever computer equipment they could find. "We'd stand on the loading dock [of the computer science building] and try to snag computers as they came in," Page recalled in a 2000 interview with *Technology Review.* "We would see who got 20 computers and ask them if they could spare one." To connect all the disparate computers they received, they used the Linux open-source operating system, which they found fast, reliable and, most important, free.

Soon, however, Brin and Page were running out of computing power. "Eventually, it became kind of unscalable, and we decided we could get better resources if we started a company," Brin told interviewer Terry Gross on *Fresh Air* in 2003. So, the two borrowed $15,000 from three credit cards to invest in a terabyte, or a million megabytes, of hard disks (for data storage). Then they wrote up a business plan very similar to the one Google still uses.

Initially, Page and Brin tried to shop their idea to the crop of search engine companies that had sprung up in the mid-1990s, such as InfoSeek and Excite. But, at the time, most of these firms were interested in following the path of thriving Internet companies like Yahoo! and America Online, billing themselves as one-stop Web portals and offering myriad services. When Page and Brin met with David Filo, one of the founders of Yahoo!, and also a Stanford alum, he strongly encouraged them to continue their work and form their own company. (Google would later become Yahoo!'s search engine.)

Page and Brin heeded Filo's advice and struck out on their own, taking leave of their doctoral programs, packing up the computer equipment they had amassed in their dorm rooms, and moving into the garage of a friend of Brin's, Susan Wojcicki. "My reaction was, 'OK, good luck, the rent's $1,700 a month and don't forget to separate the recycling,'" Wojcicki (now a Google executive) told the *Observer* in 2004. Far from becoming the professors their families had expected them to be, in the fall of 1998, Brin and Page were dropouts, living in rented rooms in Menlo Park, California, and working out of a garage. They were also on the cusp of becoming wildly successful.

> **Google is not a conventional company. We do not intend to become one.**
>
> *—Page in "'An Owner's Manual' for Google Shareholders," April 2004*

## GOOGLE, INC.

In the course of their work, BackRub the project had become Google the company. Google is a play on the word *googol* which is the number one followed by one hundred zeroes. "We were looking around at large numbers and thought that it would be fun to use something really large, 'cause we wanted to index, you know, all the information in the world," Page told *Fresh Air* in 2003. According to one legend, an early employee misspelled the name Googol when registering the site, infuriating the company's founders. In any case, in 1998, the Web was already awash in Websites, and Goog*ol*.com had already been registered. Goog*le*.com had not.

One of their earliest investors was Andy Bechtolsheim, a founder of Sun Microsystems. Bechtolsheim was impressed by Brin and Page's business pitch, and, being in a rush, quickly wrote a $100,000 check to "Google, Inc." Unfortunately, Google, Inc. had not yet been created; in order to cash the check, Brin and Page

*Larry Page (left) and Sergey Brin. (Photo courtesy of Google)*

had to incorporate. On September 7, 1998, Google was officially born.

Ultimately, Brin and Page raised $1 million in seed money from family, friends, and so-called angel investors. They soon traded their cramped quarters in Wojcicki's garage for new offices above a bicycle store on University Avenue, near the Stanford campus. By June 1999, they had raised $25 million in venture capital from established Silicon Valley venture capital firms Sequoia Capital and Kleiner Perkins Caufield & Byers. The meeting with Sequoia and Kleiner Perkins took place on a Ping-Pong table Google's founders were using for a conference table—evidence of their age. Brin and Page were just 24 and 25 years old.

## HOW'D THEY DO THAT?

From the beginning, Google's home page was an exercise in simplicity, much as it remains today. While other sites boasted high-end conceptual design and compli-

cated motion graphics, Google's site was mostly white space, displaying only the Google logo and a single blank rectangle waiting to be filled with a query. The Web design was more a matter of practicality than aesthetics—the company did not have a Webmaster on staff—but, even as Google grew, the founders kept the site's appearance virtually the same. "We have stayed true to [the design] because we realized it helps people get their searches done faster," Brin explained in the 2000 *Technology Review* interview. "They don't want to hang out on a home page when they want to get information quickly."

In many ways, the simplicity of Google's home page was misleading. The powerful new technology behind the Google search engine made it different from all other search engines. Using an algorithm dubbed PageRank—after Larry Page—Google searches not the Web itself, but a regularly updated index of the Web stored on one of Google's many servers. The index is pieced together using traditional search technology known as Web crawlers, or "spiders," which catalog thousands of Web pages per second by following links. Once the index is created, PageRank analyzes the links on each page, which determines how important a page might be to a user.

In many ways, PageRank functions like a popularity contest, based on the social intelligence built into the way Web pages are designed. As the *San Jose Mercury News* explained in 2004, "Google interprets a link from Page A to Page B as a 'vote' by Page A for Page B." If Page A is itself a popular site (linked to by many other sites), its "vote" for Page B carries more weight.

By 2000, a critical year for the company, Google was answering more than 18 million queries per day. By then, the company's search technology had been licensed by Yahoo! and various other Web portals. Since the beginning, in addition to providing search technology for other sites, Google had been making money selling banner advertisements at the top of its search results, which were targeted to the subject of the search being performed. In 2000, Google launched a technology called AdWords, which displays additional targeted, unobtrusive, text-based advertising alongside search results—advertising revenues skyrocketed. Advertisers purchase keywords that determine when their ads show up as a "sponsored link" to the right of the search results. The more traffic a sponsored link receives, the higher it moves up the list. (By 2003, the ads were bringing in more than $600 million in revenue.)

The advent of AdWords coincided with what one technology pundit called a "critical mass" of Websites online. Indeed, by 2000, even beginner Internet users could find a great deal of useful information online. Stories of people "Googling" prospective employers and potential dates sprang up in major newspapers and magazines throughout the country; *to Google* had entered the lexicon. A January 2001 headline in the *New York Observer* proclaimed, "Don't Be Shy Ladies—Google Him!"

## NO FRIENDS OF DR. EVIL

Internet users soon learned how to manipulate Google for their own profit. By the early 2000s, a veritable cottage industry of "search engine optimizers" had cropped up, aimed at improving a Website's ranking on Google, as well as on other search engines, like Yahoo! For Google results, programmers would create a "link

farm," which tricked PageRank into pushing the site to the top of Google's search results. In response, Brin and Page changed the algorithm to include technology that not only makes manipulating PageRank more difficult, but can identify sites with optimizers and blacklist them.

Such actions are reflective of Brin and Page's now-famous credo, "Don't Be Evil." Even though search engine optimizers rank fairly low on the "evil" spectrum, Google's efforts to thwart them demonstrate that the "No Evil" credo touches almost every aspect of the business. Advertising is the key area. Brin and Page make the sometimes difficult, sometimes subjective decisions about who can and cannot advertise. Google, for instance, does not take advertisements for guns or hard liquor, which has angered the National Rifle Association and the liquor industry, but it does take advertisements for wine. "Do Good"—a corollary of "Don't Be Evil"—means that Google offers many nonprofit organizations free advertising.

The founders are quick to point out that such decisions relate to advertising, not search results. The search results are pure—determined by PageRank and other secret technologies—and cannot be manipulated by sales dollars. That is a point of pride for Google, whose founders take seriously the responsibilities associated with being the most-used search engine. Indeed, as Jonathan Zittrain, codirector of the Berkman Center for Internet and Society at Harvard Law School explained in a 2003 *New York Times* article, "They're the traffic cop at the main intersection of the information society. . . . They have an awesome responsibility."

In the spring of 2001, Brin and Page brought another key executive on board to help shoulder that responsibility. Eric Schmidt, then the CEO of Novell, a Provo, Utah–based computer company, and former chief technology officer of Sun Microsystems, became the company's chairman. Industry pundits joked that Schmidt, nearly twenty-five years older than both Brin and Page, was brought in as "adult supervision." With two decades of experience in the computer industry, Schmidt helped strengthen Google's core business—contributing to financial oversight and international sales. In August 2001, Schmidt was named chairman and CEO, while Page became the president of products, and Brin became president of technology. They have since led the company as a triumvirate, although Schmidt still defers to the founders on certain issues.

"Evil is whatever Sergey says is evil," Schmidt once quipped.

## DESTINATION: GOOGLE

By the end of 2001, Google's index had grown to more than 3 billion Web pages; in addition, since Schmidt had come aboard, Google had begun posting regular profits, even in the midst of the Internet's economic downturn.

The site soon began to add features: in February 2001, Google purchased the DejaNews Usenet service and began offering Google Groups, a massive archive of postings to Usenet, the online discussion forum; later that spring, Google debuted a test version of Google News, a service that scours the Web for news stories and makes them searchable by topic; in late 2001, the site added a beta version of Froogle, a search engine for shoppers, which helps consumers locate products and services on the Web.

With each addition, technology pundits began to speculate that Google was reinventing itself as a Web portal and moving to compete with established portals like Yahoo!, America Online (AOL), or the Microsoft Network (MSN). The pundits were wrong. Portals are based on the notion that the longer a person stayed on the site using various services, the more money could be raised from advertising. Brin and Page had founded Google on the exact opposite premise: Google aimed to put Internet users on other sites, quickly and efficiently. That was, according to Page, the whole point.

Nonetheless, in 2004 Google announced a new service aimed directly at competitors like Yahoo! and MSN—free Google email. Gmail, as it is called, is a free email service subsidized by advertising. What Gmail had that Yahoo! mail and MSN's Hotmail services did not was generous storage—a full gigabyte of email storage for each user. Hotmail, for instance, limited its free email accounts to two megabytes.

By March 2004, even in its beta version, Gmail was stirring things up. Yahoo! and Hotmail quickly announced storage increases for their free accounts. Among Internet privacy advocates, Gmail was maligned

*It was really the merger of studying the Web and data mining. The marriage of those two yielded Google.*

—*Brin in* Internet World, *June 2001*

because of the connection it created between an email's text and the related advertising that ran alongside it. Some suggested that soon Gmail would fell prey to the Patriot Act, which, they claimed, could be used to review (and subpoena) private messages for political content. Brin has often countered that Gmail poses no higher security risk than any other Web-based email program, which all employ message-scanning technology, if not for advertising, then to filter spam and detect viruses.

As the debate raged on, what became abundantly clear was that Google had grown into such a powerful online brand that whenever it moved in new directions, the company could make its competitors nervous: Gmail threatened Hotmail and Yahoo! mail, Froogle threatened e-commerce giant Amazon.com, and even Google's one-click advertising access threatened the online marketplace of eBay. Indeed, by April 2004, Google had grown so powerful that when the company announced, after months of speculation, the date of its initial public stock offering, the technology industry quivered with anticipation from Wall Street to Silicon Valley.

## GOING DUTCH

When Page and Brin filed papers with the Securities and Exchange Commission (SEC) for Google's initial public stock offering (IPO), they included a seven-page letter, "'An Owner's Manual' for Google Shareholders." It was an unusual move for an SEC filing, but not entirely unprecedented; the letter was modeled after the ones that business mogul Warren Buffet would include in his annual reports. In the letter, Page set out to describe how Google would function as a public company—outlining the leadership triumvirate and the company's "Don't Be Evil" credo. "Searching and organizing all the world's information is an unusually important task that should be carried out by a company that is trustworthy and interested in the public good," he wrote. Page also called for shareholders to understand Google's business ethics. "We may do things that we believe have a positive impact on the world, even if the near-term financial returns are not obvious."

The letter also alerted the public to Google's unusual plan for the IPO—a Dutch auction. The Dutch auction system, in which anyone who cares to can bid for the stock, puts small investors on equal footing with institutional investors. Some saw Google's choice as snubbing Wall Street in favor of the everyman. Indeed, Merrill Lynch, the nation's largest retail stock brokerage firm, passed on underwriting the Google IPO, stating that, because of the low commissions earned in stock auctions, "it wasn't worth the trouble." (In the end, Google was underwritten by Morgan Stanley and Credit Suisse First Boston; commissions were roughly 3 percent, less than half of a regular IPO.) Others claimed that sheer greed was motivating the firm, as more of the money involved in the IPO would go back to Google itself.

As the date of the IPO neared, speculation on the pros and cons of Google going public were rampant. Some believed that an IPO would ruin Google, making executives and employees more focused on the bottom line than on the company itself. Others were banking on Google's IPO to have a ripple effect on the entire Internet economy. Such concerns were only amplified when news emerged that Google had failed to register 23.2 million "buddy" shares—stock given to employees and suppliers, or when the SEC delayed the IPO because of an interview Brin and Page gave to *Playboy* magazine, which was published during the mandatory "quiet period" before stock is sold when company executives are not supposed to speak to the press. (The interview was conducted a week before Google filed with the SEC.)

Finally, on August 13, 2004, Google's Dutch auction began. Five days later, when the auction closed, the stock price had been cut from an estimated $121.50 to $85 per share, and the number of available shares dropped from about 25.7 million to 19.6 million, which dismayed some investors and generated much less money than expected for Google. Nevertheless, on August 19, when Google's stock began open trading on the NASDAQ, the price immediately surged; shares were worth $108.31 when the closing bell rang on Friday, August 20.

Google's IPO made history as the largest-ever offering of stock by an Internet company. Seemingly overnight, Google was worth more than Ford Motor Company. Brin and Page, who each owned roughly 16 percent of the company, were instant billionaires, worth nearly $3 billion apiece. While generally considered a success, the IPO did not, as hoped, immediately jumpstart an economic boom for the Internet industry. But the long-term effects have yet to be seen.

From Stanford dorm rooms to the Googleplex—as the company's headquarters are known—Google has become one of the most dominant forces on the Internet. As it continually grows, Google, which first significantly emerged *after* the Internet boom, has come to challenge the Internet mainstays like MSN and AOL. It is responsible for one of every three searches on the Web, and has become one of the most visited sites on the Net. In the wake of its IPO, Google is also one of the most closely watched Internet companies, with analysts and pundits alike following Brin and Page's next move.

## FURTHER READING

### In These Volumes

Related Entries in this Volume: Yang, Jerry and David Filo

Related Entries in the Chronology Volume: 1999: Google Officially Goes Live; 2004: Google Holds Initial Public Offering

Related Entries in the Issues Volume: E-business; Linking and Deep Linking; Privacy

### Works By Sergey Brin and Larry Page

"Letter from the Founders: 'An Owner's Manual for Google Shareholders.'" *New York Times,* April 4, 2004.

### Articles

Achenbach, Joel. "Search for Tomorrow." *Washington Post,* February 15, 2004.

Boyd, Robert S. "The Guts of Google: Crawling, Indexing, Sorting." *San Jose Mercury News,* August 18, 2004.

Deutschman, Alan. "Googling for Courage." *Fast Company,* September 2004, http://www.fastcompany.com/magazine/86/google.html (cited September 29, 2004).

Elgin, Ben. "Google: Whiz Kids or Naughty Boys?" *Business Week,* August 19, 2004, http://www.businessweek.com/technology/content/aug2004/tc20040819_6843_tc120.htm (cited September 29, 2004).

Elgin, Ben, Jay Greene, and Steve Hamm. "Google: Why the World's Hottest Tech Company Will Struggle to Keep Its Edge." *Business Week,* May 3, 2004, http://www.businessweek.com/magazine/content/04_18/b3881001_mz001.htm (cited September 29, 2004).

Flynn, Laurie. "2 Wild and Crazy Guys (Soon to Be Billionaires), and Hoping to Keep It That Way." *New York Times,* April 30, 2004.

Gaither, Chris. "For Google, Going Public Is Far From a No-Brainer." *Los Angeles Times,* April 26, 2004.

"Googlemania: The Complete Guide." *Wired* 12.03, March 2004, http://www.wired.com/wired /archive/12.03/ (cited September 29, 2004).

Gross, Terry. "Interview: Larry Page and Sergey Brin Discuss the History and Workings of Google." *Fresh Air,* October 14, 2003, http://freshair.npr .org/day_fa.jhtml?display=day&todayDate =10/14/2003 (cited September 29, 2004).

Hochman, David. "In Searching We Trust." *New York Times,* March 14, 2004.

"How Google Works." *The Economist,* September 16, 2004, http://www.economist.com/science/tq /displayStory.cfm?story_id=3171440 (cited September 29, 2004).

Kopytoff, Verne. "Growing Up Google." *San Francisco Chronicle,* August 25, 2000.

———. "Google Goes Forth Into Great Beyond— Who Knows Where?" *San Francisco Chronicle,* May 2, 2004.

Levy, Steven, and Brad Stone. "All Eyes on Google." *Newsweek,* April 12, 2004.

Markoff, John, and G. Pascal Zachary. "In Searching the Web, Google Finds Riches." *New York Times,* April 13, 2003.

Salkever, Alex. "The Keyword at Google: Growth." *Business Week,* October 26, 2001, http://www .businessweek.com/bwdaily/dnflash/oct2001 /nf20011023_3541.htm (cited September 29, 2004).

Sappenfield, Mark. "A Success Story that Taps Silicon Valley's Geeky Roots." *Christian Science Monitor,* April 30, 2004.

Saracevic, Alan. "The Boy Wonders Not All Googly-Eyed . . . Well, Not Yet." *San Francisco Chronicle,* April 30, 2004.

Sheff, David. "*Playboy* Interview: Googly Guys." *Playboy,* May 2004.

Stross, Randall. "What Is Google's Secret Weapon?" *New York Times,* June 6, 2004.

Swisher, Kara. "Beneath Google's Dot-Com Shell, A Serious Player." *Wall Street Journal,* January 21, 2002.

Vise, David A. "A Peek Inside Google's Prospectus." *Washington Post,* July 28, 2004.

———. "Tactics of 'Google Guys' Test IPO Law's Limits." *Washington Post,* August 17, 2004.

Vogelstein, Fred. "Can Google Grow Up?" *Fortune,* December 8, 2003.

———. "No Love Lost for Google." *Fortune,* August 23, 2004.

Walker, Leslie. "Gmail Leads Way in Making Ads Relevant." *Washington Post,* May 13, 2004.

Waters, Richard. "Idealists Bound for Reality." *Financial Times,* October 25, 2003.

## Websites

The Anatomy of a Search Engine. Brin and Page's original paper on search engines for Stanford University, http://www-db.stanford.edu/ ~backrub/google.html (cited September 29, 2004).

Gmail Is too Creepy. Anti-Gmail Website, http:// gmail-is-too-creepy.com/ (cited September 29, 2004).

Google. Google's main Website. The "About Google" site features some press materials and company history, as well as the "Zeitgeist" page, which lists popular Web searches, http://www.google.com (cited September 29, 2004).

Google, Inc. SEC Filing. Google's SEC filing, including "'An Owner's Manual' for Google Shareholders," http://www.sec.gov/Archives/edgar/data/1288776 /000119312504073639/ds1.htm (cited September 29, 2004).

Google IPO Central. Unofficial site related to the Google IPO, http://www.google-ipo.com/ (cited September 29, 2004).

Localized Google Search Result Exclusions. Harvard professors Jonathan Zittrain and Benjamin Edelman's site about sites that are excluded from Google results, http://cyber.law.harvard.edu/filtering /google/ (cited September 29, 2004).

The Unofficial Google Fan Club. Unofficial site containing news and commentary about Google, http://www.ugfc.org/ (cited September 29, 2004).

# Steve Case (1958–)

## FOUNDER, AMERICA ONLINE

**Steve Case is best known** for bringing the Internet to the masses. In the late 1980s, he helped turn a fledgling online computer-gaming service into America Online (AOL), which quickly became the largest Internet service provider (ISP) in the United States. Touted as the easiest way to get online, AOL gained millions of customers by appealing directly to the general public, rather than to technophiles or the computer elite. AOL introduced millions of people to email and instant messaging. In 2000, Case orchestrated the largest business merger in American history, buying the media conglomerate Time Warner with AOL stock, and was hailed as a genius. When the Internet bubble burst shortly afterward and AOL Time Warner's stock plummeted, he quickly became one of the Internet's most scorned executives. Since he stepped down as chairman in 2003, Case's reputation continues to ebb and flow with the tides of the Internet economy.

## UPPER CASE, LOWER CASE

Steve Case is a fourth-generation Hawaiian, born and raised near Honolulu. His great-grandfather moved to the islands from Kansas in the late 1800s, and the Case family grew to become one of the state's more prominent white families. Steve's grandfather was a chief accountant at Grove Farm, one of the largest sugar plantations on the island of Kauai. (Years later, Steve, by then a computer-industry billionaire, purchased Grove Farm.) Steve's father, a lawyer, was born and raised on the island of Oahu, where he met Steve's mother, an elementary school teacher. They raised Steve and his three siblings in Makiki Heights, a Honolulu suburb.

Steve began his entrepreneurial career early on, opening up a limejuice stand at age six, along with his seven-year-old brother Dan. (The limejuice was squeezed from limes from the family's lime tree and cost two cents per cup.) From the start, the two boys were dedicated business partners and so inseparable that they earned the nicknames Upper Case, for Dan, and Lower Case, for Steve. Later ventures included Case Enterprises, a so-called international mail order company that sold greeting cards and seed packets door-to-door, and the Aloha Sales Agency, which offered an advertising circular that the two Cases dubbed Budget Boosters that they distributed along their shared newspaper route. The division of labor in each Case Enterprises endeavor foreshadowed their later career paths—Steve was the idea man, Dan provided capital. As Steve recalled in a 2000 *Newsweek* interview, "Even if it was five bucks to start some business, I never had any money. . . . So Dan would give me five bucks and suddenly own 50 percent of my idea."

The Case children all attended the Punahou School, a private preparatory school in Honolulu. Steve was described as a relatively shy, inwardly focused student, and very creative. He wrote album reviews for the school paper, convincing record companies to send him freebies and concert tickets by claiming he wrote for a leading paper read by Hawaiian teenagers. He continued to pursue his musical interests when he left for Williams College in Massachusetts. A student by day, Case enjoyed a brief stint as a new wave rock singer in two amateur groups by night. He satisfied his entrepreneurial bent by promoting concerts, marketing limo rides from the airport to campus, and selling gift baskets of fruit.

Case graduated with a degree in political science in 1980. After failing to gain admission to several MBA programs, Case took a job with Procter & Gamble's marketing department in Cincinnati, Ohio. For two years,

> *Windows is the past. . . .*
> *In the future, AOL is the*
> *next Microsoft.*
>
> —*Case to the* New York Times, *July 1999*

*Steve Case. (Reuters/Corbis)*

he worked on promoting the Lilt home-permanent kit and a hair-conditioning treatment in the form of a towelette. The ad copy for the towelette, Abound, read, "Towelette? You bet!" Case considered it—and much of his short time at Procter & Gamble—a total disaster. In 1982, after two years, he took a job as the manager of new pizza development for the Pizza Hut division of PepsiCo, in Wichita, Kansas. The Pizza Hut job entailed traveling the country and trying out new pizza topping combinations. (Case's most notable contribution to the Pizza Hut menu was the pineapple topping.) As he traveled from city to city, Case kept himself busy with a new Kaypro computer, an early portable personal computer.

Case read *The Third Wave* (1980), the futurist Alvin Toffler's book about the information age, which described The Source, an online community. Intrigued, Case paid $100 to subscribe to The Source, and, after much struggle, finally got online with his Kaypro and a 300-baud modem. Via The Source, Case could peruse bulletin boards and chat with the handful of others who were online in the early 1980s. "There was something magical about the notion of sitting in Wichita

and talking to the world," Case said in a 1997 interview with *Time*.

Online, Case began to learn more about The Source's founder, William von Meister. In 1982, von Meister had also founded Control Video Corporation, which featured GameLine, an innovative service that delivered popular Atari computer games over telephone lines. (Case pointed out in a 1995 *Wired* interview that GameLine was "almost exactly what Sega's now doing with the Sega Channel.") Dave Case, Steve's elder brother, had, by then, become a successful venture capitalist with the firm Hambrecht & Quist, which represented the board of Control Video Corporation. In 1983, Steve's brother invited him to a Consumer Electronics show in Las Vegas and convinced von Meister to offer Steve a position as a marketing executive. Shortly thereafter, the younger Case moved to Vienna, Virginia, to start his new job.

Within two years, Atari had folded and von Meister was forced out of the company. Case and his coexecutives, entrepreneur Jim Kimsey and Marc Seriff, the engineer of the group, were left to build the business anew.

## QUANTUM LEAP

In 1985, Case helped refashion Control Video Corporation into Quantum Computer Services, Inc., an online computer service for owners of the Commodore 64 personal computer, then one of the most popular PCs in the United States. The initial press release described their service, dubbed Quantumlink or Q-link for short, as "useful, affordable, easy to access, and entertaining." Q-link offered home users easy-to-use software featuring color graphics, point-and-click user interfaces, and sound. Within two years, Case had secured a deal with Apple Computer for a similar service dubbed Apple Link. Similar deals with computer manufacturers Tandy Corporation and IBM followed, adding new capital and ever-wider distribution. Even as Commodore headed into steep decline in the late 1980s, Quantumlink continued to prosper.

Then, in 1989, the company shifted gears, pulling its various services under one name—America Online. In a 1998 *Washington Post* interview, Case recalled, "We wanted to see America Online as our own brand, sort of private label brand, and started adding features, content, context, community and commerce and so forth to make it a more engaging service." The fledgling company competed against CompuServe and Prodigy, two other private online services boasting millions of subscribers. Case, who became president of the company in 1990, kept America Online's focus on simplicity—making getting online easy for people. "From day one, we thought the people aspect was central to this and really the soul of the medium," Case said in a 1996 *Wired* interview.

In 1991, with 120 employees and 30,000 subscribers, Case was named CEO of America Online. He was briefly stripped of that title right before the company's initial public stock offering, for fear that his age—a mere 33 years old—would scare off potential investors. Renamed CEO following the $66-million stock offering in 1992, Case set America Online on an aggressive competitive course—a growth-at-any-cost trajectory that included "carpet bombing" AOL disks across the country through direct mail and by binding them inside magazines. When Prodigy angered its consumers by raising rates, America Online lured away customers by cutting its fees and placing ads that urged users to "jump" from Prodigy to America Online. Though still trailing its competitors by hundreds of thousands of users, America Online was gaining fast.

In May 1993, Case met with Bill Gates, head of Microsoft. With its modestly successful MSN service, Microsoft was then the third-largest online service provider, behind CompuServe and Prodigy, and the meeting was ostensibly about possible business opportunities for the two companies. Famously, Gates told Case, "I can buy 20 percent of you or I can buy all of you. Or I can go into this business myself and bury you." Unfazed, Case replied that his company was not for sale. It was a bold statement for a CEO whose business had only 200,000 members, but his confidence paid off. By August of the following year, AOL had signed its millionth customer. Five years later, Case would say, "In the future, AOL is the next Microsoft."

## GROWING PAINS

AOL grew to 4 million users by late 1995 and 8 million by January 1996. By then, it had transformed itself from a private online network into a full-fledged Internet

> *I still remember that excitement 13 years ago when I first connected to an online service. I thought it was magical then, I still think it's magical today.*
>
> —Case to Wired,
> September 1995

service provider, with local dial-up access numbers in more than 800 cities in the United States and Europe. One of every three Internet users used AOL. The company's astounding growth and popularity was attributable, in large part, to AOL's ease of use. Its constantly expanding customer base also offered users unprecedented access to other users. By 1996, AOL hosted nearly 7,000 chat rooms and handled more than 11 million email messages per day. If AOL's simplicity convinced people to get online, the people they found on AOL kept them there.

Nevertheless, AOL had its share of problems. The company was routinely derided as a pedestrian, rigidly controlled version of the Internet, first within tech circles, then by the mainstream as well. (At one point, AOL was mocked as "the K-mart network.") Its stock price began to plummet in 1996, amid speculation about the company's demise. Users came and went at astonishing rates—a process called "churn" in the ISP industry. Web companies that AOL had purchased in 1995, including BookLink, NaviSoft, and Web-Crawler, languished. In August 1996, a nineteen-hour system-wide blackout caused by an internal error made headlines all over the world.

Case's December 1996 decision to switch to a $19.95 flat-rate pricing plan (following in the footsteps of its increasingly popular competitor, EarthLink) brought a tidal wave of new customers that quickly overloaded the network. The ensuing system outages and busy signals earned AOL the nickname America On Hold and prompted a class action lawsuit charging AOL had misled consumers about rates and phone fees. (The case was settled in 1998, with AOL doling out $2.6 million to subscribers.) AOL also faced Securities and Exchange Commission investigations of its accounting practices.

No matter how many steps AOL took back, it always seemed to manage another step forward. In 1997, AOL took over former competitor CompuServe's 2.6 million customers, along with WorldCom's telephone and data networks. Case oversaw an investment of $35 million in a new data network to handle the company's still extraordinary growth—up to 1 million users per month by the late 1990s. In April 1999, AOL bought Netscape, the company that launched the first widely successful commercial Web browser, for $4.2 billion in AOL stock options (10 percent of the company at the time). The subscriber base soared to 17 million members, and the stock price hit an all-time high of $93.81 in mid-December 1999.

Earlier that year, the company had sealed several high-profile deals: with DirecTV to put AOL on television, with Bell Telephone to put AOL on high-speed Internet lines and with Gateway, Inc. to develop Web-enhanced devices. In September, Case also began to quietly pursue an ambitious deal with Time Warner, Inc., one of the country's leading cable operators and media conglomerates. The potential for growth was enormous. "Our goal is to establish AOL as a more important part of tens of millions of people's everyday life," Case told the *New York Times* in a 1999 interview. "And to do that, we have to move beyond the PC in the den."

## THE TIME BOMB

In the fall of 1999, Time Warner was the nation's second-largest cable operator and one of the most powerful media companies in the world. When, in October, Case called Jerry Levin, Time Warner's CEO, the two executives began to talk synergy—how Time Warner's media content and AOL's online communities could combine to create something new, bold, and powerful, something befitting the blossoming Internet age. Over the next year, as the talks become more serious—and as AOL's stock continued to soar—the discussions switched to the thornier issues of costs and control.

On January 6, 2000, over dinner at Case's Virginia home, Levin, Case, and two other advisers sealed the deal. After dinner, Case instant-messaged top executives who eagerly awaited the verdict on their computer screens. "It's done," he typed. The merger was announced shortly thereafter.

> *I am stepping down, but not walking away. . . . I still believe the merger was a good idea for AOL and Time Warner even though I understand the conventional wisdom at this moment is otherwise. When all is said and done, I believe people will see the logic of bringing these companies together.*
>
> —Case to the Washington Post, January 2003

At the time of the merger, AOL's stock was worth more than twice as much as Time Warner's, despite Time Warner's $26 billion in revenues in comparison to AOL's $4 billion. In the end, AOL purchased Time Warner for roughly $165 billion in a stock swap based on closing price, which, for AOL, on the day before the agreement, had been an impressive $73.75 per share. For that, AOL received ownership of 55 percent of AOL Time Warner, and Case became chairman. The estimated worth of the new company was $350 billion.

Although he was chairman, Case spent much of his time and energy behind the scenes, leaving other executives, such as AOL's Robert Pittman, to take on the more visible roles during the integration. When the stock market took its fateful downturn, and as AOL Time Warner struggled to find a workable business model, Case's absence from the spotlight began to be conspicuous. Within the new company's first year, top executives, including Jerry Levin, began to resign. Once hailed as the most spectacular merger in American history, AOL Time Warner was quickly becoming one of the biggest debacles instead.

By October 2002, Case was the only original senior executive left at AOL Time Warner, and speculation about his resignation was rampant. (Indeed, many shareholders blamed Case for the poor state of affairs and openly wished for him to resign.) By the end of the year, the value of the company's stock had dropped to $15 a share, a decline of more than 75 percent. The promises of synergy that originally drove the merger were left unfulfilled. When, in January 2003, AOL Time Warner announced that it had lost nearly $100 billion in the previous year, many placed the blame directly on Case.

Case, in turn, announced in January that he would step down as chairman in May 2003. In an interview with *CNN* shortly after his announcement, Case explained, "I was the architect of the merger. I was the chairman of the company. The company has not done well. It's certainly not done anything anybody would have expected when we did the merger. So it's not at all surprising that some of that anger and disappointment would be directed at me." In the month following his announcement, Case, once lambasted by angry shareholders, enjoyed a mild reprieve, evolving from a business scoundrel into scapegoat in some pundits' eyes. AOL was not the only company with a falling share price—the entire stock market had plummeted by almost two-thirds. In many ways, Case could be seen as doing right by AOL shareholders, turning inflated stock into something of real worth. Some pundits argue that Levin, not Case, should be blamed for the disastrous results, for it was Levin who had undersold his company.

Case made his final appearance at AOL Time Warner's annual meeting on May 16, 2003, where he formally stepped down as chairman. Given a standing ovation, Case said, "I think it is far more important for us to look to the future than the past. While the last few years have been difficult and disappointing for us all, and while nobody is happy about the stock price, I believe brighter days are ahead." Case remains on the company's board and has turned his attention to philanthropy, through the Case Foundation, and other business pursuits.

During that same annual meeting, Dick Parsons, the former Time Warner president who took on Case's role as chairman, reminded the audience of Case's legacy, that he "led a revolution that introduced the Internet and connectivity to this country and the world"—a fact that angry shareholders often forgot. Indeed, by building on his formative experience trying to get online in the early 1980s, Case was able to envision a company that would bring the Internet to millions of average people by making it simple and easy. As the AOL Time Warner disaster slips into memory, few can deny the impact of AOL—from the ubiquitous phrase "You've Got Mail" to the wild popularity of instant messaging—on contemporary American culture.

## FURTHER READING

### In These Volumes

Related Entries in this Volume: Dayton, Sky; Gates, Bill

Related Entries in the Chronology Volume: 1989: Quantum Computer Services Is Reborn as America Online; 1997: America Online's Offer of Unlimited Access Leads to Class-Action Suits; 1998: America Online Announces Deal to Buy Netscape; 2000: America Online and Time Warner Announce Merger

Related Entries in the Issues Volume: E-business; Online Communities

### Books

Ashby, Ruth. *Steve Case: America Online Pioneer.* Brookfield, CT: Twenty-First Century Books, 2002.

Klein, Alec. *Stealing Time: Steve Case, Jerry Levin, and the Collapse of AOL Time Warner.* New York: Simon & Schuster, 2003.

Leibovich, Mark. *The New Imperialists.* Upper Saddle River, NJ: Prentice Hall, 2002.

Munk, Nina. *Fools Rush In: Steve Case, Jerry Levin, and the Fall of AOL Time Warner.* New York: HarperBusiness, 2004.

Swisher, Kara. *aol.com: How Steve Case Beat Bill Gates, Nailed the Netheads, and Made Millions in the War for the Web.* New York: Times Business, 1998.

———. *There Must Be a Pony in Here Somewhere: The AOL-Time Warner Debacle and the Quest for the Digital Future.* New York: Crown Business, 2003.

### Articles

Ahrens, Frank, and Alex Klein. "Energized Steve Case Emerges From the Shadows." *Washington Post,* January 17, 2002.

Angwin, Julia, and Martin Peers. "Boss Talk." *Wall Street Journal,* January 17, 2002.

"Case Study." *Newsweek,* January 24, 2000.

"A Conversation With . . . Stephen M. Case, CEO of America Online." *washingtonpost.com,* December 9, 1997, http://www.washingtonpost.com/wp-srv /business/longterm/conversation/case/case.htm (cited September 28, 2004).

Cortese, Amy, Amy Barrett, Paul Eng, and Linda Himelstein. "The Online World of Steve Case." *BusinessWeek.com,* April 15, 1996, http://www .businessweek.com/1996/16/b34711.htm (cited September 28, 2004).

Gunther, Marc. "Understanding AOL's Grand Unified Theory of the Media Cosmos." *Fortune,* January 8, 2001.

———. "What Does AOL Want? Growth, Growth and More Growth." *Fortune,* July 23, 2001.

———. "The Mess at AOL Time Warner." *Fortune Magazine,* May 13, 2002.

Gunther, Marc, and Stephanie N. Mehta. "The Internet Is Mr. Case's Neighborhood," *Fortune Magazine,* March 30, 1998.

Hansell, Saul. "Now, AOL Everywhere." *New York Times,* July 4, 1999.

Harmon, Amy. "How Blind Alleys Led Old Media to New." *New York Times,* January 16, 2000.

Kinsley, Michael. "The Case for Steve Case." *Washington Post,* February 7, 2003.

————. "The Case Against the Case Against Steve Case." *Slate.com,* February 6, 2003, http://slate .msn.com/id/2078227 (cited September 28. 2004).

Kirkpatrick, David D. "Man in Middle of AOL Deal Is Becoming Odd Man Out." *New York Times,* July 27, 2002.

Leibovich, Mark. "From Suburban Roots to Global Ambition." *Washington Post,* June 4, 2000.

Leonard, Andrew. "Steve Case: Brilliant Visionary or Fumbling Clod?" *Salon.com,* January 14, 2003, http://archive.salon.com/tech/col/leon/2003/01 /14/case/ (cited September 28, 2004).

Lieberman, David. "Case, Levin Look to Past and Future." *USA Today,* January 15, 2001.

Maney, Kevin. "He's Not Your Typical CEO, but Steve Case Knows His Cards." *USA Today,* January 12, 2000.

Manjoo, Farhad. "Saving AOL." *Salon.com,* October 15, 2002, http://archive.salon.com/tech/feature /2002/10/15/save_aol/ (cited September 28, 2004).

Munro, Neil. "Building a Case." *National Journal,* July 31, 1999.

Nollinger, Mark. "America, Online!" *Wired* 3.09, September 1995, http://www.wired.com/wired /archive/3.09/aol.html (cited September 28, 2004).

Ramo, Joshua Cooper. "How AOL Lost the Battle but Won the War," *Time,* September 22, 1997.

Roberts, Johnnie L. "How It All Fell Apart." *Newsweek,* December 9, 2002.

Rose, Frank. "Keyword: Context." *Wired* 4.12, December 1996, http://www.wired.com/wired /archive/4.12/ffaol.html (cited September 28, 2004).

————. "You've Got Mayhem." *Wired* 8.09, September 2000, http://www.wired.com/wired/archive/8.09 /aol.html (cited September 28, 2004).

Sanders, Edmund, and Sallie Hofmeister. "Last Man Standing." *Los Angeles Times,* September 22, 2002.

Swartz, Jon. "Case Study—A Look at Mr. America Online." *San Francisco Chronicle,* February 22, 1999.

Swisher, Kara. "Steve Case Tries to Hold a Place On-line." *Washington Post,* August 27, 1995.

————. "You've Got Time Warner!" *Wall Street Journal,* January 11, 2000.

Vise, David A. "Case Plays to Win AOL's 'Survivor'." *Washington Post,* December 6, 2002.

Vogelstein, Fred. "The Talented Mr. Case." *U.S. News & World Report,* January 24, 2000.

Yang, Catherine. "Q&A With AOL's Steve Case." *Business Week,* March 27, 2000, http://www .businessweek.com/2000/00_13/b3674020.htm?scri ptFramed (cited September 28, 2004).

## Websites

America Online. Main America Online Website, http://www.aol.com/ (cited September 28, 2004).

America Online, Inc. America Online's corporate site, with a detailed AOL timeline, http://www.corp .aol.com/ (cited September 28, 2004).

Steve Case's Home Page. Case's personal home page, with links to his speeches and his philanthropic endeavors at the Case Foundation, http://stevecase .aol.com/ (cited September 28, 2004).

# Vinton Cerf (1943–)

## DEVELOPER OF TCP/IP, FOUNDER OF THE INTERNET SOCIETY

**For his role in developing** TCP/IP—the basic communications protocol that underlies the Internet—Vinton Cerf has been hailed as one of the Internet's founding fathers. Though he eschews that particular title, Cerf is undoubtedly one of the most influential figures in Internet history. From his work with TCP/IP in the 1970s and MCI Mail, one of the first email networks to connect to the Internet, in the 1980s, as well his involvement in founding some of the Internet's leading organizational bodies, including the Internet Society (ISOC), Cerf has been at the forefront of the Internet's evolution for nearly three decades. In addition to leading the telecommunications giant MCI into the next era of the Internet technologies in the 1990s, Cerf has continued to guide the growth of the Internet as chairman of the Internet Corporation for Assigned Names and Numbers (ICANN); with the Interplanetary Internet project, he has begun mapping the course for the Internet to breach the Earth's atmosphere and connect with outer space.

## PAGING DR. STRANGELOVE

Born in New Haven, Connecticut, Cerf and his family moved in1946 to California's San Fernando Valley, just north of Los Angeles. Although he sustained substantial hearing loss because of his premature birth, Cerf was an excellent student. Cerf was years ahead in his math classes and determined, after reading a children's book called *The Boy Scientist,* to become a scientist himself. By the time he entered Van Nuys High School in 1958, Cerf had established himself as a budding math whiz—and somewhat of a dandy, wearing jacket and tie to class and carrying a briefcase. "I wasn't so interested in differentiating myself from my parents," he told *CNET News.com* in a 1997 interview. "[B]ut I wanted to differentiate myself from the rest of my friends just to sort of stick out."

At age 15, Cerf met his first computer: a friend of Cerf's father, himself an executive for a defense contractor, was working on the Defense Department's computing experiment dubbed SAGE (semi-automatic ground environment), and invited Cerf to see the computer in his lab outside Los Angeles. Like the majority of punch card and paper tape computers of the day, the SAGE computer was mammoth—filling nearly three rooms with cathode-ray tubes. In a 1996 interview with *Business Week,* Cerf described it as "very Dr. Strangelove."

Two years later, Cerf joined his best friend and fellow Van Nuys high school computer whiz, Steve Crocker, at the University of California–Los Angeles (UCLA), where a professor had given Crocker special computer lab privileges. Each weekend, the two boys worked on a Bendix G-15 computer, feeding it programs on paper tape. By then, Cerf told *Forbes* in 1997, "The bug had bit, and I was infected with computers."

After graduating from high school in 1961, Cerf entered Stanford University on a full scholarship from North American Aviation, then his father's employer. He majored in math, but took as many computing-related courses as he could. Once he realized he would never be a world-class mathematician, Cerf decided, after graduating in 1965, to take a job as a systems engineer with IBM.

## FINDING ARPA

Cerf's two years with IBM gave him a taste for working with operating systems. Primarily, he worked on a time-sharing computer system known as Quiktran out of IBM's Los Angeles Data Center. (Quiktran was based on the FORTRAN programming language.) By 1967, however, Cerf wanted to go back to school to pursue an advanced degree in computer science. Cerf's high school friend, Steve Crocker, who was already working on a doctorate in computer science at UCLA, convinced Cerf to interview with his thesis adviser, Gerald Estrin.

In 1967, under Estrin's supervision, Cerf began graduate work. Initially, he worked on a project (known as the Snuper Computer project and funded by the Pentagon's Advanced Research Projects Agency [ARPA]) in which one computer was programmed to analyze the actions of another. By then, UCLA's computer science department had already been tapped as one of ARPA's key academic research labs, thus Estrin's project—as well as that of other key computer science professors, Leonard Kleinrock among them—was extraordinarily well-funded and equipped with cutting-edge machinery.

While at UCLA, Crocker and Cerf began consulting for Jacobi Systems, a small computing firm located in Santa Monica that did work for the navy. Simultaneously, ARPA was developing ARPANET, an experiment in computer networking and communications. In June of 1968, ARPA sent out its Request for Proposals to develop computers that could manage network traffic on the ARPANET. Crocker and Cerf sent in a proposal on behalf of Jacobi Systems, but eventually lost the bid to Massachusetts-based Bolt, Beranek & Newman (BBN), whose computers came to be known as interface message processors (IMPs).

Later that year, both Crocker and Cerf were able to join ARPA's networking effort under the auspices of UCLA professor Leonard Kleinrock, a pioneer in packet-switching theory, a process of dividing data into small batches of information to enable it to be transmitted more readily through a data network. Kleinrock had proposed to ARPA that he develop a Network Measurement Center, where scientists would analyze the behavior of the new network.

As Kleinrock's research assistant, Cerf wrote programs that would record the behavior of data that had been sent onto the packet-switching network—how long a computer took to acknowledge receipt of a packet, how many packets were lost, and so on—then analyze the behavior statistically. When the IMPs from BBN arrived in September 1969, Kleinrock also began

data analyses on the packets moving between the IMP and his own Sigma-7 computer. Cerf worked alongside Crocker, who studied the host-to-host protocols, as well as another future computer luminary (and, oddly enough, also a Van Nuys High School alumnus), Jonathan Postel.

By early 1970, the first four nodes of the early ARPANET—at UCLA, the University of California Santa Barbara (UCSB), Stanford Research Institute (SRI), and the University of Utah—were up and running. Shortly thereafter, scientists from BBN—most notably Robert Kahn—arrived at UCLA to test the performance of the network. Kahn worked closely with Cerf, who was then lead scientist at the Network Measurement Center. Their intense, two-week collaboration on this phase of the ARPANET would cement their friendship and, in a few years, open the doors for one of the most important eras of Internet development.

Cerf had earned his master's degree from UCLA in 1970 and his doctorate in 1972. By March 1972, the ARPANET was deemed ready for public demonstration—an event slated for October of that year at the Washington Hilton Hotel, at the first International Conference on Computer Communication. Robert Kahn led the effort. Cerf, who, by June of that year, had accepted an assistant professorship at Stanford University for the fall, stayed on at UCLA to help coordinate the demo, which was deemed successful. "I would consider it a watershed event," Cerf explained in a 1990 oral history interview with the Smithsonian. "[I]t made packet switching real to people other than the ones who were involved in designing and building the ARPANET."

> *The reason I get this 'father' label is partly because our American culture needs heroes, and partly because I stuck with the program and managed it.*
>
> —*Cerf to* Forbes ASAP, *October 1997*

## THE INTERNET HAS TWO DADDIES

Cerf began teaching courses on operating systems in Stanford's Computer Science and Electrical Engineering department in November of 1972. By then, Kahn had left his job at BBN for a post at ARPA's Information Processing Techniques Office (IPTO), then led by Lawrence Roberts. Kahn was hired to develop packet-switching architectures like the one used for the ARPANET for use with satellites and mobile radios. He then began to tackle the problem of designing an open architecture that could accommodate all these networks—creating a way for the ARPANET, a packet radio net, and a packet satellite net to connect, so that military operations in the field could communicate seamlessly among the three. Kahn called this concept, "internetting."

Internetting soon became an IPTO program in its own right, and by spring of 1973, Kahn had contacted Cerf at Stanford about it. The project already involved scientists at SRI International and BBN, as well as others working on the packet radio and satellite networks based on another ARPA program, the ALO-HAnet, originating at the University of Hawaii. The struggle was to get these separate nets talking to each other. Kahn felt Cerf was the man to help him do it.

The existing communications protocol of the ARPANET, developed in December 1970, was known as network control protocol (NCP). Cerf had worked closely on NCP during his UCLA days. NCP functioned fairly well in a closed-circuit network like the ARPANET, which used telephone lines, but lacked the flexibility to accommodate the various architectures and needs of telephones, radios, and satellites simultaneously.

Over the next six months, Cerf and Kahn worked on developing this new kind of protocol. That September, they presented a draft of their work at a special meeting of the International Networking Group (INWG) in England. (Cerf had been named chair of the INWG in October 1972, shortly after the first ARPANET demonstration.) The revised paper appeared as an article, "A Protocol for Packet Network Intercommunication," in the May 1974 issue of the *IEEE Transactions on Communications*. Then, the labor began to make theory into reality.

Much of the initial design work of what came to be known as transmission control protocol (TCP) was done in Cerf's Stanford computer lab. "You know the story: The devil is in the details—and he was there,"

*It's not right to think of the Internet as having only one father. . . . It has at least two, and in reality thousands, because of the number of people who have contributed to what it is today.*

—Cerf to Forbes
ASAP, October 1997

Cerf recalled in a 1997 *Forbes* interview. By December 1974, however, the first specification for TCP was published, which allowed for work on honing the protocol to begin elsewhere, primarily at BBN and University College in London.

As work on TCP continued, Kahn and Crocker began to urge Cerf to join them at DARPA. (It had been renamed the Defense Advanced Research Projects Agency in 1972.) After a year or so of prodding, Cerf finally agreed, seeing DARPA as an opportunity to focus more on his own research instead of teaching. He arrived at DARPA in 1976 as a program manager in charge of the Packet Radio, Packet Satellite, Internet and Network Security research programs—all programs started by Kahn.

Primarily, however, Cerf's mission was to get the Internet up and running by perfecting TCP and lobbying for its widespread adoption. A key advance came in 1978, when TCP became TCP/IP. Originally, Cerf and Kahn had established one protocol—TCP. As the development of TCP continued through the mid-to-late 1970s, however, it became clear that another, simpler protocol was needed. Thus, Internet protocol (IP) was created: IP simply addressed and forwarded packets through the network, while TCP handled the more complex tasks of reliable delivery, data flow, and lost packets.

By 1979, the military and various research institutions began to embrace these developments; Cerf led efforts to extend TCP/IP for each new network system. To help organize the efforts, Cerf formed a number of groups, including the International Cooperation Board, the Internet Research Group, and the Internet Configuration Control Board (ICCB). (In an interview for the Computerworld Honors program, Cerf admitted naming the ICCB to sound "as boring as possible," so that no one would elect to serve on it, leaving him and Kahn to stack the group with key figures in TCP/IP's development.)

The following year, TCP/IP became the official protocol for the military, and, by 1982, the decision had been made that all networks connected to the ARPANET would switch from NCP to TCP/IP. On

*Vinton Cerf. (Associated Press)*

Though he already had all the technical skills necessary to build a digital email center from his work at UCLA, Stanford, and ARPA, Cerf still described designing MCI Mail as "like climbing Mt. Everest in just nine months." MCI Mail was much more than a simple email program; in addition to send and receive functions, it was programmed to print documents and send them via fax or the U.S. Postal Service. By September 1983, the system was up and running, and within nine months, MCI Mail boasted more than 100,000 subscribers.

After four years at MCI, Cerf grew bored. The company had stopped pouring millions of dollars into the development and expansion of MCI Mail, and Cerf yearned to return to research. Though he would later coordinate the historical connection of MCI Mail to the Internet, in 1986 Cerf decided to leave the company. Over the next decade, MCI Mail, once the cutting edge of commercial digital communication, waned in popularity. Cerf explained, in a 2001 oral history interview, "It was a system that was, frankly, ahead of its time." (MCI-WorldCom shut down MCI Mail on June 30, 2003.)

## BACK ON THE INFORMATION SUPERHIGHWAY

After leaving MCI, Cerf rejoined Robert Kahn as a vice president of Kahn's new nonprofit organization, the Corporation for National Research Initiatives (CNRI). Kahn had left DARPA in 1985, with a mandate from the National Science Foundation to conduct research into Internet technologies that would boost the nation's information infrastructure—a concept that later became known as the information superhighway.

Cerf coordinated research into Internet infrastructure. In 1988, Cerf led the effort to connect commercial email services to the Internet, lobbying the government to abandon its previous policy of keeping the Internet a research- and military-only system. By June of 1989, after a year or so of programming, Cerf established a connection between MCI Mail and the Internet via an experimental relay. Soon, CompuServe, another commercial email system, followed suit, and email networks that formerly could not communicate with each other could now do so through the structure of the Internet.

CNRI would go on to develop various early applications for the Internet, including Web browsers, intelli-

January 1, 1983, after nearly eight years of development, the transition to TCP/IP transformed the ARPANET and the modern Internet was born.

## A MOUNTAIN OF MAIL

Cerf did not stay at ARPA long enough to witness the official transition. In November of 1982, an executive at MCI approached him to head the company's efforts to build a digital post office. Cerf, his interest piqued by both the concept and the increase in salary, accepted.

As vice president of MCI's Digital Information Services, Cerf led the engineering of MCI's digital post office—dubbed MCI Mail. Though MCI Mail was not the first commercial email service, it was intended to make email accessible to the average person, bringing this new form of communication out of research labs and into offices and homes everywhere.

gent agents, and digital libraries. It also organized the research, government, and commercial interests in the Internet. In June 1991, CNRI announced that it would be forming the Internet Society (ISOC). In addition to educating the public about the Internet, the ISOC expected to provide vital support for the Internet Engineering Task Force (IETF). The IETF, one of the Internet's primary technical bodies, grew out of the Internet Configuration Control Board, which Cerf founded while still at ARPA. Cerf served as the first president (January 1992–January 1995) of the ISOC.

After eight years with CNRI, Cerf was once again tempted into the commercial realm by MCI—by the same executive, Bob Harcharik, who had hired him in 1982. Harcharik had contacted Cerf in late 1993, urging Cerf to come back to help MCI get into the Internet business. By 1994, Cerf was back in California as MCI's senior vice president of data architecture. In the decade he has since spent at MCI, Cerf's primary task has been to continue development of the company's Internet backbone, which he has expanded to be one of the largest carriers of Internet traffic in the world.

## MOONLIGHTING

In addition to his work at MCI, Cerf has used his clout to guide the Internet safely into the future. In November 2000, Cerf replaced Esther Dyson, another Internet luminary, as chairman of the Internet Corporation for Assigned Names and Numbers (ICANN), the international non-profit organization charged with managing issues involving Internet infrastructure. By the time Cerf was tapped for the position, ICANN—at the time, still less than two years old—was mired in controversy. Cerf's appointment marked a departure from organizational squabbles toward a focus on technical goals. As Dyson told *USA Today,* in 2001, "Vint is the ideal person to carry this on. He's got a great reputation, and he's been around these issues a long time."

By the time of the annual ICANN meeting in October 2003, the transition Cerf had been called on to carry out seemed to be under way. As head of the nineteen-member board, Cerf had refocused the organiza-

*Three networks made the original Internet. . . . So maybe two planets make the original interplanetary Internet. Maybe toss the moon in there. Make it three.*

*—Cerf to* USA Today, *January 15, 2001*

tion on key technical matters, such as the effort to move from IPv4 (the fourth iteration of IP, developed at DARPA in 1978) to IPv6, as well as issues such as Internet security.

In February 2004, NASA announced that the first link in an interplanetary communications network had been established on Mars as part of NASA's Mars Exploration Rover effort. Since 1998, Cerf and Adrian Hooke, a scientist as NASA's Jet Propulsion Laboratory (JPL), have been investigating the promise of the Interplanetary Internet—a wireless communications network built upon Internet-like protocols, galactic in proportions. (Cerf had come up with the idea in the early 1990s, when he realized that it had taken twenty years for the Internet to take off and wondered where the Internet should be in twenty more years.) With funding from DARPA, he and Hooke began efforts to define a new standard communication protocol, dubbed IPN, for Interplanetary. By 2004, IPN was in its fourth iteration.

Cerf's work on IPN is not unlike the early days of the ARPANET—the effort has been wracked by slow hardware and technical obstacles, and few outside the government and JPL are even aware of it. Today—as then—Cerf remains optimistic. "It is not science fiction anymore," he told *Government Technology* magazine in 2000. "It may take 30 or 40 years to actually get there, but as each mission is launched, it will carry a little bit of this backbone system and leave it in place."

As Cerf plows ahead with IPN, he has not shirked the responsibilities of a founding father of the Internet. Indeed, for more than three decades of tireless work, shaping and guiding the growth of the Internet, Cerf has been recognized—receiving the University of California National Medal of Technology in December 1997, alongside Robert Kahn, and the Charles Stark Draper Prize in 2001, alongside Kahn, Leonard Kleinrock, and Lawrence Roberts. (He was also named one of *People's* "25 Most Intriguing People," in December 1994.) In 2004, on the thirtieth anniversary of the publication of his and Kahn's paper on TCP/IP, Cerf finds himself coming full circle, fathering the next generation of Internet protocols.

## FURTHER READING

### In These Volumes

Related Entries in this Volume: Kahn, Robert; Postel, Jonathan; Roberts, Lawrence; Taylor, Robert

Related Entries in the Chronology Volume: 1967: Plans for ARPANET Are Unveiled; 1969: The ARPANET Is Born; 1972: ARPANET's Public Debut; 1974: "A Protocol for Packet Network Intercommunication" Is Published; 1979: DARPA Establishes the Internet Configuration Control Board; 1983: Internet Is Defined Officially as Networks Using TCP/IP

### Works By Vinton Cerf

with Robert E. Kahn. "A Protocol for Packet Network Interconnection." *IEEE Transactions on Communications* COM-22, no.5 (May 1974).

with Barry M. Leiner, et al. "The Past and Future History of the Internet." *Communications of the ACM* 40, no. 2 (February 1997).

### Books

Hafner, Katie. *Where Wizards Stay Up Late: The Origins of the Internet.* New York: Simon & Schuster, 1996.

Richards, Sally. *FutureNet: The Past, Present and Future of the Internet as Told By Its Creators and Visionaries.* New York: John Wiley & Sons, 2002.

### Articles

Barrett, Amy, and Andrew Reinhardt. "The 'Father of the Net' Has a Problem Child." *Business Week,* September 30, 1996, http://www.businessweek.com/1996/40/b3495102.htm (cited September 16, 2004).

Brandt, Anthony. "Legends: Vint Cerf." *Forbes ASAP,* October 6, 1997.

Evenson, Laura. "Present at the Creation of the Internet." *San Francisco Chronicle,* March 16, 1997.

Harris, Blake. "Building the Net of the Future." *Government Technology,* May 2000, http://www.govtech.net/magazine/visions/may00visions/cerf/cerf.phtml (cited September 16, 2004).

Hill, G. Christian. "Cerfing the Web." *Wall Street Journal,* June 17, 1996.

Kopytoff, Verne. "Q and A: Web Pioneer Looks Ahead." *San Francisco Chronicle,* November 27, 2000.

Mackintosh, Hamish. "Father Knows Best." *The Guardian,* March 4, 2004, http://www.guardian.co.uk/online/talktime/story/0,13274,1161365,00.html (cited September 16, 2004).

McKelvey, Tara. "The Papa of Protocol." *USA Today,* January 15, 2001.

Sullivan, Ben. "Cyber-Daddy." *Los Angeles Daily News,* February 28, 1999.

Wingfield, Nick. "Still Netting After All These Years." *CNET News.com,* March 3, 1997, http://news.com.com/2009-1082_3-233721.html (cited September 16, 2004).

### Websites

Cerf's Up. Cerf's personal and professional Website, with links to various publications, http://global.mci.com/us/enterprise/insight/cerfs_up/ (cited September 16, 2004).

Computerworld Honors Program International Archives: Vinton G. Cerf Oral History. Transcripts of an oral history interview with Cerf, conducted as part of the Computerworld Honors Program, http://www.cwheroes.org/oral_history_archive/vinton_g_cerf/oralhistory.pdf (cited September 16, 2004).

ICANN. Online home of ICANN, with updated news on ICANN developments, http://www.icann.org (cited September 16, 2004).

Internet Society. Online home of the Internet Society, with extensive writings on the history of the Internet, some penned by Cerf himself, http://www.isoc.org/ (cited September 16, 2004).

InterPlanetary Internet. Online site of NASA's InterPlanetary Internet Project, with links to the latest protocol specifications as well as the latest related news, http://www.ipnsig.org (cited September 16, 2004).

Smithsonian Oral and Video Histories: Vinton Cerf. Transcripts of an oral history interview of Cerf, conducted as part of the Smithsonian Institution's Computer History Collection, http://americanhistory.si.edu/csr/comphist/vc1.html (cited September 16, 2004).

# Sky Dayton (1971–)

## FOUNDER OF EARTHLINK

In 1994, Sky Dayton, then a twenty-three-year-old owner of a hip Los Angeles coffeehouse, founded EarthLink, an Internet service provider (ISP). Dayton quickly built EarthLink into the largest independent ISP in America, second only to America Online. Five years later, at the height of the Internet boom, Dayton founded a second company, an ill-fated Internet start-up incubator, eCompanies. When eCompanies began to founder, Dayton focused his energies on one promising pet project—Boingo Wireless, a company that develops and sells wireless networking technology, also known as "wi-fi." Since then, Dayton has touted wi-fi as the next Internet frontier, and his enthusiasm, coupled with his hippie-libertarian, surfer persona, earned him the nickname, the maharishi of the wireless Internet.

> The Internet is going to go from the wires, the ground, the poles, into the air. The Internet is going to be as ubiquitous as oxygen.
>
> —*Dayton to* Business Week, *June 2004*

## SKY'S THE LIMIT

Dayton was born in New York City, the son of a sculptor father and a poet mother. His parents were iconoclasts—members of the Church of Scientology who lived, as Dayton described in a 2002 interview with *Wired,* a "kind of hippy, kind of edgy in the 1970s" life. They were the kind of parents who, upon finding Dayton drawing on the walls as a toddler, let him continue to use the wall until he had created an elaborate, aeronautically inspired mural.

The Daytons left New York when Sky was still a toddler, settling in Los Angeles, where, aside from a short stint at Sky's grandfather's home in northern California, he spent much of his youth. His grandfather, an IBM engineer, who introduced Dayton to his first computer, a Timex Sinclair. Dayton was just nine years old.

Back home in southern California, Dayton got an early start on his entrepreneurial career. Still in grade school, he began washing windows for his family for a few cents per pane, then recruited new business by posting flyers throughout his neighborhood. At age 10, he and a friend wrote a business plan for a candy store. "Unfortunately, our funding committee—our parents —shot down the deal," he told *Entrepreneur* magazine in 1998.

Dayton attended the Delphian School, a boarding school in Sheridan, Oregon, which drew from the teachings of the Church of Scientology. He had inherited many of his parents' artistic gifts and dreamed of being an animator. (During one internship, Dayton helped create the Elephant Man skeleton used in the music video for Michael Jackson's song, "Leave Me Alone.") When he graduated from the Delphian School in 1988, Dayton applied to the prestigious California Institute for the Arts but was not accepted. Dayton decided to forgo college and turn, instead, to his true forte—business.

Although he lied about his computer skills to get his first job, in the graphics department of a Burbank, California, advertising firm, Dayton quickly rose, becoming head of the department in less than three months. In 1990, having left one advertising firm's graphics department for another, he set out on his own. His childhood dream of owning a candy store came partially true in the form of Café Mocha, a Melrose Avenue coffeeshop that he cofounded with a friend. He was just 19 years old.

Café Mocha hosted poetry readings and discussion salons, and it quickly established itself as a trendy West Hollywood hangout. Dayton soon opened another coffeehouse, Joe. In 1992, the indefatigable entrepreneur also opened a boutique computer graphics firm geared to the entertainment industry. At twenty-one, with three businesses under his belt, all seemed to be going well for Dayton—until he tried to get online.

## SERVICE WITH A SMILE

In an oft-repeated story about the founding of Earth-Link, Dayton stars as an everyday frustrated consumer. When the World Wide Web arrived in the early 1990s, and the Web browser shortly thereafter, Dayton wanted to get connected. He searched for a week to find an ISP and needed an additional eighty hours to get a working Internet connection. "It was utter hell," he told the *Los Angeles Daily News* in 1998. "I got no sleep, I didn't go to work." Dayton was tech savvy, having grown up around computers—how much more difficult would it be for those who were inexperienced? No one would get online, he believed, if it was going to be this hard. He felt he could do better.

Over the next year, Dayton shopped around the EarthLink business plan, which consisted of a disheveled stack of photocopies marked "CONFIDENTIAL" with a Staples stamp. Soon, he had drummed up $100,000 from a handful of angel investors. "I hadn't heard the words *venture* and *capital* in the same sentence, and I actually believed that $100,000 was a *ton* of money," Dayton wrote in a 2002 essay for *Forbes*. One of the investors was Reed Slatkin, considered to be a cofounder of EarthLink although he did not manage any day-to-day aspects of the company. (Slatkin would later become famous for bilking friends, including Dayton, out of millions of dollars with fraudulent investments.)

EarthLink formally incorporated in 1994; the company's first customer signed on in July. Back then, it was a bare-bones operation. Dayton rented a 600-square-foot office in Pasadena and made do with used furniture from a payroll company that had gone bankrupt. The entire ISP ran on ten 14.4 modems, two used Sun workstations and a two-line phone—one line for sales, the other for tech support. Dayton himself did both. "I talked to every customer myself in those early days, and hired and added complexity only when it was deserved," he wrote in *Forbes*. "Building EarthLink was a fight—a scratching, clawing, tooth-and-nail battle, and I'm glad for it."

Less than a year after the company's founding, Dayton introduced EarthLink Software, one-step technology that made his ISP one of the easiest and most direct routes onto the Internet. In August 1995, EarthLink signed a contract with UUNET Technologies and went national, serving nearly one hundred cities. Three months after going national, EarthLink stunned the fledgling ISP industry by introducing something that is now considered utterly commonplace—a flat usage fee for unlimited access. (America On-Line, the industry leader, still charged by the hour.)

The fee structure was not Earth-Link's only revolutionary strategy. Dayton eschewed the telecom approach to the ISP industry, which focused on building vast networks and physical infrastructure. He had no interest in routing connections around the world; later, he predicted, companies that did would be stuck with a telephone-based network when ISPs inevitably switched to cable or satellite. Instead, he leased networks from well-established network backbone providers like PSINet and UUNet, and focused on applications and, more important, the customers themselves. "I wanted to set up a company that would recognize the human characteristics of the Internet and service that need," Dayton told *Government Technology* magazine in 1999.

From Dayton's point of view, EarthLink's goal was to create not just Internet access, but an entire Internet experience. That experience included user-friendly, one-step software, educational booklets, and bimonthly user newsletters. From the beginning, Dayton made a long-term commitment to provide twenty-four-hour-a-day,

> *In '94, when we launched EarthLink, we had to tell people why they needed to connect to the Internet. It was not obvious. Yahoo didn't exist. Amazon didn't exist. eBay didn't exist. All these great things you can do on the Internet were just dreams in some entrepreneur's head.*
>
> —Dayton to the Los Angeles Business Journal, January 2002

*Sky Dayton. (Associated Press)*

seven-days-per-week, friendly and competent customer service—including "guardian angels," EarthLink employees who sought out dormant users and helped them get back online. In 1997, half of the company's 700 employees offered tech support; by 1999 that number had grown to 800 of a 1,300-person staff. Keeping customers happy was the goal: "Retention is the single most important factor for profitability in this business," Dayton said in a 1997 interview with *Forbes.*

EarthLink not only kept its customers, it attracted more and more by the day. EarthLink had the fourth largest subscription base—behind AOL, Microsoft, and AT&T—and was the fastest-growing independent ISP in the country. By October 1997, EarthLink had passed the 200,000-member mark. By June 1998, EarthLink's customer base had grown to 500,000. Then, in a deal with the long-distance giant Sprint, EarthLink gained

an additional 130,000 customers, as well as $24 million in cash and $100 million in credit.

As EarthLink inched ever closer to the top of the ISP industry, the company positioned itself as the anti-AOL. In Star Trek terms, Dayton became fond of saying, "If EarthLink is the Rebel Empire, then AOL is the Borg." In February 1998, EarthLink began running "Get Out of AOL Free" advertisements, where AOL defectors could call a toll-free number to get a step-by-step "Guide for AOL Graduates," which featured an illustrated robot taking the training wheels off his bike. Dayton established a no-set-up-fee agreement for former AOL customers, complete with easy change-of-address messages. EarthLink's competitive advantage—simple, old-fashioned customer service, updated for the Internet age—was what kept those customers with EarthLink. By the late 1990s, EarthLink's customer base had swelled to 1.5 million. "Everyone knows our secret sauce," Dayton told *Government Technology* in 1999. "We publish it to the world. We tell people the main ingredients that make EarthLink work so well. Why are we the only guys who are able to execute on it?"

By February 2000, after a $4-billion merger with independent ISP MindSpring Enterprises, EarthLink became the second-largest ISP in the country. With more than 3 million customers, EarthLink moved its headquarters from California to Georgia. Dayton, still in his 20s, stepped down as acting chairman and moved on to his next Internet venture.

## DRUNK ON eKOOL-AID

With the successful MindSpring merger behind him, Dayton turned his attention to the question, "Can start-ups be mass-produced?" It was a question he had discussed regularly with former Disney Internet chief Jake Winebaum, a like-minded entrepreneur Dayton had met several years earlier at an Internet round-table discussion sponsored by the *Los Angeles Times.* The two men were both avid snowboarders, and, while on the ski lifts at Whistler Mountain in British Columbia, they had discussed various start-up ideas, never settling on just one. At one point, the two discussed a joint EarthLink-Disney venture they dubbed Cornice—but the plans never came together.

In May 1999, over Mexican food, an idea finally clicked. Winebaum, in a 2000 interview with *Forbes,* recalled, "About five minutes into the first burrito, I said, 'Why don't we create a company that can create companies? We can develop this one idea and then do many other ideas.'" In the span of a couple of bites, the two men were in business.

Their brainchild, eCompanies, was incorporated in June 1999. eCompanies would be an Internet start-up incubator associated with a venture capital fund, fashioned after IdeaLab, one of the first and most successful Internet incubators. Founded in 1996, IdeaLab had launched ventures like eToys.com, CarsDirect.com, and

*I don't see myself as a programmer. I'm a user. . . . I'm that person struggling to get connected, that person who feels their oxygen supply has been cut off whenever I can't get connected. I live that.*

—Dayton to Wired, October 2002

Goto.com. Within two months, Dayton and Winebaum raised more than $130 million from sources like George Soros, Kohlberg Kravis Roberts, Goldman Sachs, EarthLink, Disney, and others.

Dayton called eCompanies' business plan "a sort of Henry Ford approach to building a company." Using the same basic building blocks, particularly with respect to marketing and infrastructure, eCompanies could churn out start-ups in a matter of 90 to 180 days from concept to company. Their expectations for the company were as ambitious as their timeline. Indeed, one wall of their Santa Monica, California, headquarters featured the image of a money truck as an incentive—as Dayton explained, a money truck was a whole lot bigger than a moneybag.

According to *Wired* writer Brendan I. Koerner, by 2001 eCompanies had become "the poster child for new economy excess." Few of its companies, such as eMemories and eParties, had proved profitable. In fact, most of the companies held by the $160-million venture fund performed poorly. Even eCompanies' mildly successful venture Business.com was nonetheless ridiculed because Dayton had paid an unprecedented $7.5 million to a cybersquatter who held that domain name. In just over two years, eCompanies had gone from Internet incubator to Internet incinerator, so called because it had burned through so much cash.

"I admit I drank my share of Kool-Aid during the craze," Dayton wrote in a 2002 *Forbes* essay. When eCompanies' day was clearly coming to an end, Dayton turned the business into a holding company for his few surviving prospects, Business.com and a fledgling wireless ISP business, Boingo.

## GOING BOINGO

Seven years after Dayton had struggled to get online, he had an eerily similar experience with wireless computing. In 2000, he was given the equipment to set up a wireless ("wi-fi") Internet network in his home. Although setting up the network was rather difficult, Dayton realized, as he had back in 1993, that he was witnessing the opening of a new Internet frontier. As with EarthLink, Dayton decided to build an entire company inspired by his own experience. He launched Boingo Wireless in December 2001; in 2002, Dayton told the *Los Angeles Business Journal* that he hoped the name, Boingo, "connoted the freedom you get when you're able to connect to the Internet wherever you are and being able to bounce from location to location."

Dayton amassed a group of EarthLink alumni to build the company, and, still gun-shy from the financial whirlwind of eCompanies, began with a comparatively modest $15-million budget. The company would build short-range, high-speed wi-fi networks in places where wired people congregate—from airports and hotels to parks and cafes. Individuals are charged $75 per month, or $7.95 for one-time access, while places that set up Boingo networks take part in a revenue-sharing plan that includes $1 for each connection and $20 for each new customer.

On January 21, 2002, Boingo began serving 400 wi-fi "hot spots" across the United States, including airports in Texas and California, and hotels run by the Four Seasons, Hilton, Marriott, Sheraton, and Radisson. Slowly, Boingo began to succeed where other wi-fi ventures, including those of AT&T, Intel, and IBM, had failed—bolstered, in part, by a major investment by Sprint PCS. By June 2004, Boingo boasted more than 75,000 locations in twenty-nine countries. Still, Dayton admits, wi-fi is in an awkward stage. But he eagerly reminds detractors that both Internet service, in the early days of ISPs, and cell phone service, began in the same patchwork style.

By 2004, Dayton, not yet thirty-five, could claim three Internet companies. Still youthful and tireless, his entrepreneurial zeal is matched only, perhaps, by his love of surfing and snowboarding. Indeed, writers profiling Dayton rarely neglect to mention a skateboard stowed beneath his desk, a recent surfing injury, or the surfer-infused patois that insinuates itself into his business and technical discussions. Although he often works from eight in the morning to midnight, six days per week, Dayton is known to slip out of Boingo's sixty-five-person office when the swell is up in Malibu. Many credit him with bringing a new sensibility to California's technology center—a more laid-back, yet still ambitious, counterpoint to Silicon Valley.

## FURTHER READING

### In These Volumes
Related Entries in this Volume: Case, Steve
Related Entries in the Chronology Volume: 2000: The Dot-Com Crash
Related Entries in the Issues Volume: Online Communities; Wireless Internet

### Works By Sky Dayton
"When Capital Corrupts." *Forbes.com.* March 25, 2002, http://www.forbes.com/asap/2002/0325/019.html (cited September 19, 2004).

### Articles

Baker, Stephen. "Q&A with Boingo Wireless' Sky Dayton." *Business Week,* June 21, 2004.

Bloom, David. "Electronic Midas Touch." *Los Angeles Daily News,* June 9, 1998.

Caulfield, Brian. "Sky Dayton." *Internet World,* April 1, 2000.

Harris, Blake. "Owning the Enterprise." *Government Technology,* February 1999, http://www.govtech.net/magazine/visions/feb99vision/dayton/dayton.phtml (cited September 9, 2004).

"In Hot Pursuit of the Wi-Fi Wave." *Business Week,* April 29, 2002.

Keough, Christopher. "The Next Tech Adventure." *Los Angeles Business Journal,* January 21, 2002.

Koerner, Brendan I. "The Long Road to Internet Nirvana." *Wired* 10.10, October 2002, http://www.wired.com/wired/archive/10.10/boingo.html (cited September 19, 2004).

Lipschultz, David. "Earthlinking." *Forbes,* August 11, 1997, http://www.forbes.com/1997/08/11/feature.html (cited September 19, 2004).

McGarvey, Robert. "Sky's The Limit." *Entrepreneur,* January 1998, http://www.Entrepreneur.com/article/0,4621,227957,00.html (cited September 19, 2004).

O'Shea, Dan. "Sky Dayton Is Smarter Than You, Richer Than You and Younger Than You." *Wireless Review,* April 1, 2002.

Shim, Richard. "Will Patience Pay Off?" *CNET News.com,* http://news.com.com/2008-1082-5077691.html (cited September 19, 2004).

Sullivan, Ben. "Glowing Sky." *Los Angeles Business Journal,* October 28, 1996.

Vrana, Debora. "California Dealin'." *Los Angeles Times,* June 5, 2000.

Zerega, Blaise. "Spitting 'Em Out." *Forbes ASAP,* April 3, 2000, http://www.forbes.com/asap/2000/0403/036.html (cited September 19, 2004).

## Websites

Boingo Wireless. Online home of Boingo Wireless, with links to press clips and other info about Boingo and the wireless industry, http://www.boingo.com/sky.html (cited September 19, 2004).

EarthLink. EarthLink's main site. The "About EarthLink" section includes a brief history of the company, including the MindSpring merger, http://www.earthlink.net/ (cited September 19, 2004).

Sky Dayton's Home Page. Dayton's personal home page, with links to a biography, press clips, and surfing pictures, http://home.earthlink.net/~sky/ (cited September 19, 2004).

# Dorothy Denning (1945–)

## COMPUTER SECURITY EXPERT

**Since the 1970s,** Dorothy Robling Denning has become one of the most important figures in national computer security. After programming vital security software for the Internal Revenue Service (IRS), the FBI, and the U.S. Navy in her early years, Denning went on to research network-security issues for cutting-edge companies, including SRI International and Digital Equipment Corporation. She also wrote hundreds of articles and several books. Since the 1990s, she has become the "go-to" expert for both the press and government for her assessments of data security, encryption, information warfare, and cyberterrorism. A truly independent mind, Denning has, at various points in her career, championed the cause of hackers and campaigned for software to defeat them; she has testified both for and against the interests of federal surveillance agencies and high-tech industry. In an age of cyber-warfare, the soft-spoken Midwesterner—one of very few women in her field—has become America's unlikely secret weapon.

## MORE THAN A MATH TEACHER

Growing up in Grand Rapids, Michigan, Dorothy Robling planned to become a math teacher. In the 1950s it was a logical career choice for intelligent women who were good at math. She majored in mathematics at the University of Michigan, where she earned her bachelor's degree in 1967, and her master's degree in 1969. At the university, Fred Haddock, an astronomy professor, introduced Robling to working with computers, which she used to calculate Doppler shifts in stars, instead of working with a calculator.

In 1969, Robling began pursuing a doctorate in computer science at Purdue University. During a seminar on operating systems, Robling first encountered the topic that would come to define the rest of her career—computer security. (She later married the professor giving the seminar, Peter Denning.) For her doctoral thesis, Denning chose to research secure information flow—how to keep classified information stored on a computer system safe from unauthorized users. Her work was the basis of the now famous lattice model of secure information flow, which was later adopted by federal agencies, including the Central Intelligence Agency, the IRS, and the FBI. Denning was awarded her Ph.D. in 1975.

Denning spent the next eight years at Purdue as an associate professor, primarily researching security models for statistical and multilevel databases, but also exploring the field of cryptography. She also wrote a book that is often considered essential reading on the subject, *Cryptography and Data Security* (1982). In 1984, Denning went to work as a senior staff scientist for SRI International, a nonprofit scientific research center based in Menlo Park, California. She began tackling the issue of intrusion detection—methods for detecting hackers and other unauthorized users in a computer network. By the time she left SRI in 1987, she had developed one of the first real-time intrusion detection systems now used by the U.S. Navy and other government agencies to guard classified data. Her other research drew her deeply into security programs that were so mathematically complex that they were too difficult and slow to be truly useful. Denning felt her work lacked practical application. That year, she became the principal software engineer at Digital Equipment Corporation's (DEC) System Research Center, in Palo Alto, California.

## DELVING UNDERGROUND

As part of DEC's then-new distributed systems lab, Denning learned of the emerging security threats presented by the Internet. She also discovered the hacker underground. While at DEC, Denning accepted an invitation to be interviewed by a former member of the

Legion of Doom, the infamous 1980s hacker group, for the 'zine *W.O.R.M.* She then began reading hacker publications and interviewing the hackers about their values and practices. In her paper, "Concerning Hackers Who Break into Computer Systems," which she presented at the National Computer Security Conference in October 1990, Denning argued that hackers were innocent "explorers" and that the computing world should strive to work with them. In turn, hackers urged those in their community to read Denning's work.

Denning secured her reputation in the hacking underground, however, by coming to the defense of Craig Neidorf, a University of Missouri pre-law student, in the spring of 1990. Neidorf was the twenty-year-old editor and publisher of *Phrack,* a 'zine for hackers. In issue 24 of *Phrack,* Neidorf published parts of a document that explained Bell South's E911 (Enhanced 911) emergency system. In February, through the dual efforts of the Secret Service and University of Missouri campus police—acting under Operation Sun Devil, the government code name for the two-year crackdown on hackers—Neidorf was indicted by a grand jury on charges that included wire fraud and interstate transportation of stolen property. By June of that year, he faced ten felony charges and a prison sentence of up to sixty-five years.

*The United States v. Craig Neidorf* was a highly publicized case. The newly formed Electronic Frontier Foundation, the first civil rights organization for cyberspace, came to Neidorf's aid, securing him legal advice. Many in the Internet world followed the case closely to see how far Operation Sun Devil could go. Denning served as Neidorf's expert defense witness. In pretrial hearings, she easily disproved the basis of the government's case by showing that the information published in *Phrack* could not be used to break into the telephone systems and disrupt 911 service. The defense also proved that the information published in *Phrack* was already available to the public for legal purchase, which meant Neidorf had not revealed secret or classified information by publishing it. On July 27, four days after the trial began, the government dropped its case and the judge declared a mistrial.

Just as Denning had secured her place among cyberlibertarians, her point of view shifted. "All of a sudden," she recalled in a 2002 interview with *CIO Insight,* "it really hit me. A lot of security research had no relevance to what hackers were actually doing. We had been addressing the obscure threats, not the real ones." In 1991,

Denning left DEC to return to academic research, this time as chair of Georgetown University's computer science department. At Georgetown, located in Washington, D.C., Denning moved beyond programming and research and into the realm of national security policy.

## THE CLIPPER CHICK

In 1992, Denning began commuting between her Georgetown office and the National Security Administration's (NSA) offices outside Baltimore, Maryland. Her assignment was to hack into skipjack, a classified computer algorithm that would form the basis of the Clipper Chip, an 80-bit, single-key encryption device to be used for telephone communications (standard encryption devices at the time were 56 bit). A handful of other experts in the field had declined the NSA's invitation to test skipjack, fearing that the Clipper Chip and the key escrow system it depended upon would give the NSA and other surveillance agencies more power to spy on U.S. citizens. (With key escrow encryption, in addition to the decryption keys held by the sender and receiver of encrypted communication, another set of keys are held by a third party—in this case, a government-related agency. With a warrant, the NSA and other security agencies could gain access to the keys in order to wiretap otherwise private conversations.)

Denning, however, was more interested in potential benefits of the Clipper Chip. "My long-held view that security served only noble objectives was shattered when I saw how criminals and terrorists used encryption and other security technologies to evade law enforcement and intelligence collection, and how individuals and organizations got locked out of their own data," she said in her 1999 acceptance speech for a National Computer Systems Security Award.

After three weeks of testing skipjack while it withstood the NSA's arsenal of automated attacks, Denning gave it high marks and became one of its staunchest supporters. Civil libertarians, hackers, and those who, less than a year before, believed Denning to be a champion of civil rights in cyberspace now called for boycotts of her books, branding her the "Wicked Witch of the East" and the "Clipper Chick" on listservs and in chat rooms. Her apparent belief in the benevolence of the NSA led many to believe she was a dupe; kinder critics claimed she was well intentioned, but naïve. Denning countered such remarks by saying that ten years prior, she would have never thought herself capable of taking her current position. "I was raised to have no trust of the NSA and a certain amount of distrust of the FBI," she told the *Los Angeles Times.* Her research into the criminal use of encryption taught her otherwise. She added, "I would not be behind this if I thought it was going to compromise privacy."

In April 1993, the Clinton administration introduced the Clipper Chip as part of the Escrowed Encryption Initiative (EEI), which was passed into law in 1994. Soon after, however, the Clipper Chip code was broken, and the EEI, a volunteer program, fell into disuse.

## SAFE AND SECURE

In 1996, the Clinton administration again brought up the issue of key escrow—now renamed key recovery—as part of a larger effort to change the law on encryption export controls. Because of its military origins, encryption technology was classified as a munition and therefore was stringently controlled. For years, industry leaders had lobbied for relaxation of the export controls on encryption technology, claiming that the export laws suffocated e-commerce. Government agencies, including the CIA and FBI, feared that loosening export controls would put powerful encryption technology in the hands of criminals. Instituting a key recovery system, while at the same time easing export controls, seemed to be a compromise.

In the House, Representative Bob Goodlatte of Virginia introduced the Security And Freedom through Encryption (SAFE) Act, which encouraged the use of strong encryption and allowed the export of generally available American-made encryption products. In the Senate, Senators John McCain of Arizona and Robert Kerrey of Nebraska sponsored the Secure Public Networks Act (S. 909), which, while loosening some encryption export controls, called for the use of a key recovery system that would allow federal agencies, under court order, to access "spare keys" held by third-party key-recovery centers. (Unlike the Clipper Clip legislation, the third party here would be a private-sector agency.)

*I have a very strong reaction to things that are false. When I see something that's false, it doesn't matter what it is, I will often respond to it.*

*—Denning to* Wired, *September 1996*

Each piece of legislation, particularly the Secure Public Networks Act, had voluble detractors. Industry leaders complained that criminals could simply bypass the U.S. key-recovery system by using strong encryption obtained overseas. They also claimed that the industry was going to be deprived of up to $60 billion annually by the year 2000 because of export controls. Civil libertarians feared the government would be able to eavesdrop on its citizens indiscriminately. FBI Director Louis J. Freeh testified that not passing the Secure Public Networks Act would ultimately devastate the agency's ability to fight crime and prevent terrorism.

Denning, who had, since the Clipper Chip debacle, remained a supporter of key-recovery systems, weighed in on the debate with a study conducted with William Baugh, Jr., of Science Applications International Corp. Surprisingly, she sided, to some degree, with industry leaders, which raised the already heated debate to another level. Their study cited at least 500 criminal cases worldwide in which encryption had been used. Law enforcement agencies had been able to prosecute many of them without a key-recovery system in place, either through weaknesses in foreign encryption programs or through human error. While affirming that criminal use of encryption was growing by anywhere from 50 to 100 percent each year, Denning now doubted that a key-recovery system was the cure, claiming it might become "incredibly burdensome and costly, and involve a certain amount of risk to the people whose data is encrypted."

While neither the Secure Public Networks Act nor the SAFE Act passed during the 105th Congress, beginning in 1998, the Clinton administration relaxed encryption export controls, and, in January 2000, abandoned the key-recovery-system program altogether.

Denning later conceded that she had been misguided in her original support of key escrow. "Of course, I didn't convince anybody," she told *Information Security Magazine.* "In the end, the way encryption policy was handled was right. We were right to liberalize it." Meanwhile, open criticism of Denning had begun to die down. After speaking at the sixth annual

Computers, Freedom, and Privacy Conference in March 1996, she commented, "I wasn't hissed at this year, so I guess that's progress."

## GEO-ENCRYPTION

Even as she was embroiled in the encryption export debates of the early- to mid-1990s, Denning pioneered another new security technology—geo-encryption. First addressed in Denning's February 1996 paper, "Location-Based Authentication: Grounding Cyberspace for Better Security," and later patented by her in 1998, geo-encryption is a process in which data are encoded so that they remain scrambled until reaching their location, as determined by the satellite global positioning system (GPS). In short, location is the key word. Not only could geo-encryption be used to transfer secure data, Denning suggests that the unique location signature defined by the GPS system could theoretically be used to locate hackers and other cybercriminals. In addition, because documents can also be time-, date- and location-stamped, geo-encryption could be used to trace the origins of fraudulent transactions, libelous remarks, or death threats.

The applications for such technology seemed endless and, as Denning initially suggested, could be of particular interest to the Department of Defense, businesses, and the medical profession. In times of war, geo-encryption could be used to send messages from the Pentagon to commanders in the field without fear of interception. Medical records could be sent across the country for a second opinion with no fear of privacy leaks. Business meetings could be encoded and decoded hundreds of miles away. Security is guaranteed because the location signature is nearly impossible to forge.

In 2001, movie executive Mark Seiler and MapQuest founder Barry Glick approached Denning with a plan to marry geo-encryption with entertainment distribution. Although skeptical at first, Denning eventually cofounded GeoCodex, an Arlington, Virginia–based startup, with Seiler and Glick. "I find the challenge of dynamically encrypting streams of video or classified data by using GPS satellites intellectually intriguing," Denning told *CIO Insight* in 2002. While teaching at Georgetown full-time, Denning advised on the development of technology that would allow movie studios to distribute directly to movie houses and video-on-demand customers, using a geo-encryption chip that could cost as little as $10. Though the company does not expect widespread use of its technology for several years, GeoCodex has also undertaken two other joint ventures that involve the transmission of classified government data and medical records, and Richard Clarke, one of the government's top cyberterrorism advisers, showed early interest in Denning's technology.

> *I gradually came to see that commercial systems would never be perfectly secure. They would always have weaknesses. Even in the domain of cryptography, the race will never end between the code makers and code breakers.*
>
> —Denning in acceptance speech, National Computer Systems Security Award, October 1999

## CYBER-WARFARE IN THE 21ST CENTURY

Geo-encryption was not Denning's only tool against cyberterror sought by the U.S. government. In the wake of the September 11 terrorist attacks, Denning was called upon to assess the possibilities of cyberterror being used against the United States and to join the Task Force on Cyber Threats for the Future, a part of the Homeland Defense program located at the Center for Strategic and International Studies in Washington, D.C.

By September 11, cyber attacks in the midst of war and political conflict had become increasingly common. During the war in Kosovo in 1999, NATO computers came under the fire of email bombs and denial of service attacks, and, after the United States accidentally bombed the Chinese Embassy in Belgrade, Chinese hackers bombarded U.S. government Websites. Similar defacements and denial of service attacks have also been part of the ongoing Israeli-Palestinian conflict and the struggle over Kashmir on the India-Pakistan border. Indeed, a group calling themselves the Dispatchers launched attacks against hundreds of Websites, including the Iranian Ministry of Interior and the

Presidential Palace of Afghanistan. Similarly, GForce Pakistan defaced U.S. Websites with pro–Bin Laden messages.

As news of these attacks—as well as speculation about hypothetical attacks against nuclear power plants or water systems—spread through the media, the public became increasingly alarmed. A National League of Cities survey showed that, by 2001, fear of cyberterrorism ranked as high as fear of biological and chemical weapons. Denning and her fellow experts, however, could dismiss such attacks as "hacktivism"—online activism that causes more nuisance than real harm. Cyberterrorism, Denning contended, would be defined as attacks that led to death or bodily injury or sabotaged energy grids, contaminated water supplies or that wreaked massive havoc in financial markets—thus defined, Denning asserted, cyberterrorism did not exist.

While alarmists in the government continued to claim, in a phrase first uttered by Representative Lamar Smith of Texas, that a "mouse can be just as dangerous as a bullet or a bomb," Denning remained sanguine. "I don't lie awake at night worrying about cyberattacks ruining my life," Denning told the *Pittsburgh Post-Gazette*. "Not only does [cyberterrorism] not rank alongside chemical, biological or nuclear weapons, but it is not anywhere near as serious as other potential physical threats like car bombs or suicide bombers." Although she acknowledges that cyberterror could pose a threat in the future, she is quick to point out that, during the U.S. Naval War College's simulated war game, Digital Pearl Harbor, it was determined that, to be successful, cyberterrorists would need $200 million, access to sensitive intelligence, and more than five years of planning.

America's cyberwarrior—so named by *Time* magazine in November 2001—has built her career on the concept of security, at first at the level of programming and, more recently, at the level of national policy. In so doing, she has been recognized with innumerable awards, including the 1999 National Computer Systems Security Award and the 2001 Augusta Ada Lovelace Award for her lifetime achievements from the Association for Women in Computing. Currently, she is a professor in the Department of Defense Analysis at the Naval Postgraduate School, in Monterey, California, where she is affiliated with the Center on Terrorism and Irregular Warfare and the Center for Information Systems Security Studies and Research.

## FURTHER READING

### In These Volumes
Related Entries in this Volume: Diffie, Whitfield; Gilmore, John

Related Entries in the Chronology Volume: 1990: Operation Sun Devil; 1993: Escrowed Encryption Initiative Is Introduced By the Clinton Administration; 2001: The Internet and the September 11 Attacks

Related Entries in the Issues Volume: Crime and the Internet; Cyberterrorism; Encryption; Hackers; Security

### Works By Dorothy Denning
*Cryptography and Data Security.* Reading, MA: Addison-Wesley, 1982.

*Information Warfare and Security.* New York: ACM Press, 1999.

"The Limits of Formal Security Models." Award acceptance speech at National Computer Systems Security Award Conference, October 18, 1999. Available online at http://www.cs.georgetown.edu/~denning/infosec/award.html (cited September 16, 2004).

With Peter J. Denning, eds. *Internet Besieged: Countering Cyberspace Scofflaws.* New York: ACM Press, 1998.

With Herbert S. Lin, eds. *Rights and Responsibilities of Participants in Networked Communities.* Washington, DC: National Academy Press, 1994.

### Articles
Butler, Rhett, and Andrew Goldstein. "Keeping the Hackers at Bay: The Cyberwarrior." *Time,* November 26, 2001.

Epstein, Keith. "How Geo-Encryption Makes Copyright Protection Global." *CIO Insight,* April 1, 2002, http://www.cioinsight.com/article2/0,3959,1516,00.asp (cited September 16, 2004).

Hotz, Robert Lee. "Scientist Says She Can't Crack the Clipper Chip." *Los Angeles Times,* October 4, 1993.

Levy, Steven. "Clipper Chick." *Wired* 4.09, September 1996, http://www.wired.com/wired/archive/4.09/denning.html (cited September 16, 2004).

"The Myth of Cyberterrorism." *Pittsburgh Post-Gazette,* December 8, 2002.

Radcliff, Deborah. "The Security Sentinels." *Computerworld,* April 8, 2002.

### Websites
Center for Democracy and Technology: Encryption. Encryption primer from the CDT, http://www.cdt.org/crypto/ (cited September 16, 2004).

Center for Strategic and International Studies: Transnational Threats Initiative. Online home of the Cyber Threats of the Future Task Force of the Homeland Security Department, http://www.csis.org/tnt/taskcyber.htm (cited September 16, 2004).

CITW: Center on Terrorism and Irregular Warfare. The Naval Postgraduate School's Center for cyberterrorism, http://www.nps.navy.mil/ctiw/index.html (cited September 16, 2004).

Dorothy Denning's homepage. Denning's professional site contains links to many of her most important publications, http://www.cs.georgetown.edu/~denning/ (cited September 16, 2004).

International Association for Cryptologic Research. The online home of the IACR, founded by Denning, http://www.iacr.org/ (cited September 16, 2004).

# Whitfield Diffie (1944–)

## CRYPTOGRAPHER

**Bailey Whitfield Diffie's** name does not appear on the computer screen every time people purchase books, airplane tickets, or other goods and services online, but he made secure e-commerce possible. In the mid-1970s, Diffie developed the basic concept behind public-key encryption, the security function that underlies the majority of encrypted transactions online, including bank transactions and credit card purchases. Diffie's landmark discovery helped bring encryption out from the shadows, enlarging the cryptographic community from cloistered government mathematicians to thousands of cryptographers, from amateur hobbyists to professionals, and "crypto anarchists" worldwide. In the mid-1990s, Diffie began to be active in public policy, fervently opposing one of the Clinton administration's proposed encryption measures (the Clipper Chip) and strongly advocating the relaxation of encryption export controls. His unflagging support of strong encryption as a way to ensure privacy and security in the digital age has earned him the moniker, "the father of Internet-computer security."

## FINDING CRYPTO

Diffie first discovered cryptography in the fifth grade of his primary school in Queens, New York City. His teacher, Mary Collins, spent an afternoon explaining simple cryptosystems—how secret codes could be made, how they could be deciphered, and how codes could be broken. The young Diffie was immediately fascinated and asked his father, a professor of history at the City College of New York, to check out all the cryptography books in the City College library. Diffie read all the books, but his interest quickly waned. By high school, he thought of himself as a mathematician. When Diffie enrolled at the Massachusetts Institute of Technology (MIT), he declared as a math major.

Although Diffie proved to be an adept student of mathematics, he found MIT to be stifling, socially and politically. The New York City native, who had grown up in a left-wing community of "red-diaper babies," found his political kin while studying for two summers at the University of California–Berkeley. He embraced the burgeoning student movement and the leftist, anti–Vietnam War politics of that campus. When he graduated from MIT in 1965, the self-described "peacenik" avoided the military draft by accepting a job at Mitre Corporation, a Massachusetts systems engineering company that did work for the Defense Department.

Diffie worked for Mitre out of MIT's Artificial Intelligence Lab, where he fell in with a community of early hackers—hackers in the original and purest sense of the word: individuals who wrote and tinkered with software. There, his interest in cryptography, forgotten for nearly fifteen years, was rekindled through conversations with his boss, Roland Silver, about privacy, security, and the National Security Administration (NSA). Diffie later told CNN that the politics of the time led him to be "very conscious of the needs of the individual for protection from the state."

While at Mitre, someone told Diffie, mistakenly, that NSA encrypted the telephones inside its own buildings (NSA actually ran its phone lines through a shielded conduit.) At the time, NSA was practically the sole source of cryptographic use and cryptanalysis in the United States. Diffie wondered how NSA could accomplish such a thing—encrypting hundreds, if not thousands, of individual phone conversations—and then expanded his thinking to the entire North American telephone system. How could two people who have never talked before have a spontaneous conversation that was also encrypted and therefore secure? Handling the sheer number of cryptographic keys to unlock each conversation would be overwhelming for any computer, he believed.

The question lingered in the background when Diffie began working under John McCarthy, one of the pioneers in artificial intelligence, at Stanford's Artificial Intelligence Lab, in 1969. (Ironically, the project was funded by NSA.) In November of that year, McCarthy presented a theoretical paper on home shopping—what is presently known as e-commerce. It led Diffie to ponder "paperless offices" and the problem of authenticating transactions. What, Diffie wondered, would replace the one-of-a-kind signed document that was the basis for most transactions? It was the first step in his quest to develop digital signatures.

Over the next several years, McCarthy encouraged Diffie's exploration of cryptography. Diffie read *The Codebreakers* (1967), David Kahn's massive, meticulously documented history of codes and ciphers. "I read it more carefully probably than anyone had ever read it," he told *Wired*. "It's like the Veda—in India if a man loses his cow, he looks for it in the Veda. In any event, by the spring of 1973, I was doing nothing but cryptography." That year, Diffie took to the road in his Datsun 510, in search of what little information about cryptography was available in the public domain.

## THE ITINERANT CRYPTOGRAPHER

In the mid-1970s, cryptography was almost entirely the province of NSA. Hundreds of NSA mathematicians held the keys—literally and figuratively—that Diffie sought, but all the information was classified, locked inside the impenetrable NSA headquarters in Fort Meade, Maryland. In the course of his two years on the road—during which he met his future wife, Mary Fischer, who began traveling with him—Diffie pieced together what he could. "I went around doing one of the things I am good at, which is digging up rare manuscripts in libraries, driving around, visiting friends at universities," Diffie said in a 1992 interview. Given the secrecy of NSA, however, his journey often took on a sinister edge. Recalled Fischer, "It was really cloak-and-dagger—people who didn't want to talk to him, people who put their

coats over their faces, people who wanted to know how the hell he'd found their names, people who had secrets, clearly, and were not about to share them."

In the summer of 1974, Diffie and Fischer visited the IBM Laboratory in Yorktown Heights, New York, one of the only significant nongovernmental groups working on cryptography in the country. (The lab developed the data encryption standard [DES], which the U.S. government used until the code was cracked in the 1990s.) There, Diffie spoke with Allen Konheim, who told him to look up Martin Hellman, a professor of electrical engineering, at Stanford University in California.

When Diffie returned to California that fall, Hellman agreed to a half-hour appointment. "There was an immediate meeting of the minds," says Hellman. The meeting lasted into the night; the working relationship lasted far longer. Hellman hired Diffie as a part-time researcher in late 1974. "When I made my first talk at Stanford after I got back," recalls Diffie, "Hellman described me in the flyer as an 'itinerant cryptographer.'" From then on, the two men teamed up to tackle one of the primary problems in cryptography—key distribution: how can two people share the key to a code without danger of the key falling into a third party's hands?

> *The Internet is so valuable as a communication mechanism that people and corporations cannot afford not to use it. But a lot of the Internet is out of people's control and it's only cryptography that makes it safe.*
>
> —*Diffie to* The Irish Times, *September 2001*

## THE KEY

The answer came in May 1975. Diffie and Fischer were taking care of John McCarthy's house, which was just up the hill from where Martin Hellman lived. McCarthy's house had a good workstation and a high-speed connection to Stanford's Artificial Intelligence Lab. All Diffie had to do was watch the house, tend to McCarthy's daughter, and think. He spent his days alternating between chores and research. For a while, he felt as though he had become "a broken-down researcher" at the age of thirty-one. Then Diffie had a breakthrough.

Diffie had been thinking about military "identification friend or foe" systems and ways to combine those systems with one-way functions, a mathematical technique. First, he thought he had developed an authentication system that provided a true digital signature—an issue he had pondered since McCarthy's paper on home shopping. Weeks later, Diffie realized he had also answered the key-distribution question that had first occurred to him at MIT—how two people who had never communicated before could have an encrypted conversation.

Briefly, Diffie's idea was to split the key into two: a public key and a private key (hence public-key encryption). Messages encrypted with a person's public key could only be decrypted by his or her private key. So, if John wanted to send Jane a secure message, he would encrypt his message using Jane's public key, which was available for all to use. Jane could then decrypt John's message with her private key. Jane would use John's public key to encrypt her reply, which John would decrypt with his own private key. Unlike single-key encryption, Jane and John—who may or may not have ever contacted each other before—could communicate without having to exchange the same secret key. In addition, the private key could then also act as a signature, because messages written in the private key could only come from one place; thus, the public key could verify the origin of the message and could, accordingly, be used as a digital signature.

Diffie walked down the hill to Hellman's house and spent almost an hour explaining his discovery. Then, the two men set out to prove the feasibility of Diffie's concept and to write up their findings. An early version of the Diffie-Hellman paper found its way to Ralph Merkle, a graduate student at UC–Berkeley. Merkle collaborated with Diffie and Hellman, along with other scientists who had made similar inroads into the key-distribution problem, including Steve Pohlig, a Stanford graduate student, and Peter Deutsch. Although Diffie is often lauded as the father of public-key encryption, he is quick to point out that each of the men in this working group played a role in its early theoretical development.

In truth, Diffie was not the first to solve the problems of key distribution and digital signatures, he was simply the first to make his solution publicly available. In April 1969, James Ellis, a British physicist, began to tackle the problem for the top-secret Government Communication Headquarters in the United Kingdom. Ellis also developed a concept of public and private keys, but he called his discovery "non-secret encryp-

tion." He lacked the mathematical skills to implement nonsecret encryption, so it sat until, four years later, Clifford Cocks, a recent graduate of Cambridge, developed algorithms based on Ellis's work. However, Ellis's concept was kept secret, unlike public-key encryption. At a computer security conference in San Francisco, held in January 1998, Diffie publicly acknowledged Ellis's work, which predated his own. The NSA also claims to have known about public-key cryptography in the 1960s.

In November 1976, Diffie and Hellman published "New Directions in Cryptography" in the journal *IEEE Transactions of Information Theory.* (Merkle published a similar article in *Communications of the ACM* in 1978, titled "Secure Communications Over Insecure Channels.") The first line of the Diffie-Hellman article read, "We stand today on the brink of a revolution in cryptography."

Indeed, they did. In 1977, three MIT scientists—Ron Rivest, Adi Shamir, and Leonard Adleman—put the Diffie-Hellman scheme to work in an encryption process that became known as RSA for Rivest, Shamir, and Adleman. Public-key encryption—and modern-day nonmilitary cryptography—was born.

In 1978, Diffie became manager of secure systems research for Northern Telecom's lab in Mountain View, California. There, he developed a secure phone system that, though it was never implemented, provided the design for Packet Data Security Overlay, which enabled end-to-end encryption between packet data networks. In 1991 Diffie became a security expert for Sun Microsystems and once again became a champion of cryptography, this time in the realm of public policy.

## ON CAPITOL HILL

When the Clinton administration introduced the Clipper Chip in April 1993, Diffie was at the helm of the movement against it. The Clipper Chip was a security device that could encrypt telephone communications. While opponents took issue with many aspects of the Chip itself—mainly that it was based on a secret, classified algorithm—the primary problem with the plan, in many people's eyes, was its key-escrow system. A second set of keys would be held in "escrow" by a third party (in this case, a government-related agency) so that the NSA, the CIA, or the FBI could, with a warrant, eavesdrop on supposedly secure communications. Critics likened key escrow to a system in which the government declared that every citizen leave a front door key with the local police precinct. This "backdoor" to the Chip enraged civil libertarians, grassroots Internet activists, and industry leaders alike.

Diffie had been concerned about government-managed encryption systems since the mid-1970s, when DES was proposed as the U.S. government encryption standard. "I did not understand how those people dared either standardize a secure system or standardize a nonsecure system," says Diffie, "because if it was secure—since they were primarily an intelligence agency—they would be afraid they wouldn't be able to read other people's traffic. If it was not secure, since they had certified it for the use of U.S. government organizations, they risk having a tremendous black eye if it were broken."

With the Clipper Chip, nearly twenty years later, Diffie's critiques were even more acute. "The big thing we've gotten away from in contemporary cryptographic technology is the vulnerability that grows out of having to maintain secret keys for longer than you actually need them," Diffie explained. "Two of the most damaging spy scandals of the last twenty years in the U.S. . . . resulted from the fact that keys existed for longer than they needed to exist, and somebody got a chance to siphon some of them off. If you use public key correctly, particularly in interactive channels like telephones, you can avoid having this hazard. The keys exist only in the equipment, only for the duration of the call, and after that, they go away . . . Key escrow is just rescuing a dreadful vulnerability." Indeed, the vulnerability lay in the fact that key registries would require those who used encryption to trust a third party—the exact situation Diffie designed public-key encryption to avoid.

Diffie testified before the Senate Judiciary Subcommittee on Technology and the Law, chaired by Senator Patrick Leahy, in May 1994. He spoke at length about the history of cryptography, dating back to the 1790s, the threats to privacy and security the use of a key-escrow system entails, and key escrow's impact on industry and the worldwide development and deployment of encryption technologies. By November of that year, Diffie, with his long, flowing hair, was a familiar presence in Washington, having testified three times before Congress and participated in a blue-ribbon panel on the future of cryptography. The lucidity and strength of his argument, and the general media frenzy around the Clipper Chip technology, helped land Diffie on the cover of the *New York Times Magazine* that year, facing off with NSA director, Vice Admiral John McConnell.

As the encryption debate evolved over the next several years, Diffie continued to speak against key-escrow–based systems and on behalf of loosening export controls on strong encryption. Testifying for the Promotion of Commerce Online in the Digital Era Act of 1996, Diffie said, "It is a pleasure, at long last, to be able to appear and testify in favor of something." He repeated the sentiment when testifying for the Computer Security Enhancement Act of 1997.

In September 1998, the Clinton administration began to relax certain export controls on encryption. Diffie, along with the cryptography community at large, saw this as a small step in the right direction; in January 2000, export controls were further relaxed. In February of that year, Diffie was called upon by President Clinton for an Internet Security Summit to assess the security risks revealed by several significant hacker attacks against prime Websites, including eBay, Yahoo!, and sites related to Internet security companies as well as the Transportation Department. (One hacker attacked the source of public-key encryption itself, by redirecting visitors from the RSA Website to a site in South America, which showed the message "Trust us with your data! Praise Allah!") Diffie took the opportunity to call for increased security while emphasizing that the route to in-

> *If you look at security in human society, the two most important mechanisms I can think of are personal recognition and private conversation. Cryptography is essential to transplanting those mechanisms—that people don't even think of as security—into a purely digital world.*
>
> —Diffie to Newsbytes, *October 1997*

creased security was strong encryption. "This is a matter of the recognition of the importance of computer security, increasing the priority of computer security features . . . among the features that are delivered with an operating system, with a computer environment."

## IN THE SUN

In April 2002, Diffie became Sun Microsystem's first chief security officer. Less than a year later, he was appointed a Sun Microsystem fellow, the most senior role in Sun's engineering wing. Such appointments recognized Diffie's steadfast work in security, both in advocating for stronger and better standards for security in the industry and in researching key security issues as technologies develop.

While encryption is not a core business for Sun, it is still a key aspect of Diffie's work. He has been called upon to comment on the slow growth of the public-key infrastructure (PKI), the framework necessary for public-key encryption to function on a widespread basis. The public-key encryption algorithms developed by RSA provide the basis for secure e-commerce. PKI, however, goes beyond authenticating business or financial transactions. It would provide a worldwide infrastructure for secure, encrypted communications. Diffie claims that the two institutions large enough to have initially launched PKI—AT&T and NSA—eventually helped stall the initiative because, in the case of AT&T, it was disbanded, or, in the case of NSA, of difficult policy issues. Like the telephone, he explained, PKI needed widespread adoption to work. "When just 10 percent of the population had one, it wasn't very useful," Diffie said, of telephones. "By the time 99 percent had one, they were crucial."

In addition, Diffie has championed the next generation of public-key encryption, called Elliptic Curve Cryptography (ECC), which, by using advanced elliptic-curve mathematics, could make encryption work more quickly, more cheaply, and on more devices. (PKI, for instance, requires significant computer hardware.) In 2002, he announced that Sun's research about ECC would be contributed to the OpenSSL Project, an open-source (freely distributed) implementation of the secure sockets layer protocol, a dominant security protocol on the Web.

Diffie also continues to appear in the political spotlight. In 2002 he testified in opposition to the transfer of the National Institute of Standards and Technology, the federal technology agency, into the Homeland Security Department, citing his concern that "the 1990's involvement of law enforcement and national security in the development of civilian computer standards delayed the deployment of secure computer systems, moving the division . . . would be detrimental to homeland security."

Still, Whitfield Diffie is, perhaps, best known for bringing the once arcane field of cryptography out into the open. When Diffie "split the key," as the basis of public-key encryption is sometimes called, he not only answered the two questions that had plagued him since the 1960s—digital signatures and key distribution—he paved the way for the development of e-commerce as we know it. His continuing work in the realm of encryption and individual privacy make him a powerful force in shaping the future of cryptography and security online.

## FURTHER READING

### In These Volumes

Related Entries in this Volume: Denning, Dorothy

Related Entries in the Chronology Volume: 1977: The RSA Encryption System Is Invented; 1991: Pretty Good Privacy Is Released on the Internet; 1993: Escrowed Encryption Initiative Is Introduced By the Clinton Administration; 2001: The Internet and the September 11 Attacks

Related Entries in the Issues Volume: Crime and the Internet; Cyberterrorism; E-commerce; Encryption; Hackers; Security

### Works by Whitfield Diffie

with Susan Landau. *Privacy on the Line: The Politics of Wiretapping and Encryption.* Cambridge, MA: MIT Press, 1998.

with Richard Parkinson. *Cracking Codes: The Rosetta Stone and Decipherment.* Berkeley: University of California Press, 1999.

### Books

Kahn, David. *The Codebreakers: The Story of Secret Writing.* Rev. ed. New York: Scribner, 1997.

Levy, Steven. *Crypto: How the Code Rebels Beat the Government—Saving Privacy in the Digital Age.* New York: Viking, 2001.

### Articles

Arnd, Weber. "Enabling Crypto: How Radical Innovations Occur." *Communications of the ACM,* April 1, 2002.

Kerr, Deborah. "Public Key Mystery." *Computerworld,* September 9, 1996.

Levy, Steven. "The Prophet of Privacy." *Wired* 2.11, November 1994, http://www.wired.com/wired /archive/2.11/diffie.html (cited September 16, 2004).

———. "The Open Secret." *Wired* 7.04, April 1999, http://www.wired.com/wired/archive/7.04/crypto .html (cited September 16, 2004).

———. "How They Beat Big Brother." *Newsweek,* January 15, 2001.

## Websites

The Atlantic: "Homeland Insecurity." September 2002. Web-only primer on how public-key encryption works, http://www.theatlantic.com/issues/2002 /09/mann_g.htm (cited September 16, 2004).

EPICArchive—Cryptography Policy. The Electronic Privacy Information Center's reports and resources on the encryption policy debate, http://www.epic .org/crypto/ (cited September 16, 2004).

Export of Cryptography in the 20th Century and the 21st. Abstract and link to the full technical report tracking the history of export controls on cryptography from the 1980s into the 21st century, written by Whitfield Diffie and Susan Landau for Sun Microsystems, http://research.sun.com /techrep/2001/abstract-102.html (cited September 16, 2004).

Interview with Whitfield Diffie (1992). Lengthy interview between Franco Furger and Whitfield Diffie, edited by Weber Arnd, http://www.itas.fzk .de/mahp/weber/diffie.htm (cited September 16, 2004).

Matt Blaze's cryptography resources. Comprehensive cryptography resources amassed by Matthew Blaze, the research scientist who cracked the code to the Clipper Chip, http://www.rsasecurity.com/ (cited September 16, 2004).

RSA Security. In 1983, RSA began marketing its algorithm based on Diffie's concept of public-key encryption, http://www.rsasecurity.com/ (cited September 16, 2004).

Washingtonpost.com Special Report: Deciphering Encryption. Concise explanation of the encryption debate, from how encryption technology works to the policy in Washington, D.C., http://www .washingtonpost.com/wp-srv/politics/special /encryption/encryption.htm (cited September 16, 2004).

Dr. Whitfield Diffie. Diffie's professional site at Sun Microsystems, http://research.sun.com/people /diffie/ (cited September 16, 2004).

# John T. Draper (a.k.a. Cap'n Crunch) (1943?–)

## PROGRAMMER; PHREAK; HACKER

**Mention Cap'n Crunch** to most people and they think of breakfast cereal, but to hackers, and to people in the radical underground of the early 1970s, Cap'n Crunch was the legend who cracked the code to AT&T's internal circuits. The phone "phreaks" of the 1970s were the original hackers, typically young men who used various tools, including a toy whistle found in Cap'n Crunch cereal boxes, to navigate their way through the phone system. Though Cap'n Crunch, known in daily life as John T. Draper, was not the first phreak, he is undoubtedly the most famous, revered in computer communities as the forefather of hacking.

## PLUGGING IN

Cap'n Crunch's story—and the story of "phreaking" itself—begins with a gifted blind kid from Tennessee with perfect pitch. Joe Engressia, blind since birth, had been tinkering with the telephone system since childhood. By age seven, he had learned to dial a phone number without touching the dial, by simply tapping the hook switch at the proper intervals. By age eight, he had discovered that he could trick the phone circuits into dropping charges for long-distance calls by whistling into the receiver at precisely 2600 Hertz. Engressia brought this knowledge with him when he entered college at the University of South Florida in the late 1960s. He studied math and made a name for himself—"The Whistler"—by whistling dozens and dozens of free long-distance phone calls for his friends.

In 1969, when the telephone company discovered Engressia's exploits, it issued him a harsh warning, which was matched by a severe reprimand from his university. News spread. Soon, a handful of fellow "phreakers," mostly teenaged boys from across the country, many of them also blind, became aware that there was another just like them. One by one, they contacted Engressia, and the roots of an emerging phreaker underground took hold. (Engressia was arrested repeatedly for phone fraud over the next several years; he legally changed his name to Joybubbles and achieved his lifelong dream of working as a troubleshooter for the telephone company.)

Those without perfect pitch found other ways to crack open the telephone system. Some discovered that the high E above middle C on an electric organ did the trick. Others used flutes or tape recorders. A few discovered that the toy whistle that came as a prize in boxes of Cap'n Crunch cereal could, when altered by closing one of the three holes, produce the tone that functioned like an "open sesame" to the inner workings of Ma Bell. While a handful of early phreakers, like Engressia, came upon the tone by accident, most found the key in a 1954 article published in a technical journal by a Bell Telephone Laboratories engineer, which casually listed all of the exact frequencies used in Bell's multifrequency tone system, including how each of the frequencies functioned.

Dozens of contradicting accounts tell of Draper's entrée into the phone phreaking universe. Some stories claim that Draper had figured out the access tone on his own, without the whistle, while serving as a U.S. Air Force technician in Alaska (or in Europe, or in Vietnam), and that he helped (or charged) his fellows officers to call the United States using the 2600 Hertz tone. (Draper was honorably discharged from the air force in 1968.) Other stories claim Draper himself was the first to discover the key held deep inside the Cap'n Crunch whistle.

By Draper's own account, which itself contains inconsistencies, about a year after he left military service he received a phone call from someone introducing himself only as Denny. At the time, Draper was studying engineering at a community college in the San Francisco Bay area and was working as a technician for National Semiconductor. The initial point of Denny's phone call was to talk about ham radios, as Draper was an enthusiastic hobbyist, but the topic soon turned to Denny's other interest, telephones. (How Denny found Draper is unclear; they had never met.) At first, Draper was not interested. Nevertheless, Draper visited Denny's house, where he discovered not a thirty-year-old man, as he had expected, but a blind teenager and three of his blind friends. The boys showed off their phone knowledge in exchange for some of Draper's technical expertise. They asked him to build a multifrequency tone device, called an M-Fer, that could produce all the necessary tones to master the phone system. At the time, the boys were using an electric organ. When Draper returned home, it took the seasoned technician less than an hour to fashion his first "blue box."

## THE LITTLE BLUE BOX

Draper's first M-Fer was a primitive version of what was to come. M-Fers, which later came to be known as blue boxes, worked by mimicking the twelve tones used to run the telephone company's trunking system—ten separate tones for the numbers 0 through 9, plus a KP key (Key Pulse), which alerted the system that a phone number was to follow, and an ST key (Start), which told the system to process the call. When reproducing the tones, the acceptable margin of error was roughly thirty cycles per second; within that range, the phone company's circuits could not differentiate between a human whistle, an organ, a tape recording, a blue box, or its own equipment.

All of this was possible because of a cost-saving decision, made by AT&T in the early 1950s, to run voice and signaling on the same circuit. Anyone with access to the proper tones could have the entire phone system in the palm of his hand (early phone phreaks were, almost without exception, male).

Draper soon figured out patterns in phone numbers and delved deeper into the phone system, using trial and error to gain access to far more than a simple free long-distance call. At the time, all area codes contained either a 0 or 1 in the middle and began with any number other than 0 or 1 (212 for New York, 415 for San Francisco, and so on). Draper discovered that dialing 121 gave him access to the internal phone system: 121 gave him a local operator, 131 gave him information, 141 gave him a special routing operator for his area, who could give him codes for operators around the

world. In time, he discovered testing numbers (two special sequential phone numbers used to test equipment that, if dialed, connected any two people in the world for free) and other quirks of the AT&T system. Draper used his growing list of codes like a map. All the while, he built bigger and better M-F devices.

His work was spurred on by the historic 2111 conference of 1970. By dialing KP-604-2111, any phreak with a blue box could drop in on a conference call of his peers. 604 was a Vancouver area code, and by dialing 2111 the phreaks gained access to an unused Canadian Telex testing trunk, which gave them a permanently open line. Members—most notably Draper himself— could swap access codes and technical advice, gossip, and show off. Draper detailed the specific codes he discovered and bragged about his love life. On one occasion, he called himself from around the world—M-F-ed into Tokyo, heading west to India, then Greece, south to Pretoria, South Africa, across the ocean to South America, back to London, across the Atlantic to New York, and finally ringing the phone right next to him in California. The delay was twenty seconds. He could barely hear himself speak.

It was a small and motley crew of phreaks from all over the country, many of them blind teenagers, who found friendship and solace in this precursor to virtual communities. Draper later recalled there being one phreak in Long Island, one in Memphis, three in Los Angeles, three in Seattle, one in San Rafael, five in San Jose, adding that operators and technicians from as far away as Tokyo also took part. They went by names like Martin Freeman (code for an M-Fer), Frank Carson (code for "free call"), the Cheshire Cat and the Midnight Skulker (a.k.a. Mark Bernay). Indeed, it was at the historic 2111 conference that John Draper's infamous alter ego was born. On the suggestion of a fellow phreaker, Draper, who had unimaginatively used "John" as his handle, adopted a new pseudonym. "Why not call me Cap'n Crunch?" he said. The conference ended on April 1, 1970, when the telephone company discovered the dormant trunk, but Draper's nickname stuck.

## PHREAKING ACROSS AMERICA

The legend of Cap'n Crunch, well known at the time among the coterie of phreakers but virtually unheard of by the public at large, was solidified by an article in *Esquire* by Ron Rosenbaum, "Secrets of the Little Blue

Box." In the early 1970s, *Esquire* had been approached by a phreak who sought vengeance against AT&T, which had foiled his plans to sell nearly $300,000 worth of blue boxes to an organized-crime ring in Las Vegas. Initially, the idea was to include a 45-rpm recording of the blue box tones, bound into the magazine along with the story, which would give the entire nation keys to AT&T's kingdom. *Esquire's* lawyers vetoed the recording, but approved the story, which ran in October 1971.

Rosenbaum called the phreak at the center of the article, the one seeking revenge, "Al Gilbertson." But ultimately Cap'n Crunch (and, to a lesser degree, Joe Engressia) proved to be the real star. It was Cap'n Crunch who could send his voice flying around the globe before coming through the receiver on the other end; Cap'n Crunch who had built a complex M-Fer from parts pilfered from his job at National Semiconductor; Cap'n Crunch who could, he warned, bring down the entire telephone system with the help of only two other phreakers. He bragged about his prowess, telling Rosenbaum, "My ears are a $20,000 piece of equipment. With my ears I can detect things [AT&T] can't hear with their equipment." And, perhaps most important for generations of hackers to come, he revealed to Rosenbaum the allure of phreaking. "The phone company is a system, . . ." said Cap'n Crunch. "If I do what I do, it is only to explore a system. Computers, systems, that's my bag. The phone company is nothing but a computer." He added, "Ma Bell is a system I want to explore. It's a beautiful system, you know."

Soon after *Esquire* hit the newsstands, phreaking exploded across the nation. Although the record never appeared in the magazine, premade tone recordings made their way into the public from the phreaker underground. Tech-savvy readers also sought out the Bell Systems technical journal and built blue boxes of their own. Indeed, two students, Steve Wozniak and Steve Jobs, sneaked into the library at Stanford's Linear Accelerator Center (SLAC) to do just that. Frustrated, they sought the advice of Cap'n Crunch himself, contacting him through KPFA, Berkeley's free-speech radio station.

Wozniak and Jobs were taken aback when the mythic Cap'n Crunch arrived at Wozniak's UC–Berkeley dorm room. Expecting a super-engineer, they found, instead, a wild-haired madman with horn-rimmed glasses. Draper introduced himself simply by saying, "It is I," then began explaining his vast knowledge of Ma Bell's systems. Wozniak's first call was to the pope, "to confess," remembers Draper. Soon after, Wozniak and Jobs had sold enough blue boxes to finance a year of college.

In June 1972, eight months after *Esquire* published Rosenbaum's article, *Ramparts* magazine published schematic designs for blue boxes for all to see. By then, however, Draper had already been arrested for wire fraud.

> *True hackers don't learn from books. They work with very little information and go out and find things on their own, instead of learning it from someone.*
>
> —From "Cap'n Crunch FAQs," www.webcrunchers .com/crunch (cited September 16, 2004)

## WHERE THE PHREAKING NEVER STOPS

According to Draper, he was on his way home from a FORTRAN programming class in May 1972 when he was nabbed by "men in suits" at a 7-11 parking lot near his home in Mountain View, California. He received five years' probation. Most believe the publicity from the *Esquire* article brought him to the attention of authorities.

Draper told authorities he hadn't phreaked in months, which may have been at least partly true, as his interests had turned to computers themselves. At the time, Draper was working as a programmer at Call Computer, a time-sharing computer services company, hacking into his own programs to make them better. He was also a member of the People's Computer Club (PCC), a group of programmers and hackers who met on Wednesday nights in a storefront in Menlo Park to talk tech and eat potluck dinners. (The PCC eventually became the famed Homebrew Computer Club, where Steve Wozniak first displayed the Apple I.)

According to Draper, two years into his probation he was befriended by a man named Adam Bauman. The pair bonded over phones and computers. Draper eventually committed the White House Toilet Paper Crisis phreak with a friend of Bauman's in Los Angeles. (This was a prank in which, as Draper tells it, he and Bauman's friend called the White House, allegedly reaching President Nixon by using his code name, Olympus, and said "We have a crisis here in Los Angeles! . . . We're out of toilet paper sir!") Within a period of months, however, Bauman led Draper to his second arrest, in 1974, by entrapping him on an FBI-tapped pay phone outside of the PCC storefront. (Others claim Draper's entrapment story is false, and that Cap'n Crunch never stopped phreaking.) In October 1976, Draper was sent to a minimum-security prison in Lompoc, California.

"Jails are the perfect venue for transferring hacking knowledge," Draper admits. "Inmates have a lot of spare time on their hands, and a patient teacher can teach just about anyone anything, given enough time." While at Lompoc, Draper held informal classes three evenings per week on the fine art of phreaking. He used his technical savvy to win favor among his inmates, unearthing the phone numbers of ten unlisted pay phones to give to other inmates. He made himself invaluable—and assured his safety— by giving lessons on detecting bugs and traces, building blue boxes, and altering FM radios to use for communication. Draper later boasted that thirty to fifty ex-cons were released into the outside world with enough information to defraud Ma Bell hundreds of times over.

Toward the end of his sentence, Draper entered the work-furlough program, during which he wrote the preliminary design for EasyWriter, the word processing program that eventually came bundled with the first IBM PC in 1981. During the day, Draper wrote code in an early form of FORTH, one of the first extensible (designed to allow the addition of new features) computer languages, at a computer at the recording studio where he worked; at night, he debugged the program line by line in his prison cell. Within twenty-four hours of his release, in February 1977, Draper was president of his own software company, Cap'n Software, Inc., headquartered on Telegraph Avenue in Berkeley, California. Soon after, he also became the thirteenth employee of Apple Computer.

Later that year, Draper was arrested a third time, having violated his probation by associating with "known phone phreakers." He was sentenced to two months at the Northampton State Prison in Pennsylva-

nia. There, he gave false technical advice to a fellow inmate he suspected of being an informer and, in return, was beaten so badly that he was left with several broken vertebrae.

## FROM BLACK HAT TO WHITE

Despite Draper's legal problems, the success of Easy-Writer brought him an unaccustomed amount of acclaim and fortune. By the mid-1980s, he had a winter home in Honolulu in addition to his Bay area apartment. He drove a Mercedes 240D through the streets of Berkeley. By several accounts, he was a millionaire.

Over the next decade, however, Draper's life slowly unraveled. He bounced from job to job and city to city, programming here and there, but never staying in one place for long. In 1996, Draper showed up at The Loft, an artists' collective in San Diego. There, he found a high-speed Internet connection and a group of like-minded individuals (some nearly half his age) who shared his interests in raves, electronic music, and rave culture. When that community fell apart, Draper was homeless. He spent some time in Tijuana, Mexico, before heading east to Florida to work on Website development. Along the way, in Texas, the sole manuscript of his autobiography was stolen from his car while he slept. From Florida, Draper landed across the globe in India. In late 1999, he left the Goa coast to return to Fremont, California.

> *I'm not a bad guy . . . [b]ut I'm being treated like a fox trying to guard the hen house.*
>
> —Draper to the New York Times, *2001*

Back in the Bay area, Draper sought to rehabilitate his tarnished reputation and put his hacking abilities to good use. In November 1999, he and several partners founded ShopIP, a security software and consulting firm. Over the next year, Draper developed an advanced firewall-IPS system he dubbed CrunchBox. When it was released in 2001, he told the *New York Times* that ShopIP was "a real change in direction" for him, one that "made [him] realize that [he] could pay back society for [his] deeds in the past." CrunchBox 3.3, released in May 2003, consists of a "packet filtering firewall, an intrusion prevention system and an automatic response module," running on a secure OpenBSD operating system on a Pentium III. Fifty-seven years old when it came out, Draper still has the hacker spirit. In an interview with the British newspaper, the *Guardian,* Draper said he felt sorry for any amateur hacker who came up against his CrunchBox system, adding, "They are gonna be in for a nasty surprise!"

Decades after the fact, Draper still enjoys a quiet fame as the mythic Cap'n. He was paid homage in the movie *Sneakers* (1992), when Whistler, a blind hacker based on Joe Engressia, is shown eating Cap'n Crunch cereal. Although ShopIP has since gone out of business, Draper is still called upon to comment on the latest bout of cell phone hacking. Email hoaxes about false technicians stealing long-distance calls from unsuspecting customers, based on the Cap'n's antics, pop up now and again. But mostly, Draper busies himself with the painstaking task of mastering a new system—Internet security.

## FURTHER READING

### In These Voumes

Related Entries in this Volume: Mitnick, Kevin
Related Entries in the Issues Volume: Crime and the Internet; Hackers; Security

### Books

Freiburger, Paul, and Michael Swaine. *Fire in the Valley: The Making of the Personal Computer.* Berkeley, CA: Osborne McGraw-Hill, 1984.

Levy, Stephen. *Hackers: Heroes of the Computer Revolution.* New York: Doubleday, 1984.

Mungo, Paul, and Brian Clough. *Approaching Zero: The Extraordinary Underworld of Hackers, Phreakers, Virus Writers, and Keyboard Criminals.* New York: Random House, 1992.

### Articles

Besher, Alexander. "The Crunching of America: Cap'n Crunch Tries to Resurrect His Company's Image." *InfoWorld,* June 18, 1984.

Lundell, Allan, and Geneen Marie Haugen. "Merry Pranksters: Jobs and Wozniak Unearth the Secrets of Phone Phreaking." *InfoWorld,* October 1, 1984.

Markoff, John. "The Odyssey of a Hacker: From Outlaw to Consultant." *New York Times,* January 29, 2001.

Rosenbaum, Ron. "Secrets of the Little Blue Box." *Esquire,* October 1971.

———. "Pynchon and Crunch: Heroes of the Underworld Wide Web." *New York Observer,* February 5, 2001.

### Websites

Cap'n Crunch in Cyberspace. Personal site of John Draper, http://www.webcrunchers .com/crunch/ (cited September 16, 2004).

Phone Trips. Mark Bernay's personal site features audio clips of phone phreaks, http://www.wideweb.com /phonetrips/ (cited September 16, 2004).

ShopIP Information Security Systems. Draper's security software business site, featuring the CrunchBox Computer Protection Systems, http://www.shopip .com (cited September 16, 2004).

Telephone Tribute: Phone Phreaking. Telephone hobbyist Dave Massey's personal site, http://www .telephonetribute.com/phonephreaking.html (cited September 16, 2004).

Woz.org. Steve Wozniak's personal Website. Search under Cap'n Crunch, John Draper, and phreaking for more information, http://www.woz.org (cited September 16, 2004).

# Esther Dyson (1951–)

## INDUSTRY ANALYST

**Esther Dyson, a leading** technology industry analyst, is one of the most influential and powerful women in the high-tech sector. Over the years, she has been known by many names: Queen of the Digerati, Doyenne of Cyberspace, a Pied Piper, a guru, a one-woman think tank. To many insiders of the high-tech world, however, she is known simply as Esther.

Dyson was a part of the defining high-tech events of the 1990s: the emergence of ICANN, the international Internet naming group founded during the Clinton administration; the PC Forum, a yearly conference where CEOs, investors, and budding entrepreneurs rub shoulders, exchange ideas, and make or break new technology; the Electronic Frontier Foundation, the first civil rights organization for cyberspace. Her newsletter, *Release 1.0,* is read by a veritable who's who of CEOs and venture capitalists. Her book, *Release 2.0,* was touted by John Doerr, a seasoned venture capitalist, as a "cyber tour de force." Dyson helped bring high-tech to Eastern Europe, making regular circuits from Moscow to Silicon Valley to her home base in New York City. Indeed, on a map of the Internet world, Esther Dyson has been everywhere. She remains a vital voice in conversations about technology, speaking through regular columns for the *New York Times, Release 1.0,* and, most recently, her online Web diary, or blog, *Release 4.0.*

The circumstances of Dyson's upbringing make her role as a leader of cyberspace seem a given. Her British father, Freeman Dyson, has been hailed as a visionary astrophysicist; her Swiss mother, Verena Huber-Dyson, is a distinguished mathematician. (The couple divorced when she was five.) Growing up near Princeton's Institute for Advanced Studies (IAS), where Freeman Dyson worked when Esther was young, friends and neighbors included several Nobel laureates, the inventor of color television, and a handful of key architects of the hydrogen and atomic bombs. As her brother, George, is fond of noting, Esther was born in the same year that John

Von Neumann invented the programmable computer at the IAS. Indeed, one of her first playthings was the remains of an early IAS computer.

A gifted student, Dyson entered Harvard at age sixteen. Although graduating with a degree in economics, Dyson spent most of her time at the student newspaper, the Harvard *Crimson.* The rhythms of her later life began at college: early morning swims, writing, holding her own in a "boys' club," the nonstop networking. A February 1998 article in *Wired* reported that young Esther once said, in response to her father's admonishments that she never seemed to study or attend class, "Oh no, Daddy, you don't understand. You don't come to Harvard to study. You come to Harvard to get to know the right people." More than thirty years later, knowing the right people—and connecting them with each other—remains her stock in trade. "What I try to do is find worthy ideas and people and get attention for them," Dyson told the *New York Times.*

After Harvard, Dyson pursued a career in journalism. Dyson began, in 1974, as a fact checker for *Forbes,* rising quickly to reporter. Reporting for *Forbes* on start-up companies and the people who ran them was like "real-life business school," says Dyson. She covered the electronics and burgeoning computer industries and wrote an article on the economic threat posed by Japanese hardware years before others realized how strong a competitor Japan would be in the computer industry.

Although Dyson never intended to be a technology guru, by the late 1970s, she seemed to be headed in that direction. After three years, Dyson left *Forbes* to become a securities analyst on Wall Street. She covered up-and-coming companies, including the newly formed Federal Express, and got her first taste of the venture capital and high-tech industries working for New Court Securities (now Rothschild). After three years, she moved to another securities firm, Oppenheimer & Company, where she specialized in software companies and high-tech. In

1982, Ben Rosen, a former star analyst for Morgan Stanley, brought Dyson to Rosen Research to write his monthly newsletter, the *Rosen Electronics Letter,* which was aimed at the computer industry and its investors. She changed the newsletter's name to *Release 1.0*—a name that retained the core identity (and initials) of the original—and shifted its focus from personal computers to software, and style from short articles to feature-length discussions. As the new editor in chief, one of her first articles described a trip to Bellevue, Washington, where she covered an interesting new start-up, Microsoft. She suggested the fledgling company lose some of its "charm" if it planned to succeed.

In 1983, Dyson also took charge of the PC Forum, which had begun in 1977 as an afternoon session of the Ben Rosen Semiconductor Forum. When Dyson took over, the PC Forum was, as she wrote in *Release 2.0,* "mainly a showcase for PC companies preening in front of investors." Under Dyson's hand, the PC in PC Forum—which had stood for personal computer—came to stand for Platforms for Communications, a broad umbrella that covered not just regular computer hardware and software, but the industry as a whole—its struggles, its identity, and its challenges. Open only to subscribers of *Release 1.0,* the conference was neither cheap nor conveniently located: attendees had to travel to Scottsdale, Arizona.

Also in 1983, Dyson bought Rosen Research (Rosen left to become chairman of Compaq), renaming it EDventure Holdings, Inc. Armed with a must-read newsletter, an elite annual conference that functioned as a meeting ground for the computer industry, and a growing Rolodex of names and contacts, by the mid-1980s Esther Dyson had begun to secure her place among the leaders of the coming technological revolution.

## ESTHER 1.0

*Release 1.0* was Dyson's primary voice to the (rather select) masses. Since its beginnings, the newsletter attracted a small but impressive list of subscribers, with addresses ranging from Silicon Valley to Wall Street. By the late 1990s, nearly 1,800 subscribers paid $695 per year to read what Esther Dyson deemed important. As with the *Forbes* article on Japanese hardware, her ruminations and investigations were often miles ahead of the curve, featuring cutting-edge ideas and technologies, in-

*Esther Dyson. (Reuters/Corbis)*

cluding object-oriented programming in 1987 and emerging artificial intelligence technologies in the early 1990s. When the Internet took hold, she tackled such vital issues as anonymity, education, and security early on. The newsletter has steered clear of the basic diet of similar publications, avoiding stock tips and conventional wisdom, offering, instead, unconventional thinking about whatever seemed to capture Dyson's attention. Its critics claim *Release 1.0* is too technical, abstruse, erratic—even unreadable. Its champions claim it is worth every penny.

*Release 1.0* has been an industry regular for more than twenty years, save for Dyson's short departure in 1985 to head another publication, *Computer Industry Daily (CID).* In the mid-1980s, Ziff-Davis, the highly successful computer-magazine publisher that boasts periodicals like *PC Magazine* and *PC Week,* had proposed an exclusive, $1,000-a-year industry newspaper, to be delivered daily in electronic form by MCI Mail or, on paper, via overnight mail. *CID* was to be an advertisement-free paper filled with computer industry gossip and late-breaking tech news, aimed at a select group of CEOs, analysts, journalists, and investors.

Dyson was tapped to be editor-publisher in late March 1985. She spent the summer building a staff of nearly thirty and courting sources and international news services. By September 1985, after just a few months of publication (most of which was sent free to subscribers of *Release 1.0*), Ziff-Davis decided to cancel the newspaper. Many commented that *CID* may have survived had it been more like *Release 1.0*—a distillation of Dyson's industry analysis—adding that Dyson's talents were lost on managing a staff or reporters. Others blamed the slow market or vague editorial directions from on high. Dyson, only thirty-four, revived *Release 1.0* and reclaimed EDventure Holdings, including her primary moneymaker—the PC Forum.

By the 1990s, each March more than 500 CEOs, venture capitalists, and entrepreneurs would pack up and head to the Arizona desert, shelling out thousands of dollars to attend the conference. (By 2000, a ticket cost $4,500.) Most years, hundreds were turned away. According to *Upside Magazine* in March 1997, the PC Forum had become the "industry conference of consequence."

Much of the excitement was engendered by the Internet. In the early 1990s, the mood at the PC Forum was dampened by industry losses and the apparent inevitability of industry domination by Microsoft. The Internet changed the mood. In 1996, Eric Schmidt, CTO of Sun Microsystems, told *USA Today* that the Internet had given the industry "something to debate, something to hope for." Indeed, each year the PC Forum tackles a different aspect of the debate: from which metaphor best explains the Web (1997) to how to make money online (1999) to how to manage and utilize data (2002). Dyson decides the topics, the speakers, and the dozen or so "debutante" companies that are featured each year; industry players decide which way the conversation goes.

In the years since the Internet bubble burst, some of the excitement and delirium that marked earlier conferences has been replaced by what one conference-goer deemed a measured seriousness. Although the atmosphere of PC Forum remains congenial and informal,

dispatches from late-night jam sessions of a motley crew of entrepreneurs and venture capitalists performing Grateful Dead covers are long gone, replaced by relatively staid articles on business-to-business products, open source software, and radio-frequency-identification technology. Still, Dyson strives to keep the momentum up. She admonished a rather glum crowd of nearly 700 in 2001, "Don't define yourself in terms of what you thought you'd be last year, when any idiot was worth a billion dollars. Define yourself by what you'll be next year."

## INVESTING IN EASTERN EUROPE

In 1989, right as the PC Forum was on the brink of one of its most exciting eras, Dyson had become, by her own admission, bored. That spring she flew to Moscow. After three weeks, she returned to the United States invigorated by the opportunities she saw in the nascent computer and high-tech industry in Russia. (The country was not a new interest. As a child, her father frequently traveled to Russia for conferences, and she had studied the language in high school.) Within a year, Dyson had conceived and organized *REL-East* and the East-West High-Tech Forum, the Eastern counterparts to *Release 1.0* and the PC Forum, and had begun to invest time and money and use her prodigious networking skills to make them a success.

"Things in the U.S. are too easy almost," Dyson told the *Irish Times,* explaining why she was drawn to Eastern Europe. "I don't like being the tenth person to do something. I realized there was nobody noticing the computer industry in Central and Eastern Europe. Nobody was doing their public relations for them, so I took this odd combination of things I was interested in and tried to fill the vacuum." Dyson filled the information vacuum about the industry with *REL-East,* which, though irregularly published, was, unlike its costly U.S. counterpart, distributed free. She filled the investment vacuum by bringing together Eastern European entrepreneurs and Western investors at her exclusive annual conference. Dyson also made pioneering investments in dozens of start-ups in the region, such as IBS and uproar.com (both from Russia), Poland Online, and Middle Europe Networks (CompuServe's Hungarian outpost), to name just a few.

Although Dyson stopped publication of *REL-East* in the mid-1990s, the conference continued to grow and evolve. Now known simply as EDventure's Hi-Tech Forum, its focus has shifted from Eastern and Central Europe to the computer and information industries of all of Europe. The change is evident in the shift in location of the conference, from the early days of Budapest, Hungry, and Bled, Slovenia, to the international cities of Copenhagen, Denmark, and Barcelona, Spain. Dyson's efforts were recognized in 1996, when Hungary awarded her the von Neumann Medal for "distinction in the dissemination of computer culture." Indeed, she played a vital role in bringing the often aggressive and inventive entrepreneurial spirit of Silicon Valley to the emerging technology sectors in Eastern Europe at a time when they had just begun to embrace capitalism. Almost fifteen years after she first stepped foot in Moscow, Eastern Europe is still her favorite developing area. And her objective remains the same. "I'm shining a flashlight," she told *Electronic Media.* "I'm trying to help them to find their own heroes and content so that they don't have to rely on Bill Gates."

## MS. DYSON GOES TO WASHINGTON

Dyson, already a familiar face in Palo Alto, Moscow, and New York, added another stop on her regular worldwide itinerary in the early 1990s: Washington, D.C. Dyson's foray into Washington—and Washington-style politics—may have seemed surprising at the time: she was one of the most staunch and vocal cyber-libertarians of the time and yet practically apolitical. (She has never voted.) Still, Dyson had long been wary of government incursions into the growing Internet. "Whether you like it or not," she told the audience at the 1993 PC Forum, "there are people in Washington that have more control over your future than Microsoft does."

Dyson first came to Washington as a member of the Electronic Frontier Foundation (EFF), the first civil rights organization dedicated to online issues. Lobbying on behalf of the EFF, Dyson held forth on topics from encryption to pornography with top Washington offi-

cials, including President Bill Clinton. Dyson became chairman of the EFF when the organization moved its headquarters to California in 1995. President Clinton then asked Dyson to sit on his Export Council Subcommittee on Encryption, which reviewed U.S. export laws on commercial encryption products. In 1994, she was tapped by Vice President Al Gore to cochair the Information Privacy and Intellectual Subcommittee of the National Information Infrastructure Advisory Council. By the mid-1990s, the three most powerful men in American government—President Clinton, Vice President Gore and Speaker of the House Newt Gingrich (himself a cyber-libertarian)—would bend an ear when Esther Dyson had something to say.

Other Washington players were also deeply influenced by Dyson. Senator Bill Bradley, for one, encouraged Dyson to publish her insights for audiences beyond Capitol Hill, Wall Street, and Silicon Valley. In 1997, Broadway Books published *Release 2.0: Design for Living in the Digital Age,* a collection of Dyson's thoughts on the Internet's impact on our ideas about work, security, privacy, education, and governance. (The paperback edition, *Release 2.1,* for which Dyson solicited and incorporated reader comments, was published the following year.) Dyson took less than seven weeks to finish the more than 300-page book, for which she was paid $1 million. To many reviewers, her haste showed. *Wired* decried aspects of the book as "downright banal," and the formidable Michiko Kakutani of the *New York Times* dismissed much of Dyson's writing as "fuzzy, perfunctory and often willfully Pollyanna-ish." Even positive reviews that cited the depth of her insights kindly suggested that the book suffered from "thinking out loud" syndrome.

Regardless, the same year *Release 2.1* hit bookstores, Dyson rose to even greater prominence when she became the founding chairman of the Internet Corporation for Assigned Names and Numbers (ICANN), which was established by the U.S. Department of Commerce. ICANN was the Clinton administration's effort to create an international Internet organization that

> *Being a woman is a disadvantage within the system, but an advantage being on your own. . . . I might have been more mainstream if I were a guy; I might have wanted to be Bill Gates.*
>
> —Dyson to Wired, November 1993

would manage the Internet's infrastructure and domain name system.

From the start, ICANN was beset by conflict. Some were alarmed by the apparent secrecy of ICANN's operations and questioned the need for closed-door meetings. Others accused the group of misusing funds and demanded to see financial records. Still others expressed fear that ICANN was moving toward controlling content, not just domain names, and criticized the group for being corporate lapdogs, beholden to U.S. trademarks even though its reach was intended to be global. And some blame Dyson, in part, for ICANN's rocky start. Most accounts of Dyson's sometimes-difficult personality refer to an incident during an ICANN meeting, when Dyson suggested that the group create an election process that would protect them from "people who are stupid."

In August 2000, Dyson stepped down as chair to become ICANN's "best critic" as part of the At Large Advisory Committee, an independent body that advises ICANN on behalf of Internet users. Although Dyson still believes in ICANN, her support for the institution is lukewarm. As she told *New Media Age,* "ICANN is just there to fill the vacuum with something powerless, so that the vacuum doesn't get filled with something powerful."

Having stepped down from ICANN in 2000, Dyson returned to "normal life"—investing in more than three dozen companies worldwide, sitting on the board of nearly a dozen other organizations, writing incessantly, flying from meeting to meeting and spending hours each day answering hundreds of emails. She still spends nearly three quarters of her life in transit, amassing, some accounts suggest, nearly 6 million frequent flier miles. At EDventure Holdings, she still "nurses" start-ups, claiming during a PBS interview that her "maternal instinct is spent on Eastern Europe." She continues to run the PC Forum and writes much of *Release 1.0,* as she has since the early 1980s. Since January 2000, Dyson also speaks to the public through her column, "Release 3.0," published every other week for the *New York Times Syndicate,* and, more recently, through her blog, Release 4.0.

> *Our common task is to do a better job with the Net than we have done so far with the physical world.*
>
> —*Dyson in* Release 2.0, *1997*

Soon after founding EDventure Holdings, Dyson became one of the most important and dynamic figures in the high-tech industry. From Silicon Valley to Wall Street, and from Washington to Moscow, Dyson has helped chart the course of information technology. She uses her considerable influence to connect entrepreneurs with investors worldwide, to champion both emerging technologies and emerging markets, and to steer such prominent and powerful institutions as ICANN and the EFF. In doing so, she has secured a name for herself as an analyst, theorist, writer, and prophet in world of the Internet.

## FURTHER READING

### In These Volumes

Related Entries in this Volume: Barlow, John Perry; Gilmore, John; Lessig, Lawrence; Postel, Jonathan; Winblad, Ann

Related Entries in the Chronology Volume: 1998: ICANN Is Chosen by the U.S. Commerce Department as Successor to InterNIC

Related Entries in the Issues Volume: E-business; E-commerce

### Works by Esther Dyson

*Release 2.0: A Design for Living in the Digital Age:* New York: Broadway Books, 1997.

*Release 2.1: A Design for Living in the Digital Age.* Rev. ed. New York: Broadway Books, 1998.

### Articles

Borsook, Paulina. "Release." *Wired,* November 1993.

Evenson, Laura. "Esther Dyson." *The San Francisco Chronicle,* August 13, 1995.

Manjoo, Farhad. "Esther Dyson Defends ICANN." *Salon.com,* July 25, 2002, http://archive.salon.com/tech/feature/2002/07/25/dyson/ (cited September 16, 2004).

Rosenberg, Scott. "E.D. Phone Home." *Salon.com,* December 9, 1997, http://archive.salon.com/21st/feature/1997/12/09feature.html (cited September 16, 2004).

Weeks, Linton. "Small Wonder: Pixieish, Puckish Computer Wizard Esther Dyson Imposes Order on Chaos of Cyberspace with Magical Clarity." *The Washington Post,* November 3, 1997.

## Websites

EDventure Holdings, Inc. Esther Dyson's company Website, home of the PC Forum and *Release 1.0,* http://www.edventure.com (cited September 16, 2004).

Electronic Frontier Foundation (EFF). Online home of the EFF, a nonprofit civil rights group for cyberspace, http://www.eff.org (cited September 16, 2004).

The Internet Corporation for Assigned Names and Numbers (ICANN). Online home of ICANN, the international Internet naming group, http://www.icann.org (cited September 16, 2004).

Release 4.0. Esther Dyson's Web diary, http://release4.blogspot.com (cited September 16, 2004).

# Shawn Fanning (1980–)

## CREATOR OF NAPSTER

**At just 19 years old,** Shawn Fanning transformed the Internet into the world's biggest music library. His invention—Napster—was a free downloadable program that turned every computer into a server for exchanging music files called MP3s. During Napster's fifteen-month life, more than 60 million people downloaded the software, making Napster the fastest-growing at-home software program in history. Napster's popularity quickly drew the attention—and the ire—of the recording industry, which sued Napster just six months after its debut. As Napster struggled, under court order, to evolve from a haven for music piracy into a legitimate business, a host of peer-to-peer (P2P) file-swapping programs emerged. Although the original Napster has been replaced by a moderately popular online music service, the P2P movement that Napster popularized lives on.

## IT's A HARD-KNOCK LIFE

Shawn Fanning grew up in a rough, working-class neighborhood south of Boston, Massachusetts. His was a childhood marked by struggle—Shawn's biological father left his mother, just 17 years old at the time, soon after getting the news of her pregnancy; the man she later married could not always provide for the growing family. "Money was always a pretty big issue," Shawn told *Business Week.* "There was a lot of tension around that." At one point, Shawn and his four siblings were put into foster care. When his mother reunited with her husband, the family moved to Harwich Port, a small, quiet town on Cape Cod.

Shawn escaped troubles at home by playing sports, earning the nickname Napster on the basketball court for the messy head of "nappy" hair he tucked under a baseball cap. He found a father figure in his uncle John. John, head of NetGames, a small computer gaming company, rewarded Shawn's good grades—$100 per A—and showered him with gifts, including a purple

BMW. John was also the one to notice young Shawn's aptitude for computing. During Shawn's sophomore year at Harwich High, John bought him an Apple Macintosh 512+. "I saw this as a way for him to work his way out of his situation," John later recalled in a *Business Week* interview. John also paid for an additional phone line so that Shawn could connect to the World Wide Web—in the early 1990s, a relatively new phenomenon—and to other like-minded hackers and hobbyists through Internet Relay Chat (IRC), one of the Internet's earliest chat services.

Shawn's main form of escape quickly changed from sports to computers. Every summer, Shawn worked at NetGames, learning the finer points of programming from his uncle's college-age employees. Between his junior and senior years in high school, Shawn taught himself to program UNIX computers—a platform favored by computer science students and hackers alike. He became increasingly absorbed by programming and spent hours on IRC, eventually falling in with a loose-knit group of elite hackers who called themselves the w00w00 club. On IRC Shawn met Sean Parker, who would become a cocreator of Napster.

In the fall of 1998, Fanning enrolled at Northeastern University in Boston. Intending to major in computer science, he took courses in the field, but was bored by the elementary programming classes. He began spending much of his time on the $7,000 laptop computer his uncle had bought him.

At Northeastern, Fanning discovered MP3s. MP3—short for the Motion Picture Experts Group Audio Layer 3—is an international standard for compressing audio files. MP3 files were very small, yet retained CD-quality sound, thus they were the ideal format for listening to music on a computer and for exchanging music files over the Internet. By the late 1990s, MP3s had become so popular that, at one point, *MP3* beat out *sex* as the most popular Internet search term.

The idea for Napster dawned on Fanning after his roommate began to complain about the difficulties of finding good MP3s on the Web. Popular Internet sites, such as MP3.lycos.com and Scour.net, often contained dead links, and search engines were not much more reliable. Fanning, longing to become absorbed in a programming project, believed he could find a way to fix his friend's problem.

Ultimately, Fanning decided to create a program that combined a real-time system for finding MP3s with the sense of community he had discovered on IRC. Included would be aspects of instant messaging and file sharing, as well as a search function. He spent endless hours working through the idea. By winter, Fanning had dropped out of school and moved into his uncle's office to work on his new project full-time.

Fanning first wrote the server software, then moved on to user interface. To enable him to make his program PC-compatible, Fanning bought a book and taught himself Windows programming in a matter of weeks. Programmers at the NetGames office were amazed by Fanning's focus and concentration.

With the help of fellow hacker-cum-programmers Sean Parker and Jordan Ritter, the source code for what would become known as Napster came together in three months. Later, Fanning would claim that the urgency to complete Napster before someone else came up with the same idea kept the program clean and simple. "With a few more months," he told *Time* in 2000, "I might have added a lot of stuff that would have screwed it up. But in the end, I just wanted to get the thing out."

Fanning posted an early version of his program on IRC in June 1999, urging some two dozen hacker friends to vet the program—but to keep it to themselves. Within days, some 15,000 people had downloaded Napster.

Napster's early growth was spectacular, primarily because the program, and the idea behind it, was so simple. Once downloaded, the Napster software allowed a user to log onto Napster's server, which provided an index of all the MP3s (songs) of other users currently online. A user could then search for a particular song title. Once the title was found on the server, the Napster service would put the two computers (peers) in contact with each other, so that files could be shared directly (hence the term *peer-to-peer*). In addition, Fanning added a feature that allowed users to chat. Through Napster, Fanning and colleagues had created a central meeting ground for the hundreds of preexisting communities of music fans spread across the globe, then gave them the added and undeniable pleasure of getting something for nothing. The music-swapping revolution had begun.

## THE FAMILY BUSINESS

While Fanning programmed night and day, his uncle handled the business end of Napster.com. The two Fannings incorporated Napster, Inc. in May 1999, with the elder Fanning claiming the lion's share—70 percent—of the company. That summer, John worked to raise Napster's first round of funding. By September, he had secured at least six months' of financing and had hired Napster's first CEO, Eileen Richardson, a Boston-area venture capitalist.

Richardson immediately moved the younger Fanning and his Napster cofounder, Sean Parker, to San Mateo, California, on the fringes of Silicon Valley. The following month, the elder Fanning secured another $2 million of funding from so-called angel investors—wealthy individuals willing to support a start-up company—that bought Napster another six months to develop a workable business model. Despite its overwhelming popularity, Napster had yet to make a single cent in profit.

Although Napster did indeed provide a service, Shawn (and many others at Napster.com) saw it primarily as a music community—a virtual place to swap songs and admire each other's music collections. The struggle, then, was to figure out how to build a business on top of such a community.

For his part, Shawn imagined early on that the service would be bought out by the record industry, which would not be able to deny the power of Napster's revolutionary technology. Once it was clear that Shawn had created what is known in the computing world as a "killer app," John Fanning consulted lawyers about the legality of his nephew's program under existing copyright law. John then assured

*I never expected anything like this to happen.*

—*Fanning to* USA Today, *May 2000*

Shawn and the rest of Napster's growing staff that they would win any copyright case brought against them.

John Fanning was somewhat of an imprudent and unrealistic businessman. According one source, his ultimate plan was to earn millions during a Napster initial public offering (of stock), then leave the company before the copyright issue could be settled. His chosen CEO, Richardson, was a skilled venture capitalist with a decade of experience in the tech industry, but she had never run a company. Her inexperience, coupled with the elder Fanning's lack of business sense, set Napster on a collision course with the music industry.

## I FOUGHT THE LAW (AND THE LAW WON)

On December 7, 1999, the Recording Industry Association of America (RIAA) sued Napster on behalf of five of the biggest record labels in the country. The RIAA suit claimed that Napster was guilty of encouraging illegal copying and distribution of copyrighted materials, and asked for fines up to $100,000 per copyright-protected song exchanged via Napster.

News of the lawsuit—and the RIAA's threats against the song-swappers themselves—did little to dissuade Napster users from swapping songs. In fact, both inside and outside the recording industry, it had become abundantly clear that Napster had become far more than a service—it had become a movement, one that proved difficult to stop. While the RIAA cracked down on the company itself, university officials struggled to keep Napster users off their networks. Within months of Napster's debut, the service had eaten up more than 25 percent of the bandwidth on university networks from Washington state to Florida. By March 2000, Napster had been banned or restricted on more than 100 campuses—and still, Napster users numbered in the millions.

In April 2000, on the coattails of the RIAA lawsuit, Metallica, a heavy metal band popular in the late 1980s, brought its own, separate lawsuit again Napster, claiming that, in addition to copyright infringement, Nap-

*I don't think a day goes by when people don't recognize me. I mean, it's been good for getting girls ... but it's hard to move past that.*

*—Fanning to* Time, *October 2000*

ster's repeated transgressions violated the racketeering statutes normally used to prosecute mobsters. Two weeks later, Metallica's attorney, Howard King, brought a similar case against Napster on behalf of Dr. Dre, a rapper who was once part of groundbreaking rap group N.W.A.

For Lars Ulrich, the drummer for Metallica, the fight against Napster took on the air of a crusade against piracy in all forms. "This is not just about music," Ulrich argued in a *60 Minutes II* interview. "We are probably right now two to three months away from having a situation where you can download a 90-minute motion picture in . . . real time." In May 2000, Ulrich drove to Napster's headquarters in San Mateo to deliver personally a box with a list of 335,000 names—names of Metallic fans who had illegally downloaded music from Napster. But while Napster technicians struggled to meet Metallica's demands by kicking users off the system, within minutes the users would log back on under new aliases.

Not all pop musicians followed in Ulrich's footsteps. (Indeed, some Metallica fans aligned themselves with Napster, and soon sites like KillMetallica.com and Metallicasucks.com cropped up on the Web.) Limp Bizkit, a 21st-century alternative rock band (whose members were in grade school when Metallica topped the music charts), saw Napster as a viable means to circumvent the recording industry, which had become more and more homogenous and risk averse. Chuck D, another famous rapper, claimed Napster was responsible for providing an alternate infrastructure for new artists to find their audience and create fan bases.

While pop artists argued their case in the media, Napster and the RIAA duked it out in the courts. Going into the case, one of Napster's main arguments was the Audio Home Recording Act of 1992, which allowed for any noncommercial copying of music, along the lines of making tapes for friends. Napster could have also cited the 1984 Supreme Court decision regarding Betamax (a now-obsolete home-video standard), in which the Supreme Court declared that technology could not be outlawed just because some users put it to unlawful purposes. The Betamax precedent had been cited in an earlier case, when the RIAA sued the maker of a portable MP3 player called the Rio.

According to David Boies, Napster's lead attorney who is best known for winning the federal antitrust case against Microsoft, the center of the Napster case would be to define commercial and noncommercial use. Although any profit from the exchange of pirated music would clearly be illegal, Napster had yet to find a way to make money. Still, Boies only gave Napster a 50-50 chance at winning the judgment.

The Napster case was argued in California district court. In May 2000, Judge Marilyn Patel declared Napster to be in violation of the Digital Millennium Copyright Act (DMCA) of 1998, which had been passed to protect the music, film, and publishing industries from piracy and illegal distribution over the Internet. Napster executives had originally believed that, because Napster facilitated the connecting of computers but did not host copyrighted material itself, the service was more like America Online and other Internet service providers, which were protected under the DMCA. The biggest blow to Napster came on July 26, 2000, when Judge Patel handed down, at the RIAA's request, a preliminary injunction that would shut Napster until the coming trial unless it could remove all copyrighted music from its service in five days. Judge Patel also made clear that she thought Napster had very little chance of proving its case.

Shawn, who had attended the trial with Napster's new CEO, Hank Barry, was visibly crushed by the verdict, tears welling up in his eyes. Boies immediately vowed to appeal. Two days later, the Ninth U.S. Circuit Court of Appeals granted a stay of the injunction, and Napster remained online.

In the midst of the legal seesaw, Napster saw its usage rise more than 70 percent, as fearful users scrambled to find that one last song before Napster cut its signal forever. Outside Napster's nondescript headquarters, which had since moved to Redwood City, California, supporters cheered the stay. Inside, the mood was less jubilant, with Fanning telling *Business Week,* "I feel numb."

The court case dragged on for another year (Napster flourished, with well over 30 million users worldwide and more than 1 million new users each week). On October 9, 2000, Fanning testified before the Sen-

*Shawn Fanning. (Steve Azzara/Corbis)*

ate Judiciary Committee, convened by Senator Orin Hatch in Provo, Utah. He described Napster's birth and its ascent, citing emails that spoke to the more legal aspects of Napster. "People tell us that they use Napster to sample new music before deciding what to buy, find new artists, and house music in their computers that they already own on CD, cassette, vinyl and sometimes 8-track. We hear regularly from mothers who say they use Napster to screen the music their children are listening to and parents who say that Napster is a shared activity that helps them communicate with their teenagers." Indeed, by several accounts, CD sales had increased during Napster's reign—a claim that undermined the record industry's assertion that Napster was killing it.

Weeks after his testimony, Fanning and Napster CEO Barry announced that Napster would partner with Bertelsmann AG, the company behind the BMG record label, to develop a subscription-based Napster. Once the newly legitimate Napster was on its feet, BMG would withdraw from the RIAA suit, and use

Napster's name recognition to gain a foothold in the online music realm.

The courts, however, slowed the process. By February 2001, the appeals court found Napster guilty of "contributory and vicarious infringement" and ordered Napster to stop users from trading pirated music. (Much of its decision was based on a memo from Napster cofounder Sean Parker, in which Parker urged the company to remain ignorant of the true identities of users who were trading illegally.) The following month, Napster began to voluntarily block songs that appeared on the Top 100 record charts—though Napster users continued to flout the law by trading songs with altered names and titles. But each injunction took its toll. Music-swapping volume on Napster fell by more than 35 percent between March and April. By July, even after Napster had announced further plans to launch its legitimate subscription-based service, Judge Patel ordered the site to shut down until it was free of copyrighted materials. For the first time, Napster went offline.

## AND THE BAND PLAYED ON

By the fall of 2001, Napster had settled out of court with Dr. Dre and Metallica and agreed to pay copyright owners $26 million for past copyright infringements, plus an additional $10 million toward advances on licensing deals to come into play when Napster emerged as a subscription service. The major music labels, forced to embrace the Internet by Napster's success, raced to stake claims online. In December 2001, music industry leaders such as EMI Recorded Music and AOL Time-Warner collaborated with streaming-media giant Real-Networks to introduce a new music subscription service, MusicNet, just days before another service, PressPlay, backed by Sony Music Entertainment and Universal Music Group, arrived on the Web.

The response was tepid. The record labels, still reeling from early days of Napster, offered only limited distribution of newly released materials, which did little to pique public interest. Artists who had casually sup-

*You can't stop technology. Even if they succeed in shutting down those particular services, new services will spring up. It's the nature of the Internet.*

*—Fanning to the* Wall Street Journal, *October 2002*

ported the crackdown on piracy now railed against the subscription services, claiming that they had never given the labels permission to distribute online. And, of course, music fans who had previously downloaded free via Napster were far more interested in the next generation of Napster-alikes that had cropped up since 1999. Free downloadable music had flooded the Web via applications like Kazaa, Gnutella, and Morpheus, which had adopted an even more dispersed network for file swapping. Some programmers had seen that Napster's Achilles heel was its centralized servers, which could be easily turned off; the new P2P services operated without central servers, thus stopping them was more difficult. Nevertheless, in October 2001, the RIAA cracked down on Kazaa, the most popular of the new-breed P2P services.

While the P2P revolution took hold, Napster strained to achieve legitimacy. By January 2002, the Napster case was set aside while both sides worked to reach a settlement. The record labels had already passed on Napster's offered settlement of $1 billion, paid to the five main record labels over five years. In March 2002, Bertelsmann AG, which had already pumped nearly $85 million into Napster in preparation for a buyout, proposed to pay an additional $16.5 million to Napster investors to help the service regain its footing—terms that everyone on the Napster board of directors seemed to like, except John Fanning. To stop the sale, John sued to have two board members removed. By the time his case was thrown out of court, the Bertelsmann offer had been withdrawn. The elder Fanning left the board within days; weeks later, Shawn and other key executives stepped down. Napster sat on the brink of defeat.

Days after Shawn stepped down, however, Napster underwent an abrupt about face. Bertelsmann announced that it would acquire the company through bankruptcy proceedings and pledged several million dollars toward paying Napster's debts and to keep the company going as the paperwork was settled. Bertelsmann CEO Joel Klein announced, in a press release,

"Creating new ways of doing business is never easy, but Napster will be at the forefront of finding business models that respect copyright, reward artists, and deliver entertainment value to consumers."

The RIAA and other music industry groups opposed the sale; in September 2002, a Delaware judge blocked the deal. Most of the Napster staff was laid off; Napster.com, once a font of activity, simply read "Napster was here." Shawn left the company for good in September 2002.

Napster, however, lived on—both in name and, through services like Kazaa, in spirit. An online media company, Roxio, acquired Napster's assets during Napster's bankruptcy liquidation in November 2002 and launched Napster 2.0 in October 2003. No longer peer-to-peer, Napster.com now competes in an online music world dominated by the overwhelmingly popular Apple iTunes service. The site bears little resemblance to the original Napster, other than name and logo. Fanning worked as consultant for Roxio early on, but has since, as some see it, gone to the "dark side." As the music industry floods the country with thousands of subpoenas and lawsuits against file swappers, Fanning has been developing technology that would help the music industry track songs within file-sharing networks, in order to charge individuals for their use. Fanning hopes his technology, dubbed SNOCAP, will be music to the record industry's ears.

## FURTHER READING

### In These Volumes

Related Entries in this Volume: Glaser, Rob; Lessig, Lawrence

Related Entries in the Chronology Volume: 1999: Napster Roils the Music Industry; 2003: Music Sharers Sued by Recording Industry; 2003 (sidebar): Is File Sharing Killing Music?

Related Entries in the Issues Volume: Copyright; Peer-to-peer Networks

### Books

Alderman, John. *Sonic Boom: Napster, MP3, and the New Pioneers of Music.* Cambridge, MA: Perseus, 2001.

Menn, Joseph. *All The Rave: The Rise and Fall of Shawn Fanning's Napster.* New York: Crown, 2003.

### Articles

Ante, Spencer. "Inside Napster." *Business Week,* April 12, 2000, http://www.businessweek.com:/2000/00_33/b3694001.htm (cited September 16, 2004).

———. "Napster's Shawn Fanning: The Teen Who Woke Up Web Music." *Business Week,* April 12, 2000, http://www.businessweek.com/ebiz/0004/em0412.htm (cited September 16, 2004).

Dyson, Esther. "Napster Forcing Music Industry to Change Its Money-Making Tune." *Los Angeles Times,* August 21, 2000.

Gorov, Linda. "Hi, I'm Napster." *Boston Globe,* June 11, 2000.

Graham, Jefferson. "Caught Between Rock and a Hard Drive." *USA Today,* May 15, 2000.

Greenfeld, Karl Taro. "Meet The Napster." *Time,* October 2, 2000, http://www.cnn.com/ALLPOLITICS/time/2000/10/02/na/pster.html (cited September 16, 2004).

Kover, Amy. "Who's Afraid of This Kid?" *Fortune,* March 20, 2000.

Walker, Jesse. "Music for Nothing." *Reason,* October 2000, http://reason.com/0010/fe.jw.music.shtml (cited September 16, 2004).

Wingfield, Nick. "Napster Boy, Interrupted." *Wall Street Journal,* October 1, 2002.

### Websites

CNN.com—In-Depth Special: Napster. CNN.com's complete coverage of the Napster phenomenon, including timelines, analyses, and the Napster legal case, http://www.cnn.com/SPECIALS/2001/napster/ (cited September 16, 2004).

Electronic Frontier Foundation: Intellectual Property and P2P File-sharing Archive. Compilation of white papers, articles, and legal findings related to Napster and other P2P file-sharing services, http://www.eff.org/IP/P2P/ (cited June 17, 2004).

FindLaw Legal News: Special Coverage: Napster. Compilation of the documents and oral arguments involved in the Napster lawsuit, http://news.findlaw.com/legalnews/lit/napster/ (cited September 16, 2004).

Napster.com. Online home of the subscription-based service called Napster, http://www.napster.com (cited September 16, 2004).

Salon.com Directory: Napster. Compilation of all *Salon.com*'s coverage of Napster, http://dir.salon.com/topics/napster/ (cited September 16, 2004).

Save Napster Dot Com. Former online home of the Students Against University Censorship Website, which is now archived on this site. Savenapster.com offer links to news about P2P file sharing and the new Napster, http://www.savenapster.com (cited on September 16, 2004).

SNOCAP, Inc. Future online home of Shawn Fanning's post-Napster endeavor, SNOCAP, Inc., a file-tracking service geared toward the recording industry, http://www.snocap.com/ (cited on September 16, 2004).

# Bill Gates (1955–)

## FOUNDER OF MICROSOFT

**Bill Gates is most often** described in superlatives—the richest person in America, chairman of the largest software company in the world, either the most hated or most revered man in computing. Since founding Microsoft in 1975, at age 19, Gates has become one of the strongest driving forces in personal computing—his company has created some of the most widely used and successful computer programs available, including MS-DOS, Windows, Word, and Excel. However, Microsoft's fierce—and by several accounts, not entirely legal—competitive practices have, more than once, prompted investigation by the U.S. and other governments. Investigations begun in the mid-1990s resulted in *United States v. Microsoft*—easily one of the most closely watched antitrust trials in American history.

Throughout the trial, the media often portrayed Gates as a modern-day robber baron, much like robber barons of old, such as John D. Rockefeller. However, as occurred with Rockefeller, in time a new image of Gates began to emerge—the philanthropist. The Bill and Melinda Gates Foundation, established in 2000, is the most well-endowed charitable organization in the world, with more than $25 billion in assets. Through the foundation, Gates has arranged to give away nearly all his wealth to causes: closing the Digital Divide and eradicating disease, including malaria, tuberculosis, and HIV.

## REVENGE OF THE NERD

Bill Gates, born William Henry Gates III on October 28, 1955, grew up in Seattle, Washington, the son of Mary Maxwell and William Henry Gates II. Bill was fourth in a line of prominent Gates men: his great-grandfather was a state legislator, his grandfather was a banker, and his father was a leading Seattle corporate attorney. His mother, equally accomplished, served on the boards of various large institutions, including the United Way, and was a regent for the University of

Washington, as well as director of a bank. Throughout Bill's youth, the Gates household was alive with discussions of current events, business, and politics. Thus, the worlds of power and money were familiar to Gates when he began his career.

As a student, with his slight build, freckled face, unkempt hair, and thick glasses, Gates was the quintessential nerd. Although exceedingly intelligent, he took on the role of class clown and was considered by some teachers to be a problem student. In 1967, his parents enrolled him at the Lakeside School, an extremely rigorous and prestigious prep school in Seattle. At Lakeside Gates, not yet a teenager, befriended Paul Allen, three years his senior, with whom he would later form Microsoft.

The first turning point in Gates's young life came in eighth grade. Lakeside had raised money for a time-share on an early PDP-10 computer. "Very quickly, the teachers were intimidated," Gates recalled in an interview with the Smithsonian Institution. "So, it was sort of a group of students reading the manuals and trying things out." Gates, just thirteen, began to program.

Together with a handful of other, older Lakeside students—including Allen and another future Microsoft employee, Ric Weiland—Gates devoted much of the school year to mastering the PDP-10. In their spare time, the boys read every computer manual they could get their hands on and joined computer hobbyist email lists and even professional user groups, such as Digital Equipment Corporation's DECUS User Group. Then, when the money for the computer time-share ran out (much earlier than the school had expected, since the boys had used the computer for hours on end), Gates and his cohorts—who by then had dubbed themselves the Lakeside Programming Group—paid for time spent online.

Although its members were still in high school, the Lakeside Programming Group began to hire out its serv-

*Bill Gates. (Kim Kulish/Corbis)*

ices, in return for what was to them a most precious commodity—free computing time. Seattle's Computer Center Corporation (dubbed C-cubed) was among the first to hire Gates et al. The University of Washington and Lakeside School itself soon followed, as did Portland-based Information Sciences Inc., which hired Gates, Allen, and Weiland to write a payroll program in the computer language COBOL. Gates and Allen then formed a company, Traf-O-Data, and developed a data-processing program on traffic patterns for the Washington State Road Department. For their efforts, the two teenagers grossed about $20,000. In Gates's senior year, they did work finding and fixing computer bugs for TRW, a defense contractor. By then, talk of forming yet another software company was becoming increasingly more serious. Nevertheless, in 1973, Gates graduated and left Seattle for Harvard.

The second great turning point of Gates's youth occurred just over a year later, in December 1974. That month's issue of *Popular Electronics,* a hobbyist magazine, featured the Altair 8800 on its cover, with the headline "World's First Microcomputer Kit to Rival Commercial Models." Gates and Allen (who, by this time, had also moved to Massachusetts) believed that the Altair heralded the new age in personal computing and, possibly, a new business opportunity. They called MITS, the New Mexico–based company that built the Altair and promised a version of the computer language BASIC for its product—although the young programmers had not actually seen the Altair, much less written any code for it.

When MITS agreed to review the language, Gates and Allen immediately began a marathon eight-week code-writing session for the new Altair. Without the computer chip, they worked solely from the manual. Allen had secured a meeting with MITS and in February he traveled to Albuquerque to give MITS a demonstration. The demo went off without a hitch—much to Allen's amazement, he admitted later—and Gates and Allen, calling themselves Micro-Soft, became the proud creators of the first-ever computer language for the first mainstream personal computer. In turn, MITS became Micro-Soft's first customer. (The hyphen in Micro-Soft was later dropped.)

By March, Allen had moved to New Mexico to become director of software for MITS; Gates followed soon after. Micro-Soft worked closely with MITS, but also began licensing its BASIC programs to other manufacturers that used the same Intel chip as the Altair. After eighteen months, Allen quit MITS to join Micro-Soft full-time, and on November 26, 1976, Micro-Soft officially registered as a company with the New Mexico secretary of state.

The early days of Micro-Soft were heady and stressful, setting the tone for decades to come. Programmers worked day and night, with Gates pushing them to meet increasingly tight deadlines. The small office was peopled, primarily, by young men in their early 20s, just sixteen programmers at its peak. Gates himself did a large amount of the programming, in addition to functioning as the sales and legal departments. Gates's shrewd legal sense helped young Micro-Soft retain the rights to BASIC after a lawsuit brought by MITS that was settled in 1977. His business savvy prompted him to pen the now famous "Open Letter to Hobbyists," one of the first screeds against software piracy, in which Gates argued that pirating software simply keeps good software from being written. (Members of the free software movement assert that the success of the freely available operating system Linux, for instance, has proven Gates wrong.)

Micro-Soft grew steadily—November 1978 it had opened an office in Japan. By early 1979, the company had changed its name to Microsoft and moved to Seattle. In June 1980, Gates hired his best friend from college, Steve Ballmer, who became, in effect, Number Two at the company. This left Gates more time to guide the business end of Microsoft.

That year, Gates secured the deal that would cement his success in the software industry. Gates had purchased the rights to DOS, a new operating system that had been developed by Seattle Computer, a local company. For a $50,000 flat fee, Microsoft was free to license DOS to others as a Microsoft product. By the end of the year, Microsoft and IBM, then the largest computer company in the country, signed a contract that would create MS-DOS (Microsoft Disk Operating System). The contract included a key stipulation—instead of selling the rights to MS-DOS to IBM, as Seattle Computer had sold them to Microsoft, Microsoft would retain all rights. Gates knew that IBM products were quickly cloned, and he believed that the real money would come from licensing MS-DOS to IBM's many competitors. He was right on target. IBM PCs loaded with MS-DOS began to ship in early fall of 1981. Consumers flocked to the product, and, later, to PC clones running MS-DOS as well. The modern PC era—and the era of Microsoft—had just begun.

## LOSING THEIR CHARM

In the early 1980s, even in the midst of the growing success of MS-DOS, few in the computer industry took Microsoft, and its 20-something CEO, very seriously. After her first visit to Microsoft's Bellevue offices in 1982, Esther Dyson, a respected industry analyst, suggested that the company needed to "lose some of its charm" if it planned to succeed. Over the next decade, Microsoft did just that, assiduously. MS-DOS, which was on its way to becoming the industry standard in operating systems, opened the doors for a slew of new software applications. However, Microsoft's first major software release, Multiplan, a spreadsheet program, was no competitive match for the Lotus 1-2-3 spreadsheet program. However, its second offering, Microsoft Word, was released in 1983 and grew to surpass WordPerfect, then the primary word processing program, to become one of the most commonly used programs on a personal computer. By 1984, Microsoft had also become one of the primary software developers for Apple Computer, one of the few computer companies to develop a

*[Microsoft] cannot think like a big company or we are dead.*

—*Gates to Entrepreneur, December 1999*

non-MS-DOS operating system. A year later, Microsoft Windows—a graphical user interface (GUI) that ran as an extension of MS-DOS—hit the shelves.

The advent of Windows marks a key moment in Microsoft history. On the one hand, the immense popularity of Windows helped stoke the flames of excitement about Gates's company, which, in turn, helped Microsoft's initial public offering (IPO) on the stock market, just months later, become a resounding success. (On March 13, 1986, the company raised $61 million, and Gates became, at 31, America's youngest billionaire.) On the other hand, Windows helped reveal one of the darker truths about Microsoft: far from being an innovator, Microsoft was, in essence, a world-class copycat. BASIC, for example, had been developed at Dartmouth in the mid-1960s. Gates and Allen were simply the first to craft a version for the Altair. Similarly, Microsoft had purchased DOS from Seattle Computer, then sold it to IBM as MS-DOS. Also, many claim that Windows, the platform on which some of Microsoft's most successful software applications run, was modeled on the Apple Macintosh GUI (which, in all fairness, was based on innovations from Xerox's computer lab, the Palo Alto Research Center). It was becoming clear that Gates's great talent was not in developing groundbreaking technologies but in brokering the deals that would make those technologies a consumer success.

Post-IPO, and with a new multi-building corporate campus located in the Seattle suburb of Redmond, Microsoft still insisted that it was "the little guy" in the software world and continued to function like a startup. Programmers still worked day and night in small teams; although the company had grown to more than 1,100 employees, many still dealt directly with Gates about their code. Microsoft also continued to compete as though it were still struggling for market share. Gates would often match, dollar for dollar, any price break on any competitive software—no matter how small the company—a practice that easily quashed fledgling software companies.

Throughout the late 1980s, Microsoft continued to bring out new products—the OS/2 operating system

*Fear should guide you, but it should be latent. I have some latent fear. I consider failure on a regular basis.*

—*Gates to* Playboy, *December 1994*

(one of several products that Microsoft developed in conjunction with IBM), updated versions of Windows and Word, as well as new products like Excel and the CD-ROM. By spring 1988, Microsoft had become the premier software vendor in the country, having beaten out Lotus. Nevertheless, trouble was brewing. In March 1988, Apple filed suit against Microsoft, claiming copyright infringement of the GUI interface. Two years later, the U.S. Federal Trade Commission (FTC) began to investigate Microsoft, contending that it was in violation of the Sherman Antitrust Act; the FTC maintained that Microsoft used licensing agreements to force companies that installed Microsoft operating systems to also install Microsoft applications.

## THE LONG ARM OF THE LAW

By the early-1990s, Microsoft seemed to have settled its early legal troubles: in August 1993, a federal judge dismissed Apple's copyright complaint, a decision that was upheld on appeal; that same month, the FTC closed its four-year antitrust investigation without further action, primarily because the investigators were deadlocked about whether to prosecute (two for, two against). However, the Department of Justice (DOJ) picked up where the FTC left off. In July 1994, Microsoft entered into a consent agreement with the DOJ that required Microsoft to refrain from tying any of its software to its operating systems but did not prohibit the company from making "integrated products." Microsoft claimed victory. Many, including then-attorney general Janet Reno, saw the agreement as little more than a slap on the wrist. "Microsoft's unfair contracting practices have denied other U.S. companies a fair chance to compete, deprived consumers of an effective choice among competing PC operating systems, and slowed innovation," she announced on the day of the agreement.

Over the next year, both sides challenged the legitimacy of the consent decree, until District Court Judge Thomas Penfield Jackson upheld the decree in August 1995. (A month earlier, Gates had assured industry executives at Intel, the microchip manufacturer, that the "antitrust thing" would "blow over.") In the intervening months, the DOJ's investigative pressure forestalled Microsoft's planned $1.5-billion acquisition of Intuit, which would have given Microsoft immediate dominance in the personal financial software market. Seemingly unfazed and with the DOJ's consent, Microsoft shipped Windows 95 on August 24, 1995.

Unlike previous versions of the operating system, Windows 95 featured Microsoft's new World Wide Web graphical browser application, Internet Explorer (IE). Graphical Web browsers had, since 1993, revolutionized the way people interacted with the Internet and were quickly changing the nature of computing by bringing increasingly more people online. At the time of Windows 95's release, Netscape Navigator clearly dominated the browser market. Microsoft, somewhat late to the Internet game, placed IE in direct competition with Navigator; on December 7, 1995, just days after Netscape's stock had reached an all-time high, Gates announced a tough new Internet strategy, which included giving IE away. (Netscape cost $39, though the public could easily use free test-versions.) The browser wars had begun.

Initially, many believed Navigator to be the superior browser. Early versions of IE did not support various cutting-edge technologies such as Java, which allowed for dynamic Web page design. Still, IE's price was right and its market penetration, thanks to the immense popularity of Windows 95, was overwhelming and it prevailed. By mid-1996, Netscape's share of the market—once as high as 80 percent—had dropped to less than half. In an August 8, 1996, letter to the government, Netscape asserted that Microsoft was violating the 1994 consent decree by bundling Internet Explorer with the Windows 95 operating system; in addition, Netscape contended that Microsoft was engaging in many other illegal competitive activities. The DOJ began investigating Netscape's claims the following month.

Even as the DOJ announced the filing of an injunction against Microsoft for violating the consent decree on October 20, 1997, new antitrust allegations were flying. In August of that year, Microsoft had invested $150 million in one of its key rivals, Apple Computer—an act that piqued the DOJ's curiosity. (Later investigations revealed that the $150 million was part of a deal in which Steve Jobs, head of Apple, was given the choice between replacing Netscape with IE as the default browser on

Apple's operating systems or losing its lifeblood—Microsoft Office.) In addition to announcing the injunction, the DOJ asked the court to impose a staggering $1-million-per-day fine on Microsoft for as long as it continued to tie IE to Windows 95. (On appeal, Microsoft won the right to keep IE as part of Windows 95.) The following November, Texas became the first state to file a wider antitrust suit against Microsoft, alleging various violations of the Sherman Antitrust Act. On May 18, 1998, nineteen other states, as well as the U.S. government, followed the lead of Texas. Two months later, a handful of Silicon Valley executives testified before a Senate Judiciary Committee about Microsoft's unfair and allegedly illegal business practices.

The slow grinding of the justice system took its toll on Gates. By all outside appearances, Gates was utterly confident that Microsoft would prevail; privately, however, Gates sometimes succumbed to the pressure. Family and friends, speaking to the press, referred to Gates's "blue funk." In a monthly executive meeting in late December 1997, Gates is said to have broken down and cried, saying, "The whole thing is crashing in on me." At the same time, customer satisfaction with Microsoft was at an all-time low. In July 1998, Gates relinquished his post as president of Microsoft to Steve Ballmer. (Ballmer would, in 2000, become CEO.) In recorded pretrial hearings, Gates would be seen as uncooperative, evasive, and arrogant, denying knowledge of information contained in his own emails and questioning the meaning of such common words as *ask*. (Indeed, Gates's evasiveness was later characterized as "Clintonesque," referring to the former president's testimony on his relationship with intern Monica Lewinsky.) Many could not believe that Microsoft's legal team let Gates possibly harm the case by appearing so openly hostile; still others believed it was Gates, not the legal team, who was calling the shots.

## TAMING THE BEAST FROM REDMOND

*U.S. v. Microsoft* began on October 19, 1998, in Washington, D.C. By then, IE had gobbled up Netscape's market share, and the browser wars were all but over. Indeed, by the end of the year, America Online (AOL) would announce its purchase of Netscape for $4.2 billion in stock. Gates asserted that the buyout proved that

Microsoft had competition and was not a monopoly. Netscape CEO James Barksdale, the government's first witness, claimed otherwise. He testified that Microsoft had deliberately set out to "cut off Netscape's air supply" and recalled a June 1995 meeting in which Microsoft suggested that it and Netscape illegally divide the browser market.

As the case progressed, the DOJ outlined the myriad ways in which Microsoft's business practices earned it the moniker, "the Beast from Redmond." Prosecutors alleged that Microsoft pressured AOL to adopt IE over Netscape by leveraging the power of Windows, which by then had nearly 90 percent of the operating system market; in exchange for dropping Netscape, Microsoft would reward AOL with vital market penetration by placing an AOL icon on every Windows desktop it sold. The DOJ also put forth claims that Microsoft threatened to terminate its Windows 95 licensing agreement with Compaq, a leading PC maker, when Compaq tried to replace IE with Netscape on its Presario computers. As for Intuit, the company with which Microsoft nearly merged, the DOJ offered as evidence a damning email, written by Gates, that read: "I was quite frank with him, that if he had a favor we could do for him that would cost us something like $1 million . . . in return for switching browsers in the next few months, I would be open."

Then, the DOJ widened the case beyond the browser wars. Apple executive Avie Tevanian testified that, in addition to similar pressures to drop Netscape from the Apple desktop, Microsoft had illegally attempted to divide the burgeoning streaming media market, allegedly offering to not compete with Apple's Quicktime technology in the Macintosh market if Apple agreed not to push Quicktime into PC markets. In addition, Microsoft allegedly tried to convince Apple to help undermine the influence of Java, a platform-independent programming language offered by Sun Microsystems, by using Microsoft's Windows-only version. Executives from Sun Microsystems supported these allegations. Even Microsoft's supposed ally in the computer market, the computer chip maker Intel, testified for the DOJ: Intel's CEO, Andy Grove, testified that Gates had made vague threats about shifting Microsoft's alliances in the chip-making market if Intel continued to develop what would become the company's own operating system. In each instance, Microsoft seemed to be leveraging its power as a monopoly to protect Windows—its bread and butter—from becoming obsolete.

Microsoft's main defense against these allegations was that the software industry, unlike any other earlier industry, was so incredibly competitive that even an industry leader must fight to remain in business. The company also countered each of the DOJ's claims by asserting that each company—from AOL to Sun—made its own business decisions and that the choice of either or both IE and Windows was made because of their superiority to the competition.

In November 1999, Judge Jackson, who had presided over both the consent decree case and the wider antitrust trial, released the first phase of the court's decision: the findings of fact. Jackson found that Microsoft was indeed a monopoly (which, in itself, is not illegal) and that the company had abused its market position to quash competition. "The ultimate result is that some innovations that would truly benefit consumers never occur," Jackson wrote. In the second phase of the court's decision, the conclusions of law, issued in June 2000, Jackson decreed that Microsoft be split into two companies. Microsoft appealed this decision all the way to the U.S. Supreme Court, which remanded the case to the U.S. Court of Appeals for the District of Columbia. By September 2001, the DOJ announced that it would not seek to break up Microsoft. In January 2003, Microsoft was able to settle the antitrust case with all but two of the original twenty states, agreeing to certain restrictions on competitive practices and paying more than $1.1 billion to consumers who had purchased Microsoft products.

Once again, detractors called the settlement a mere slap on the wrist. Individual companies, such as Sun Microsystems, pursued their allegations in separate cases. As early as May 2003, industry trade groups sought to challenge the settlement. However, by 2003, several other factors had begun to wear away Microsoft's dominance in the software market, including the rather

> *I have a high enough level of visibility that people will second-guess anything I do.*
>
> —Gates to Time, *July 2000*

limited success of Microsoft's Internet initiative, .NET; increasing security problems with the company's products; and the growing popularity of one of the greatest threats—open-source software—to Windows.

## A Strong Foundation

Even before Microsoft ran so deeply afoul of antitrust laws, Gates had been giving a significant portion of his fortune to charity. (Since the mid-1990s, Gates had been the richest or second richest individual in the United States.) Given Microsoft's legal troubles, many criticized his early donations as a feeble attempt to regain public support. However, as the years passed, he and his wife, Melinda French Gates, have earned a reputation as true philanthropists.

The Bill and Melinda Gates Foundation, established in January 2000 by the merger of the Gates Learning Foundation (1997) and the William H. Gates Foundation (1994), has pledged more than $25 billion to education and global health. One of Gates's earliest programs pledged millions to connect American libraries to the Internet, especially in low-income neighborhoods on the wrong side of the Digital Divide. The Gates Millennium Scholars program, begun in September 1999, was set up to fund thousands of scholarships at elite universities for minority students pursuing degrees in the fields of math, science, or technology. Gates's global health program focuses on eradicating diseases such as malaria, tuberculosis, and HIV/AIDS, particularly in Africa, in addition to working for increased education, particularly in the arena of reproductive health, around the world. In one of the most-lauded programs, the Gates Foundation has donated more than $150 million for the Malaria Vaccine Initiative, which hopes to save the lives of millions in impoverished nations.

In all, Gates has pledged to give away 95 percent of his personal fortune. As his wealth hovers somewhere between $30 and $40 billion (depending on the price of Microsoft's stock), this makes the Bill and Melinda Gates Foundation the richest philanthropic organization in the world.

Although for much of the 1990s, Gates was suspected, in certain circles, of being the devil himself—and Microsoft, the Evil Empire—both he and his company have received a much fairer shake in the post-antitrust trial era. Not much has changed about Gates's fiercely competitive business nature or his ambitious visions for computing. Indeed, since stepping down from his post as CEO in 2000, he has, as chief software architect, refocused his energies on breaking new ground in software and technology, most notably with Microsoft's newest operating system, code-named Longhorn, which, Gates claims, will usher in a new era in integrated computing. (Longhorn is due to arrive in 2006.) As Microsoft faces new threats from open-source software and the ever-changing competitive environment of the Internet, and as Gates grows closer to retirement age, speculation abounds about the future of his empire. "For now," Gates confided in a 1999 interview with *Entrepreneur* magazine, "Microsoft is my career, and I think I have about the best job in the world."

## Further Reading

### In These Volumes

Related Entries in this Volume: Andreessen, Marc; Case, Steve; Dyson, Esther; Glaser, Rob; Torvalds, Linus; Winblad, Ann

Related Entries in the Chronology Volume: 1975: Bill Gates and Paul Allen Found Microsoft; 1976: Bill Gates Pens Open Letter to Hobbyists; 1984: AT&T Is Broken Up; 1998: Department of Justice Files Suit Against Microsoft; 1998: Linux Operating System Becomes a Cause Célèbre; 2003: SCO Group Sues IBM Over Linux; 2004: Microsoft Is Fined by European Union

Related Entries in the Issues Volume: Copyright; Hackers; Open Source; Security

### Works By Bill Gates

"Open Letter To Hobbyists" (1976), http://www.tranquileye.com/cyber/1976/gates_open_letter_to_hobbyists.html (cited September 16, 2004).

*The Road Ahead.* New York: Viking, 1995. Also available online at: http://www.roadahead.com/ (cited September 16, 2004).

*Bill Gates Speaks.* New York: John Wiley, 1998.

*Business @ The Speed of Thought.* New York: Warner Books, 1999.

### Books

Andrews, Paul. *How the Web Was Won: Microsoft from Windows to the Web.* New York: Broadway Books, 1999.

Bank, David. *Breaking Windows: How Bill Gates Fumbled the Future of Microsoft.* New York: Free Press, 2001.

Edstrom, Jennifer. *Barbarians Led by Bill Gates: Microsoft from the Inside.* New York: Henry Holt, 1998.

Gatlin, Jonathan. *Bill Gates: The Path to the Future.* New York: Avon Books, 1999.

Ichbiah, Daniel. *The Making of Microsoft: How Bill Gates and His Team Created the World's Most Successful Software Company.* Rockland, CA: Prima, 1991.

Manes, Steven. *Gates: How Microsoft's Mogul Reinvented an Industry and Made Himself the Richest Man in America.* New York: Simon & Schuster, 1994.

Rohm, Wendy Goldman. *The Microsoft File: The Secret Case Against Microsoft.* New York: Time Business, 1998.

Wallace, James. *Hard Drive: Bill Gates and the Making of the Microsoft Empire.* New York: HarperBusiness, 1993.

———. *Overdrive: Bill Gates and the Race to Control Cyberspace.* New York: John Wiley, 1997.

## Articles

"83 Reasons Why Bill Gates' Reign Is Over." *Wired* 6.12, December 1998, http://www.wired.com/wired/archive/6.12/microsoft.html (cited September 16, 2004).

Gleick, James. "Making Microsoft." *New York Times Magazine,* November 5, 1995.

Harmon, Amy. "*U.S. v. Microsoft:* The Overview." *New York Times,* November 2, 2002.

Heilemann, John. "Fear and Trembling in Silicon Valley." *Wired* 8.03, March 2000, http://www.wired.com/wired/archive/8.03/deepthroat.html (cited September 16, 2004).

———. "The Truth, The Whole Truth and Nothing But The Truth." *Wired* 8.11, November 2000, http://www.wired.com/wired/archive/8.11/microsoft.html (cited September 16, 2004).

Isaacson, Walter. "In Search of the Real Bill Gates." *Time,* January 13, 1997, http://www.time.com/time/gates/cover0.html (cited September 16, 2004).

Karlgaard, Rich. "ASAP Interview: Bill Gates." *Forbes,* December 7, 1992.

———. "On the Road with Bill Gates." *Forbes,* February 28, 1994.

"The Rise and Rise of the Redmond Empire." *Wired* 6.12, December 1998, http://www.wired.com/wired/archive/6.12/redmond.html (cited September 16, 2004).

Seabrook, John. "E-mail from Bill." *New Yorker,* January 10, 1994, http://www.booknoise.net/johnseabrook/stories/technology/email/ (cited September 16, 2004).

Strouse, Jean. "How to Give Away $21.8 Billion." *New York Times Magazine,* April 16, 2000.

Weigner, Kathleen K., and Julie Pitta. "Can Anyone Stop Bill Gates?" *Forbes,* April 1, 1991.

## Websites

Bill and Melinda Gates Foundation. Home of the Gateses' eponymous charitable organization, the most well endowed philanthropic organization in the world, http://www.gatesfoundation.org/default.htm (cited September 16, 2004).

Bill Gates Home Page. Bill Gates' Microsoft site, with links to his speeches, published articles, essays, and books, http://www.microsoft.com/billgates/default.asp (cited September 16, 2004).

Gates' Millennium Scholars Foundation. Online home of the Gateses' scholarship fund aimed at increasing the population of minorities at institutions of higher learning, particularly in the fields of math, science, and technology, http://www.gmsp.org/ (cited September 16, 2004).

Microsoft Corporation. Online home of Microsoft Corporation, with links to products, resources, tech support, and corporate history. The Press Pass section includes information on the company's history and organization, http://www.microsoft.com/ (cited September 16, 2004).

OpenLaw: The Microsoft Case. Companion Website for a Harvard course on the Microsoft case, taught by renowned Internet lawyer Lawrence Lessig. Site contains extensive links to court documents and legal analysis, http://cyber.law.harvard.edu/msdoj/index.html (cited September 16, 2004).

Smithsonian Institution: Interview with Bill Gates. As part of the Smithsonian's Computer History Collection, David Allison, Smithsonian computer division director, interviewed Bill Gates at length about his early experience with various computers and the history and growth of Microsoft, http://americanhistory.si.edu/csr/comphist/gates.htm (cited September 16, 2004).

The Time 100: Bill Gates. Gates is nominated as one of *Time*'s top twenty innovators of the 20th century, http://www.time.com/time/time100/builder/profile/gates.html (cited September 16, 2004).

*U.S. v. Microsoft.* The Department of Justice's Website on the antitrust suit against Microsoft, including a full list of documents and decrees—from the most recent developments to the settlement—reaching back through the appeals process and with links to other significant Microsoft cases, http://www.usdoj.gov/atr/cases/ms_index.htm (cited September 16, 2004).

Washingtonpost.com: Microsoft. The *Washington Post*'s special report on Microsoft and the *U.S. v. Microsoft* saga, with links to *Washington Post* articles on each development in the case, http://www.washingtonpost.com/wp-dyn/business/specials/microsofttrial/ (cited September 16, 2004).

# William Gibson (1948–)

## SCIENCE FICTION AUTHOR

**When science fiction** writer William Gibson coined the word *cyberspace* for a short story published in the early 1980s, he had no thought of defining the coming Internet Age. A full decade before the Internet burst into the mainstream, Gibson's critically acclaimed novel, *Neuromancer* (1984), introduced a generation of science fiction and other readers to ideas like cyberspace and virtual reality. The bleak vision of a networked society that marks Gibson's early novels earned him a reputation as the "Orwell of the Internet" and placed him at the head of a new movement in science fiction—cyberpunk. Gibson has since made a name for himself in Hollywood and as a futurist, commenting on today's technologically driven society.

## LIVING IN EXILE

Gibson had an itinerant childhood. Born in South Carolina, his family moved from one budding Southern suburb to another—wherever his father, a civilian contractor, found work. Gibson's father died when Gibson was six, and Gibson moved with his mother to her family's home in a small town in southwestern Virginia. For Gibson, the move was a jarring shift toward the conservative and the rural—he has described his mother's hometown as more Depression 1930s than 1950s. "The trauma of my father's death aside, I'm convinced that it was this experience of feeling abruptly exiled, to what seemed like the past, that began my relationship with science fiction," Gibson explained in a 2002 essay, "Since 1948." His interest in science fiction was also influenced by stories of his father's work on the U.S. government's secret Oak Ridge Project in Tennessee, where the first atomic bomb was constructed. In grade school, Gibson devoured books by science fiction greats, including Ray Bradbury and Isaac Asimov. These authors spoke to the outsider in Gibson. "That was the first inkling I had that there were quite a few people in the world who felt as not a part of what was going on around them as I did," he said in a 2004 interview with the *Financial Times.* By junior high, his taste had shifted to include the seminal beat writers—Jack Kerouac, Allen Ginsberg, and William Burroughs. This mix of the fantastic and bohemian would mark his writing for years to come.

Gibson remained terribly alienated from kids his age in southern Virginia. At the age of 15, he convinced his mother to send him to the Southern Arizona School for Boys, a private school-cum-dude ranch located in the desert just outside Tucson. Once there, his childhood dream of becoming a science fiction writer drifted away. Indeed, it would be another decade before he would return to writing.

In Arizona, Gibson was exposed once again to an entirely different slice of life. He described the school as "a dumping ground for chronically damaged adolescent boys," with peers that included a volatile teenage alcoholic and a boy from San Francisco who had undergone bouts of plastic surgery since age 10. During Gibson's senior year, school authorities suspected him of being part of a drug scandal—a charge he denied. Though it was the late 1960s, and Gibson boasted a distinctly antiestablishment sensibility, his most countercultural act at the time was reading Kerouac. His mother died the same year, when Gibson was 18. Her death and the false drug charge by school authorities prompted Gibson to leave school; he returned briefly to Virginia, then moved north to Toronto, Canada.

Gibson settled in among American expatriates fleeing the Vietnam-era draft, although, unlike them, Gibson was not especially political and never received a draft notice. While trying his hand at being an artist or painter, Gibson met his future wife. Once married, they moved to her hometown of Vancouver, where Gibson enrolled in an English degree program at the University of British Columbia. He earned a bachelor's degree in 1977 and, after the birth of his first child, Gibson decided to attempt to settle down.

*William Gibson. (Matthew Mcvay/Corbis)*

## THE CYBERPUNK MUSE

With no career prospects—and, in fact, what he calls "an absolute lack of enthusiasm for anything like 'career'"—Gibson stayed at home with his infant son while his wife earned her master's degree. "I couldn't leave the house, so I was kind of trapped, it seemed to me, in a situation where all I could do was write," he recalled in the 2003 *Guardian* profile. "It was a matter of options narrowing down to the typewriter. The baby would be asleep and I would go and write fiction."

For inspiration, Gibson, then twenty-seven, returned to his childhood fascination with science fiction. By then, an early love of toy robots had evolved into a more sophisticated interest in artificial intelligence and sentient machines. The basic theme of many of his works was "where do we stop and they [machines] begin?"

Gibson's first short story, "Fragments of a Hologram Rose," was published in 1977 in *UnEarth,* an underground science fiction magazine out of Boston. Gibson

introduced notions of virtual reality, tied in with a classic end-of-the-romance plot line, written in style he calls "noir naturalism." Another story followed shortly thereafter, and soon Gibson was earning a living. "Almost as soon as I'd started, it started to work," he told the *Guardian* in 2003. "I wrote and I sold. By the time I'd sold two or three stories, I was earning enough money that I couldn't actually afford to stop because I had no other income."

As he wrote, Gibson became more involved with the science fiction community. Terry Carr, a legendary science fiction editor, bought the rights to his story "The Gernsback Continuum," to include in a 1981 edition of the science fiction anthology *Universe,* and quickly became one of Gibson's unflagging supporters. At a science fiction convention held in Vancouver, Gibson met John Shirley, who subsequently helped get Gibson's stories into Isaac Asimov's highly respected *Omni* magazine. When Bruce Sterling's first science fiction novel, *Involution Ocean,* was published in 1977, Gibson sent him a fan letter, which quickly turned into a five-year correspondence. (A lifelong friendship formed as well as, in 1990, professional collaboration on the book, *The Difference Engine.*)

What Gibson shared with Shirley and Sterling was a new vision for the genre of science fiction. In the late 1970s and early 1980s, the primary thrust of most science fiction still reflected rather conservative notions of 1950s technological progress. The turbulence and the disillusionment of the 1960s and 1970s had not made its way into the genre, nor had numerous cultural developments, such as the burgeoning punk rock movement, the birth of the personal computer, and looming threats of nuclear war. As Gibson explained to the *Philadelphia Inquirer* in 2004, "The art form I loved as a kid had gone completely flat. I realized no one had tried to write a science-fiction novel as if Lou Reed and David Bowie were writing it."

Gibson, Shirley, and Sterling set out to bring a modern sensibility to science fiction. They became the leading writers in what came to be known as the cyberpunk subgenre of science fiction, characterized by a taste for technology, anarchy, and bohemia, stories with a distinctly dystopian tenor. (Of the three, Shirley was the only one who looked the part of a cyberpunk—he sang in a punk band at night, and science fiction was his day job.) Their work tapped into a previously unknown cultural intersection where geeky technophiles and surly punk rockers met. Gibson affectionately nicknamed this subculture of fans M&Ms, for "modems and Mohawks." Gibson himself knew little of modems and the other technological gadgets he wrote about; until 1985, he worked on a 1927 Hermes manual typewriter.

> *Science fiction is a very good tool for getting a handle on an unimaginable present. Forget the unimaginable future, the present is damn near unimaginable this morning.*
>
> —Gibson to the National Post, October 2003

## THE COLOR OF TELEVISION

From the beginning, Gibson was writing about future worlds where virtual reality and vast interconnected electronic networks were the norm. In part, he was inspired by watching kids at the video arcade on Granville Street in Vancouver. "I could see in the physical inten-

sity of their postures how rapt the kids were," he told the *Christian Science Monitor* in 1998. "These kids clearly believed in the space the games projected."

In his short story "Burning Chrome," first published in 1982, he sought to formalize his description of that space. He also sought to give this new space a name, a "buzz word devoid of meaning." He wrote down and crossed out several ideas, including *infospace* and *dataspace,* before finally setting on the term *cyberspace.* The word *cyber* harkened back to the 1940s, when Norbert Weiner, a famed Massachusetts Institute of Technology scientist, coined the term *cybernetics* to describe the study of control systems. Its Greek root translated into "steersman"—a concept that dovetailed nicely with Gibson's vision of cyberspace as defining much of his characters' lives.

Gibson continued to use *cyberspace* when writing *Neuromancer,* his first full-length novel, which had been commissioned by Terry Carr in 1981. At first, Gibson was terrified at the prospect of writing an entire novel—a feeling only sharpened after he went to see the seminal science fiction movie *Blade Runner* (1982). (Reviews of *Blade Runner* call it a "cyberpunk vision of the year 2019.") In an interview with the London newspaper *The Independent,* Gibson explained, "I fled the cinema after 15 minutes, deeply dismayed because it looked exactly like the pictures on the inside of my forehead. Actually, it looked better."

Nevertheless, Gibson continued to write. *Blade Runner* had captured the look and feel of Gibson's imaginary future, but in *Neuromancer,* Gibson focused on the subjective experience of cyberspace. He wrote, in a now-famous passage from the book:

Cyberspace. A consensual hallucination experienced daily by billions of legitimate operators, in every nation, by children being taught mathematical concepts. . . . A graphic representation of data abstracted from the banks of every computer in the human system. Unthinkable complexity. Lines of light ranged in the nonspace of the mind, clusters and constellations of data. Like city lights, receding.

In *Neuromancer,* published in 1984, Gibson offered readers a complete world the likes of which few had conceived of before: one populated by "data thieves" and "neural implants," by people "jacking in" to an electronic network, to "cyberspace" and the "matrix." Indeed, the book's opening line—"The sky was the color of television, tuned to a dead channel"—clearly indicated that readers were in new territory. Marianne Trench, a virtual reality theorist, explained in a 1995 interview with *CMC* magazine, "When William Gibson's visions were published, they struck sparks in the real world. . . . Never before had science fiction literature determined the way people thought and talked."

His novel was hailed by critics; the *Washington Post* called it "an amazing virtuoso performance," while literary theorists drew parallels between Gibson and Thomas Pynchon, author of *Gravity's Rainbow,* anointing them both key figures in postmodern literature. The book became the signature cyberpunk novel, and Gibson the signature cyberpunk novelist (even though other authors were more prolific). *Neuromancer* went on to win the Hugo Award, Nebula Award, and Philip K. Dick Award— top prizes for science fiction. It is the only novel to ever win all three.

By 1988, Gibson had published two more novels, *Count Zero* (1986) and *Mona Lisa Overdrive* (1988), which, along with *Neuromancer,* made up the Sprawl trilogy. (*Sprawl* refers to Gibson's vision of the East and West coasts of America, where cities have grown together to form massive urban wastelands.) His depictions of insidious multinational corporations, massive data networks, and a vast, multicultural global village earned him a reputation as prophet. As James Flint, a science fiction novelist and one-time *Wired* contributor, told *The Guardian* in 2004, "Reading Gibson in the early 1990s, you got a sense that you were glimpsing 'the truth' about where we were all headed. . . . For about five years I don't think I met anyone who hadn't read him, who didn't use his work as a crucial touchstone."

## HOLLYWOOD CALLING

The success of *Neuromancer* pulled Gibson from relative obscurity and into the mainstream. By 1994, the term

> *I invented cyberspace because I needed something to replace the aliens-and-spacecraft part of science fiction.*
>
> —Gibson to the Globe and Mail, June 1995

cyberspace had been universally embraced as shorthand for the new territory carved out by the booming Internet and the World Wide Web. Gibson's work made an impact in another cultural sector—Hollywood.

"Johnny Mnemonic," a short story first published in *Omni* magazine in 1981, was the first of Gibson's writing to make it to the screen. Optioned in 1990, it took more than five years to finance, produce, shoot, and edit, and what began as a low-budget art-house film grew into Hollywood blockbuster proportions.

The movie featured Keanu Reeves as a mnemonic courier from the near future who delivers data via a computer chip implanted in his brain. Although the screenplay—filled with hit men, bounty hunters, and all-powerful multinational corporations—hewed closely to the story, the finished product was severely edited by Sony Pictures and only vaguely resembled Gibson's adapted screenplay. Gibson told writer Daniel Robert Epstein, in an interview for the alternative online 'zine *suicidegirls.com,* "Nobody got to see the film that Robert Longo shot. . . . The tragedy with Johnny Mnemonic is that we shot an ironic broadly comic action film that at some level was supposed to be about bad science fiction movies. We were not trying to make a blockbuster mainstream adventure film starring Keanu Reeves."

Several of Gibson's other screen adaptations also foundered, including a draft proposed for *Alien 3.* (The sole element of Gibson's draft to appear in the movie were bar code tattoos located on the backs of the characters' necks.) Still, the dramatic allure of Gibson's stories proved irresistible for filmmakers. In 1998, noted director Abel Ferrera made the film version of Gibson's short story "New Rose Hotel," starring Christopher Walken and Willem Dafoe. A long-awaited adaptation of *Neuromancer* was announced in 2000 (though, four years later, production had still not begun), and in April 2004, the announcement of plans for a film version of his latest book, *Pattern Recognition,* to be released in 2006, made waves.

More important than the mixed success of his own stories, Gibson's vision played a vital role in shaping the entertainment industry in the 1990s and early 21st century—most notably with the *Matrix* trilogy. Indeed, Gibson has called the film "the ultimate cyberpunk artifact." Images and concepts from his work have also appeared in films like *eXistenZ* and *Strange Days*. In 2001, Gibson appeared as himself in the documentary *No Maps For These Territories,* an exploration of Gibson's writing and unique vision. He has also made cameo appearances on science fiction inspired television series such as *The X-Files* (for which he wrote two scripts) and *Wild Palms.* Playing himself in the *Wild Palms* episode, he is pointed out as the man who invented the word *cyberspace.* Gibson's sole line was, "And they won't let me forget it."

## TAKE TWO

After finishing the Sprawl trilogy, Gibson continued to mine the Internet world for ideas that lent themselves nicely to futuristic dramatization. His work also became more personal. In 1992, he published a poem, "Agrippa: A Book of the Dead." It was, at once, a meditation on fading memories and the loss of father and an Internet-age art piece. The poem was distributed via floppy disk. Once read, a computer program on the disk turned the words into gibberish. Hackers needed only three days to crack the encryption code—copies of "Agrippa" soon began circulating on the Internet and, since Gibson did not use email, fans sent hacked copies of the poem to his fax machine.

His next three novels, *Virtual Light* (1993), *Idoru* (1997), and *All Tomorrow's Parties* (1999), known as the Bridge Trilogy because of recurring image of a futuristic shanty town built on the postearthquake remains of the San Francisco Bay Bridge, were set in the none-too-distant future and commented on evolving aspects of the Internet Age. *Virtual Light* (set in 2005) focused on the hunt for virtual reality glasses; *Idoru* described a love affair with a completely virtual Japanese celebrity (Tokyo is one of Gibson's favored settings). Referring to *All Tomorrow's Parties,* a novel about a man who (like Gibson himself) sees emerging patterns in cyberspace's endless streams of data, Gibson told the *Toronto Star,* "I was working so close to the present it's almost an alternate present."

> *The Internet is strange. It doesn't make any money. It is transnational, beyond anyone's control. It is the great anarchist event.*
>
> —Gibson to the Financial Times, September 1994

When the terrorist attacks of September 11, 2001, took place, Gibson was in the midst of writing his eighth novel, *Pattern Recognition.* One of his only books to be set in the present, he felt he had to incorporate the attacks into the plot. "I had this very surreal, unpleasant, Kafkaesque sense of the world that I had been blithely telling interviewers was incomprehensible and catastrophic and terrifying—and suddenly it was," he told the *Los Angeles Times* in 2003. "It was like the universe had called my bluff."

Gibson's move toward the present tense coincided with the waning popularity of cyberspace, which happened when the wild and darkly glamorous Internet frontier of the 1990s gave way to its bland commercialization in the 21st century. The most famous of his literary inventions, cyberspace, is not something Gibson regrets leaving behind. "Those days are over," he said in a 2000 interview with *Science Fiction Weekly.* His vision for the modern-day equivalent? "Biopunk . . . kids in the Haight doing their own genetic manipulation."

## FURTHER READING

### In These Volumes

Related Entries in this Volume: Sterling, Bruce
Related Entries in the Chronology Volume: 1984: William Gibson Introduces the Term *Cyberspace*

### Works By William Gibson

*Neuromancer.* New York: Ace Books, 1984.
*Burning Chrome.* New York: Ace Books, 1986.
*Count Zero.* New York: Arbor House, 1986.
*Mona Lisa Overdrive.* New York: Bantam Books, 1988.
*Virtual Light.* New York: Bantam Books, 1993.
*Idoru.* New York: G.P. Putnam's Sons, 1996.
*All Tomorrow's Parties.* New York: G.P. Putnam's Sons, 1999.
*Pattern Recognition.* New York: G.P. Putnam's Sons, 2003.
with Bruce Sterling. *The Difference Engine.* New York: Bantam Books, 1990.

### Films

*Johnny Mnemonic.* Columbia TriStar Home Video, 1995.
*No Maps for These Territories.* New Video Group, 2000.

## Books

Cavallora, Dani. *Cyberpunk and Cyberculture: Science Fiction and the Work of William Gibson.* New Brunswick, NJ: Athlone Press, 2000.

Olsen, Lance. *William Gibson.* San Bernardino, CA: Borgo Press, 1992.

## Articles

Chollet, Laurence. "William Gibson's Second Sight." *Los Angeles Times,* September 12, 1993.

Darling, Peter. "Sandpapering the Conscious Mind with William Gibson." *Science Fiction Weekly,* February 7, 2000, http://www.scifi.com/sfw /issue146/interview.html (cited September 23, 2004).

Doctorow, Cory. "William Gibson Turns Humanist." *Globe and Mail,* November 30, 1999.

Evenson, Laura. "The Oracle of Cyberspace." *San Francisco Chronicle,* September 24, 1996.

Farrell, John Aloysius. "The Cyberpunk Controversy." *Boston Globe,* February 19, 1989.

Garreau, Joel. "Cyberspaceman." *Washington Post,* October 18, 1993.

Harmon, Amy. "Crossing Cyberpunk's Threshold." *Los Angeles Times,* May 24, 1995.

Heer, Jeet. "William Gibson Can't Be Bothered with the Future." *National Post,* October 20, 2003.

Hiltbrand, David. "Squinting at the Present." *Philadelphia Inquirer,* February 17, 2004.

Kirchhoff, H. J. "William Gibson's Cyber-career." *Globe and Mail,* June 2, 1995.

Lewis, Peter H. "Present at the Creation, Startled at the Reality." *New York Times,* May 22, 1995.

Poole, Steven. "Tomorrow's Man." *Guardian,* May 3, 2003.

Popham, Peter. "Cyber Face." *Independent (U.K.),* October 24, 1996.

Quittner, Josh. "Lost in Cyberspace." *Newsday,* September 14, 1993.

Ulin, David L. "Present Worries in Future Tense." *Los Angeles Times,* March 4, 2003.

Van Bakel, Rogier. "Remembering Johnny," *Wired* 3.06, June 1995, http://www.wired.com/wired/archive /3.06/gibson.html (cited September 23, 2004).

Wheelwright, Geof. "Superhighway in Need of a Route." *Financial Times,* September 6, 1994.

## Websites

EFF "William Gibson Publications" Archive. Links to several lengthy interviews with Gibson, http://www .eff.org/Misc/Publications/William_Gibson/ (cited September 23, 2004).

No Maps for These Territories. Online site for *No Maps for These Territories,* a documentary about Gibson's life and work, http://www.nomaps.com (cited September 23, 2004).

"Technoculture from Frankenstein to Cyberpunk." Extensive site of a 1997–1998 course at Georgetown University that studies works of Gibson and other cyberpunk writers, http://www .georgetown.edu/faculty/irvinem/technoculture/ (cited September 23, 2004).

William Gibson. Gibson's official site, with links to his 2003 Weblog, essays, a discussion board, and a list of Gibson's preferred links, http://www .williamgibsonbooks.com/index.asp (cited September 23, 2004).

William Gibson aleph—essential information collection. A comprehensive fan site, with links to news and books, http://www.antonraubenweiss .com/gibson/ (cited September 23, 2004).

# John Gilmore (1955–)

## ENTREPRENEUR; CYBER-LIBERTARIAN

**Since the 1980s,** cybermillionaire John Gilmore has funneled his wealth into groups concerned with the protection of Internet freedom and privacy, most notably the Electronic Frontier Foundation (EFF), the first civil rights organization for cyberspace. A lifelong member of the Libertarian Party, Gilmore has been an untiring champion of individual rights and freedoms, both online and off, and has taken on the federal government in several significant lawsuits, especially on the issue of encryption. Although he officially retired from the computer business in 1995, Gilmore continues to use his wealth and influence—and what some might call his audacity—to help shape the future.

## EMPLOYEE NO. 5

Gilmore, who spent his childhood in rural Pennsylvania and teenage years in Alabama, is a self-taught computer programmer; he does not have a college degree. He first learned about computers during his senior year in an Anniston, Alabama, high school, in a year-long data processing class where he used an IBM 1401. The town of Anniston hired Gilmore as a programmer after he graduated. In the early 1970s, he wrote mainframe email software that was used by Jimmy Carter's presidential campaign. By 1978, Gilmore had moved to the San Francisco Bay area, where he worked as a freelance programmer. He was also an early user of ARPANET (a prototype of the Internet) and Usenet, an online community of discussion groups, or newsgroups. Indeed, it was on Usenet that Gilmore learned of an opening at a fledging software company named Microsoft.

In 1981, Gilmore had a job interview with Bill Gates. Microsoft was hiring programmers for a project that involved writing a version of the UNIX operating system for a new computer being built by Andy Bechtolsheim, a graduate student at Stanford University. Bechtolsheim and three fellow graduate students—

Vinod Khosla, Scott McNealy, and Bill Joy—were in the process of founding a new company they called Sun Microsystems (named for Stanford University Networks). When Gilmore sought out Bechtolsheim as a part of researching the Microsoft job, Bechtolsheim convinced Gilmore to join Sun instead. Gilmore turned down the Microsoft job and, in 1982, became Sun employee No. 5.

Gilmore quickly became known as Sun's "little revolutionary," Bechtolsheim told the *San Francisco Examiner* in 1995. His decidedly liberal political worldview had been formed in the 1960s, but by the 1970s, he had aligned himself with a more libertarian point of view—one that valued individual liberty above all else. On the job, Gilmore was known as a maverick, someone who would show up to work in a sarong just to shake up people's expectations He would do excellent work for Sun when a project was in line with his beliefs. Bechtolsheim told *USA Today* that Gilmore would "never take on a job unless he was convinced it was the right thing to do."

After three and a half years working on the architecture, design, and implementation of Sun Workstations 1, 2, and 3, Gilmore's interest waned. By then, the start-up had become a large corporation—too large for Gilmore's tastes. "They couldn't figure out how to manage me," he told *USA Today*. "I didn't fit into any corporate niches." In 1985, Gilmore left Sun with $10,000 in his pocket, a Sun workstation, and significant stock holdings in the company. In 1986 Sun had its initial public offering and made Gilmore a millionaire. Rather than retire, Gilmore used his millions to fund projects that furthered the goals closest to his heart—freedom of speech, freedom of software, and freedom of encryption.

## FREEDOM OF SPEECH

Gilmore is credited with bringing freedom of speech to the Internet when he and fellow computer industry figures Brian Reid, of Digital Equipment Corp. (DEC), and Gordon Moffett, of Amdahl Corp., founded the alt. hierarchy of newsgroups on Usenet in 1987. In 1986, during what Internet historians call The Great Renaming, Usenet established seven top-level hierarchies to help organize the growing number of newsgroups—comp. (for computer-related group), rec. (for recreational and entertainment topics), sci. (for science-related issues), soc. (for social issues), talk. (for discussion of controversial issues), misc., and news. Traffic on these newsgroups was more or less controlled by a group of system administrators informally known as the Backbone Cabal—named for the primary "backbone" of network links that controlled the traffic for most Usenet groups. At the time of the Great Renaming, members of the Usenet community were already growing frustrated with the way the Backbone Cabal was exerting its power. Gilmore and Reid—himself a member of the Backbone Cabal—were among the frustrated.

In May 1987, Gilmore, Reid, and Moffett met at a computer industry barbecue in Mountain View, California, and the topic of Usenet came up in conversation. Gilmore was frustrated because several Backbone administrators refused to carry the rec.drugs newsgroup, which featured online posts discussing illegal drugs, even though the Usenet community had already approved the newsgroup's creation. Similarly, Reid was upset that he was being forced by administrators to use the name rec.food.recipes for his gourmet food newsgroup, plus he was concerned about the increasingly authoritarian nature of the Backbone Cabal itself. A solution to all these problems presented itself; they would create their own hierarchy—an alternet (or alternate) to Usenet, with the abbreviation alt. Moffett offered to join the effort.

Gilmore had a high-speed Internet connection at his San Francisco home, which he calls "Toad Hall," a reference to the children's book, *The Wind in the Willows*. To carry the newsgroups on the alt. hierarchy, Gilmore connected his computer ("hoptoad") to Reid's computer ("mejac") and to Moffett's computer at Amdahl. Reid, a system administrator at DEC, also linked the computers to the company's server, "decwrl." By the end of May, the newsgroups alt.test, alt.config, Gilmore's alt.drugs, and Reid's alt.gourmand were up and running. Gilmore financed the entire network until it became part of the mainstream Usenet network.

Unlike the regulations imposed by the other hierarchies, Gilmore established a comparatively simple creation policy for alt.—anyone who is technically capable of creating a newsgroup can do so. No one was officially in charge of the groups, which meant no censorship. Gilmore gave only the guidance "use common sense." In 1988, individuals bypassed Usenet administrators, who refused to create the group soc.sex, to create the group alt.sex. The alt. hierarchy—sometimes jokingly referred to as the "anarchists, lunatics, and terrorists" hierarchy, even though many of the groups are relatively staid—carries far more newsgroups than any other hierarchy on Usenet. As Gilmore would explain—in a much-repeated aphorism—"the Net treats censorship as damage and routes around it."

## FREEDOM OF SOFTWARE

Gilmore spent much of the late 1980s working with the Free Software Foundation (FSF), a Boston-based group dedicated to development of and use of free computer software, for which he wrote several programs for the GNU family of free software. Free software, also called open-source software, was pioneered by programmer Richard Stallman; it is based on the belief that software should not be copyrighted, so that everyone who uses the software is free to alter it as they see fit. It was while collaborating with the FSF that Gilmore earned the moniker of "gnu."

Through the FSF, Gilmore connected with Michael Tiemann, a fellow programmer and entrepreneur. In the late 1980s, Tiemann wanted to create a business that would offer service and support for the GNU software. It would be the first company of its kind, and many—skeptical of the free software movement itself—doubted such a company could thrive. Tiemann knew Gilmore through occasional email exchanges and brought him on board in 1989, along with another programmer, David Henkel-Wallace. Together, they formed Cygnus Support (later called Cygnus Solutions).

Cygnus—purposely named with *gnu* at its center—was an enormous success, proving that free software could be not only profitable, but significantly so. From the company's humble beginnings in the spare rooms of its founders' Palo Alto apartments, Cygnus became a full-fledged company that nearly doubled its sales every year; Cygnus won consulting contracts with big-name clients like Sun Microsystems and Northern Telcom. Indeed, Cygnus helped bring free software to the mainstream.

In addition to cofounding Cygnus, Gilmore updated and repaired the GNU Debugger (GDB) until 1993, as well as maintained MIT's free Kerberos encryption software, from 1994 to 1995. By 1995, however, Gilmore was forced out of Cygnus because of personality clashes. He remained on the board of directors until January 1997 and received another financial windfall when, in 1999, the company was sold for $675 million to Red Hat, Inc., now one of the leading providers of open-source technology.

## ON THE FRONTIER

When Gilmore left Cygnus in 1995, he turned much of his attention to the Electronic Frontier Foundation (EFF), devoting many hours to the organization. In 1990, Gilmore had been one of several members of the burgeoning online community, including EFF cofounders Mitch Kapor, former head of Lotus, and John Perry Barlow, to express deep concern about the violation of individual rights brought about during the hacker crackdown Operation Sun Devil. As Kapor and Barlow discussed forming an organization to fight government incursions, Gilmore sent them a brief email, offering support. As Kapor recalled to *USA Today*, Gilmore wrote, "I don't necessarily have Mitch's financial horsepower, but I'm supportive of what you're doing. Would $100,000 help?" Steve Wozniak, cofounder of Apple Computer also contributed a hefty sum, and the EFF was born.

Gilmore has since played an active role in championing the rights of hackers, cryptographers, and other Netizens, and has been a vital lifeline for the organization. In 1994, when the EFF suffered a major organizational rift and left the East Coast for San Francisco, the organization set up shop in the basement of Toad Hall. By then, the EFF had expanded beyond fighting cases brought on by Operation Sun Devil to other online issues, including one of Gilmore's essential concerns—access to encryption.

For much of the 1990s, Gilmore had been waiting for someone to challenge U.S. export regulations that kept strong encryption software from the hands of private citizens. In 1995, he got his chance, in the form of *Daniel Bernstein v. United States*. In the early 1990s, Bernstein, then a graduate student at the University of California–Berkeley, had written encryption software he called Snuffle. He was aware, at that time, of government restrictions on encryption technology, so he refrained from posting Snuffle on the Usenet group sci.crypt until he contacted the State Department. (Because the Internet is global, posting the software on Usenet would amount to exporting it, even without Bernstein leaving his home.) Over the course of several months, a number of government agencies told him that posting his program on the Internet would require a munitions license and that his application for a license would be denied, as Snuffle was "too secure" an encryption program.

With financial backing from the EFF, and with Gilmore himself serving as technical adviser to Bernstein's legal counsel, Bernstein brought a case against the State Department, the Department of Defense, the Department of Commerce, and the National Security Agency (NSA) in February 1995. The case stated that computer source code was constitutionally protected free speech; in essence, Bernstein's team argued that the government's insistence that Bernstein have a license to publish amounted to censorship.

In 1997, for the first time, a court affirmed that computer code could be considered speech and was therefore protected by the First Amendment. The government quickly appealed, but was met with the same decision in 1999—if code is protected by the First Amendment, then U.S. export controls on encryption were unconstitutional. Indeed, one judge wrote that government controls on encryption not only violate the rights of cryptographers to publish but also the rights of U.S. citizens to be protected by the security that encryption provides. The case was later remanded to a lower court for review and ultimately dismissed; however, the initial findings in the *Bernstein* case were a major victory for the EFF. In 2003 the U.S. government revised its encryption regulations.

## AMONG THE CYPHERPUNKS

The *Bernstein* case was not Gilmore's first tussle with the government over encryption. In June 1989, he posted a cryptography paper written by a Xerox Corporation researcher on the Internet, after NSA had requested that Xerox not publish the document. Within hours, thousands of people had downloaded copies. In September 1992, Gilmore filed suit against NSA for ignoring his request, under the Freedom of Information Act, for access to two volumes of a classified series of books written in the World War II–era by Colonel William F. Friedman, the father of American cryptography. When Gilmore later came into the possession of the two books from university libraries, government officials ordered him to return the books, on the grounds that, if he distributed the materials, he would be violating the Espionage Act, which carried a ten-year prison sentence. Gilmore challenged the government in the courts on First Amendment grounds. Shortly after learning that Gilmore had found the books himself, the government—without explanation—declassified them, and the case was dismissed.

The early 1990s was a significant time for encryption. In addition to the Friedman situation, the Clinton administration announced its plans for the Clipper Chip, an encryption device for telecommunications that, as proposed, would give NSA and other government intelligence agencies increased powers to eavesdrop on U.S. citizens. A new version of the PGP (Pretty Good Privacy) encryption technology, which barely skirted the U.S. export laws, was released. For many in the Internet community, encryption ranked high on their list of hot topics.

In the fall of 1992, Gilmore, along with Tim May, a former Intel employee, and Eric Hughes, a fellow software engineer, founded Cypherpunks, a loose-knit group of amateur cryptographers, security aficionados, and crypto-anarchists that advocated the elimination of all government regulations on encryption and the development of strong encryption products to protect individual privacy. (Officially, Cypherpunks was a forum for "discussing personal defenses for privacy in the digital domain.") Their raucous voice in the encryption debate helped land the three original Cypherpunks—albeit masked and unnamed—on the cover of the May-June 1993 issue of *Wired,* for Steven Levy's story, "Crypto Rebels."

Cypherpunks continued to grapple with encryption and other privacy-based technologies—such as re-mailers, which allow people to send anonymous email—over the next decade, communicating primarily through the Cypherpunk email list, which was hosted by Gilmore's Website, www.toad.com. Once membership had fallen to fewer than 500, Gilmore kicked Cypherpunks off toad.com. The group lives on in forums like the alt.cypherpunks newsgroup. "I definitely expect to participate," Gilmore told *Wired News,* after the split. "I just won't run it. Consider me as demoting myself from 'manager' to 'programmer'—or from 'publisher' to 'writing letters to the editor.'"

One way Gilmore has stayed in touch with Cypherpunks, particularly those outside the United States, is through the FreeS/WAN project, which he became involved with in 1996. FreeS/WAN (Free Secure Wide Area Network) brings encryption technologies to Internet software, in an effort to deter government wiretapping. Although Gilmore knew hundreds of able cryptographers in the Bay area alone who could take on the project, in order to sidestep existing encryption export controls, he had to look abroad, first to two Greek programmers, then to a group of Cypherpunks in Canada. Gilmore's wildly ambitious hope was to secure 5 percent of all Internet traffic against wiretapping by Christmas 1996.

To date, Gilmore has not succeeded, but FreeS/WAN continues. "There is no perfect security and never will be," Gilmore told the *Ottawa Citizen* in 2001. "But our goal is to increase the work factor of wiretapping, such that the government cannot effectively wiretap all the citizens. Or even a large percentage of them."

## CRACKING THE CODE

In 1998, Gilmore made a frontal assault on government encryption regulations by setting out to crack the Data Encryption Standard (DES). DES was adopted as the federal encryption standard in 1978, but for years those in the cryptography community had decried DES, which employed 56-bit keys, as insufficient and insecure.

In 1997, RSA Data Security Inc., a Silicon Valley software company that sells encryption technology (considered by many to be superior to DES), began the DES Challenge, a contest to see who could successfully crack an encrypted DES message. In 1997, the prize was claimed after five months. For the second DES Challenge, held in early 1998, the encrypted message was cracked in thirty-nine days. Both times, thousands of computers spread all over the Internet were needed to crack the encryption. When the next DES Challenge came up, Gilmore was ready to take it on—not for the $10,000 prize, but to drive home the point that the DES was insecure and that, therefore, the government should allow for stronger encryption technologies. With funding from the EFF, Gilmore and Paul Kocher, a cryptography researcher, built a homemade, unclassified computer for $220,000, which would take on the DES Challenge. In July 1998, the EFF's DES Cracker—a single computer built from more than 1,000 chips—revealed the encrypted message after just fifty-six hours. It read, "It's time for those 128-, 192- and 256-bit keys."

Upon winning the prize, Gilmore and the EFF took the government to task. In a press release, Gilmore, who claimed that the government had been lying all along, said, "When the government won't reveal relevant facts, the private sector must independently conduct the research and publish the results so that we can all see the social trade-offs involved in policy choices." According to many, the EFF DES Cracker was, in effect, the final nail in the coffin for the 56-bit DES. Gilmore has since called for the use of Triple DES.

After the Clinton administration relaxed export controls in the late 1990s and the encryption debates died down, Gilmore turned his attention elsewhere, though he has continued in his entrepreneurial and libertarian ways. He served for three years (1997–2000) on the board of trustees for the Internet Society, and currently serves on the board of ReQuest, Inc., an MP3 stereo equipment company, and CodeWeavers, Inc., a free software company. In the wake of the September 11, 2001, terrorist attacks, Gilmore has been particularly active on his freedom to travel campaign, in which he is suing the federal government and several major airlines because he was barred from boarding a commercial aircraft because he refused to show ID. The ID requirement, Gilmore asserts, violates several of his constitutional rights. Gilmore shows no signs of backing down from a fight for the rights of the individual—whether for himself or his fellow citizens.

## FURTHER READING

### In These Volumes

Related Entries in this Volume: Barlow, John Perry; Denning, Dorothy; Joy, Bill; Stallman, Richard

Related Entries in the Chronology Volume: 1975 (sidebar): The Free Software Movement; 1985: Richard Stallman Establishes the Free Software Foundation; 1990: Operation Sun Devil; 1990: The Electronic Frontier Foundation Is Established; 1993: Escrowed Encryption Initiative Is Introduced by the Clinton Administration

Related Entries in the Issues Volume: Encryption; Hackers; Open Source; Privacy

### Books

Electronic Frontier Foundation. *Cracking DES: Secrets of Encryption Research, Wiretap Politics, and Chip Design.* Sebastopol, CA: O'Reilly & Associates, 1998.

### Articles

Abate, Tom. "Little Revolutionary." *The San Francisco Examiner,* August 17, 1995.

Cave, Damien. "It's Time For ICANN to Go." *Salon.com,* July 2, 2002, http://www.salon.com/tech /feature/2002/07/02/gilmore/ (cited September 16, 2004).

Hum, Peter. "The Question After Sept. 11: Will They Be Viewed as Privacy's Champions or a Security Threat?" *The Ottawa Citizen,* December 6, 2001.

Levy, Steven. "Crypto Rebels." *Wired,* May/June 1993, http://www.wired.com/wired/archive/1.02/crypto .rebels.html (cited September 16, 2004).

———. "Courting a Crypto Win." *Newsweek,* May 17, 1999.

Markoff, John. "In Retreat, U.S. Spy Agency Shrugs at Found Secret Data." *New York Times,* November 28, 1992.

———. "U.S. Data Code Is Unscrambled in 56 Hours." *New York Times,* July 17, 1998.

Weise, Elizabeth. "A Maverick Cracks the Code." *USA Today, August 5, 1998.*

### Websites

Cypherpunks Home Page. Archived material from the original Cypherpunk mailing list, http://www.csua .berkeley.edu/cypherpunks/Home.html (cited September 16, 2004).

Electronic Frontier Foundation. Home of the first civil rights organization for cyberspace, http://www.eff.org/ (cited September 16, 2004).

FreeS/WAN. Home base of the FreeS/WAN project, which aims to secure Internet traffic against FBI wiretapping through encryption technology for open-source Linux operating systems, http://www .freeswan.org/ (cited September 16, 2004).

Free To Travel.org. Website that tracks *Gilmore v. Ashcroft,* http://freetotravel.org/ (cited September 16, 2004).

John Gilmore's home page. Personal Website of John Gilmore, with extensive links to the various projects he has supported and press clippings, http://www .toad.com/gnu/ (cited September 16, 2004).

# Rob Glaser (1962–)

## FOUNDER AND CEO OF REALNETWORKS

**Although his name is** not as recognized as those of some Internet pioneers, in the computing world, Rob Glaser is virtually synonymous with streaming media. Since 1994, his company, RealNetworks, has brought sound and video to the Internet, transforming the Web into a portal for radio and television. Real-Networks dominated the streaming media early on, but has since encountered fierce competition from Microsoft's Media Player and Apple's QuickTime streaming media platforms, as well as Napster and other peer-to-peer music-sharing services. A former Microsoft employee, Glaser is also renowned as one of the few CEOs to have faced down Bill Gates's empire time after time and survived.

## THE PROGRESSIVE ERA

Glaser grew up in Yonkers, New York, a middle-class suburb of New York City. He had a liberal upbringing—both at home and at his school, the Ethical Culture Fieldston School in nearby Riverdale, New York. His mother was a social worker specializing in inner-city children; his father owned a print shop. Both were politically active. By the time Rob was 12 years old, he was handing out leaflets in support of the United Farm Workers' grape boycott alongside both parents.

In addition to his early support of social causes, Glaser was a rabid baseball fan, loyal to the New York Mets. (His dream was to become the next Lindsey Nelson, a Mets announcer; instead, the adult Glaser is part owner of the Seattle Mariners.) As a kid, he read the 3,000-page *Baseball Encyclopedia* cover to cover, memorizing stats. In a 2001 interview with *Business Week,* he mused, "It's probably the way I got into computers and math." Glaser's first contact with computers came while he was still in grade school, where he learned the basics of programming on a mainframe computer at Fieldston. By high school, the tech-savvy Glaser and his friends

had set up a simple pirate radio station that broadcast to the Fieldston cafeteria.

After high school, Glaser entered Yale University, where his political leanings drew him to the antiwar and leftist movements of the early 1980s. He wrote a column for the *Yale Daily News,* entitled, "What's Left?" in which he railed against his conservative peers and rallied support for labor movements and disarmament groups. At night, he shed his daytime role as student-activist and programmed games for his fledgling computer company. In 1983, after only four years, the hard-working Glaser graduated with three degrees—bachelor's degrees in both computer science and economics, and a master's degree in economics.

Upon graduation, Glaser had many options. He contemplated getting another graduate degree or going to work for Hewlett-Packard. Finally, he decided to start work with a then little-known software company in Seattle, Washington, called Microsoft. Just 21 years old when he started, Glaser was a fast-rising star at Microsoft, becoming the company's youngest vice president when he was promoted to lead the Multimedia Systems Group. His upward trajectory was marred only by some colleagues' reactions to his management style, which has been reported to be highly demanding and sprinkled with profanities.

By the time Glaser decided to leave, at age 30, he had worked at Microsoft for a decade. According to some accounts, Glaser left when control of the multimedia group was wrested from him by one of Microsoft's resident visionaries, Nathan Myhrvold. Glaser himself maintains that his departure was amicable, and that he sought to funnel his Microsoft millions into his passion for social causes. "I wanted to put up my periscope and regain perspective on the world," he told *Time* in 1997.

Upon leaving Microsoft in 1993, Glaser joined the boards of several organizations, including the Electronic Frontier Foundation, a civil rights organization for cyberspace, and the Foundation for National Progress, which supports progressive, independent media. As Glaser pondered the next chapter in his career, he found inspiration in a new Internet technology—the Web browser Mosaic that upon its release in 1993, was taking the Internet by storm. (Mosaic was the predecessor of Netscape Navigator.) "I downloaded Mosaic and, immediately, all the lightbulbs went off," he told *CNN.com* in 2000. Glaser realized that a platform—Internet—for interactive television (then believed to be the next big thing, technologically) was right at his fingertips.

That fall, he and a colleague from Yale, David Halperin, brainstormed about a new kind of company that would marry their interest in social causes with their newfound interest in the Internet. It would be part PBS—educational and socially conscious—part MTV—hip, popular, and video/music-oriented, all to be delivered online. Glaser commissioned three engineers to develop the technology that would allow the transmission of real-time audio and video. What they developed would become what is now known as streaming media.

Glaser and Halperin traveled to Washington, D.C., to meet with possible investors soon after the technology was ready. In the meeting, Glaser opened his laptop and, with the click of a mouse, let the room of astonished people watch a baseball game on his computer. At the time, it was revolutionary. Months later, in February 1994, Progressive Networks was born.

## KEEPIN' IT REAL

Glaser learned, from his meeting in Washington, that investors were far more interested in the technology than in his socially conscious programming, so he initially focused on delivering existing media rather than creating his own. (Glaser kept his commitment to progressive causes by pledging 5 percent of Progressive Networks' profits to charity.) On April 10, 1995, Glaser debuted his first product, RealAudio, the first real-time

"streaming" audio system for the Web. The first shows broadcast included *ABC News* and National Public Radio.

RealAudio was an instant success, playing the primary role in launching the entire streaming-media industry. The technology behind streaming audio—and, later, video—solved a key problem in the transmission of audio files. Before streaming media, users experienced lengthy waits while audio files, which contained huge amounts of data, downloaded completely. With streaming media, the data are fed to the computer in bits—data streams—that can be played while the file continues to download.

Because RealAudio was one of the first products of its kind in an entirely new industry, Glaser did not have a clear business model to follow. Indeed, most in the Internet world saw streaming media as a luxury, rather than a necessity. Glaser chose the somewhat risky path of giving RealAudio away free to consumers (while charging new-media companies for RealAudio server software), in an attempt to capture the greatest market share. Glaser also chose to require all RealAudio users to register with the company—including name, home address, and email—a move contrary to the reigning Internet ethos that favored anonymity. In 1999, Glaser explained to *Wired,* "I was influenced by [industry analyst] Esther Dyson's argument that in a networked economy, you don't sell; you build relationships. I believed that if people were inviting us to take up a megabyte on their hard drives, they were inviting us into a relationship."

> *I think we all know there is going to be sort of this celestial jukebox. . . . I am just as impatient as hell to get there as soon as possible.*
>
> —*Glaser to the* Wall Street Journal, *April 2000*

RealAudio's next product, the Live RealAudio System, debuted in September 1995, with the live broadcast of a Seattle Mariners baseball game. By the following year, Glaser was able to use his registration procedure to contact thousands of users and sell them an upgraded RealAudio player for $29.95. RealVideo—which combined streaming audio with streaming video—followed in February 1997, with opening day appearances by pop singer George Michael and the filmmaker Spike Lee. By September, the company had changed its name to RealNetworks

*Rob Glaser. (Dan Lamont/Corbis)*

and filed for an initial public offering (IPO) on the stock market. The November IPO was a huge success, raising nearly $35 million.

In the midst of such success, however, the company struggled internally. Nicknamed Oppressive Networks by some employees, the company was racked with management problems. In just over two years, the company had had three chief executive officers and two chief financial officers. Glaser was as well known for his temper and profanity as for his business acumen and ambition, and the company reflected, and suffered from, his volatile personality. By the time RealNetworks had arrived on the scene, via its IPO, the board of directors felt it necessary to call in consultants to work with Glaser on his management style.

## A DEVIL'S BARGAIN

By the mid-1990s, most of the streaming-audio and -video content on the Web was formatted for Real products, and Real enjoyed the vast majority of the streaming-media market. However, Microsoft was in pursuit,

and Glaser was well aware of his former boss's competitive practices. When Glaser got word of Microsoft's attempt to buy another small, struggling streaming-media company, he realized that RealNetworks (still known as Progressive Networks) faced the difficult choice of trying to work with its competitor or lose the much-needed Windows market. Glaser chose to cooperate.

Glaser and Microsoft came to an agreement on July 21, 1997. For $30 million, Microsoft licensed the source code of RealAudio and RealVideo, and purchased a nonvoting minority interest in RealNetworks. The deal allowed Microsoft to bundle Real's media players with its browser, Internet Explorer, giving Windows users easy access to Real-formatted media online.

To many, the deal seemed too good to be true—and it was. Shortly after sealing the deal, Microsoft released its own free version of the Real server software, thereby undercutting one of Real's primary sources of income. "Once they did that," Glaser told *Wired* in 1999, "they were signaling, not necessarily in a rational way, that they were focused on cutting our oxygen supply."

Glaser was prepared. Within six weeks, RealNetworks released a new, improved version of its Real media player that was incompatible with the source code Microsoft had licensed. Microsoft retaliated. When RealNetworks released its RealPlayer G2 in April 1998, Glaser found that Windows rendered his newest product ineffective. Windows users trying to use G2 software discovered that their computers defaulted to the Windows Media Player.

At the time, Microsoft was in the early stages of its legendary antitrust trial. Senator Orin G. Hatch, a tenacious foe of Microsoft, urged Glaser to testify at a Senate Judiciary Committee hearing on Microsoft's anti-competitive practices. On July 23, 1998, Glaser stood beside fellow industry leaders, including Oracle CEO Larry Ellison and IBM's Jeffrey Papows, and before the committee and testified, "I believe Microsoft is taking actions that create obstacles to the freedom and openness of the Internet," and went on to describe how Windows "breaks" the RealPlayer G2 software. Microsoft countered that the problem lay in the RealPlayer software, not Windows—testimony that made waves in the business sector, as consumer confidence in Real products fell. Real's stock price plunged the day after Glaser's testimony. What could have been an early victory for RealNetworks in the nascent media-player wars turned into, at best, a draw.

The following month, however, a record number of users downloaded RealPlayer from the Internet. On September 21, 1998, upon the release of President Bill Clinton's taped grand jury testimony, his relationship with Monica Lewinsky, more than 250,000 users downloaded RealPlayer in a single day, and nearly 2 million users watched the tape via the Internet. By the end of the year, RealNetworks carried nearly 85 percent of all Internet audio and video broadcasts. Nevertheless, Microsoft's Media Player and Apple's streaming-media product, QuickTime, were making steady progress.

## DON'T PUT ANOTHER DIME IN THE JUKEBOX

In the early days of RealAudio, Glaser was given to talking about the Internet's potential to be "the world's coolest radio," and the "jukebox in the sky." By 1998, his visions of music on the Web were just beginning to become reality, ushered along, in no small part, by the growing popularity of the MP3. The MP3—shorthand for MPEG-3, or the Moving Picture Experts Group Audio Layer 3—is a standard compression format for audio files. MP3 compressed music files (songs) were small and easily shared, with minimal sacrifice of audio quality. Glaser realized that the greatest potential for music on the Web lay in the personal collections of music fans. So he decided that RealNetworks would create the technology that allowed Real users to upload their CD collections to their computers and download music to their hard drive from the Internet. In essence, it would allow users to digitally organize and manipulate their entire music library, using their computer. That product, released in May 1999, was RealJukebox.

With RealJukebox, Glaser walked a fine line. To counter the spread of pirated music, major record labels had banded together in 1998 under the name Secure Digital Music Initiative (SDMI). SDMI fought the expansion of the MP3 format, viewing it as a threat to album sales. Although he strove to work with the record companies, Glaser knew that for RealJukebox to thrive in the new world of digitized music, it had to support

*People in Silicon Valley see things unnecessarily in black and white: You either hate Microsoft or you are a vassal of them. I am saying there is a third way.*

*—Glaser to the* Wall Street Journal, *February 1998*

MP3s. Risking the ire of the recording industry, Glaser made RealJukebox (and later versions of RealPlayer) MP3 compatible. Glaser hoped by the time the recording industry could react, RealJukebox would be too ubiquitous to ignore.

Both fortunately and unfortunately for RealNetworks, Napster emerged on the scene in the same month RealJukebox debuted. The riotously successful peer-to-peer file-sharing service easily outmatched RealJukebox as the most popular way to download music from the Internet. However, while RealNetworks at least paid some lip service to the recording industry, Napster ignored such concerns completely and, thus, became the focus for the industry's wrath over pirated music. By December 1999, the Recording Industry Association of America, on behalf of its major record labels, had sued Napster. When Napster was finally ordered offline by the courts, RealJukebox was still thriving.

Once Napster proved that downloading music was immensely popular, Glaser set out to prove it could be equally profitable. In April of 2001, RealNetworks, working with EMI Recorded Music, BMG Entertainment, and AOL Time-Warner, announced a new digital music subscription service—MusicNet, the first major collaboration between an Internet company and the recording industry.

MusicNet was designed to let users download popular music with the ease of Napster, while preserving the rights of the musicians and record companies, all for a low monthly fee. However, MusicNet—and the handful of similar efforts that followed—never quite took off. Their popularity was hampered by extreme restrictions imposed by the record companies, which often made sought-after new releases unavailable. Also, the notion of subscribing to music, rather than owning it outright, did not prove to be as popular as Glaser had hoped.

The disappointing growth of MusicNet was yet another setback for RealNetworks, which was also in the throes of the industrywide recession. Between the summer of 2000 and fall of 2001, the company lost more than $150 million. Just months after MusicNet's debut,

RealNetworks was forced to lay off 15 percent of its workforce. According to an April 2002 Neilson/Net-Ratings report, Microsoft's Media Player had made significant gains in the streaming-media market, with more than 15 million at-home users compared with RealPlayer's 17 million; in addition, Media Player surpassed RealPlayer by more than half a million in the corporate realm.

## STREAMING DNA

In July 2002, to combat Microsoft's growing market share, Glaser announced Helix, a bold new streaming-media technology. The departure from Real's normal naming conventions was significant, because Helix itself differed significantly from all other Real products: it was to be open source (meaning that much of the Helix code would be kept public so that programmers worldwide could adapt and improve upon it). Before the advent of Helix, Real was sometimes criticized for engaging in proprietary practices not unlike Microsoft's own.

Real's tactic mirrored an attempt by Netscape Communications, in the late 1990s, to use the growing popularity of the open source movement as leverage against Microsoft in the browser market. But where Netscape failed and was eventually bought out by America Online, Real planned to succeed. By 2002, the open-source movement was far more powerful than it had been just four years earlier—one of its leading softwares, the Linux operating system, had been embraced by large corporations and governments worldwide, and industry analysts showed that Linux was gaining fast on the heart of the Microsoft empire, Windows.

Real's open-source initiative began with the Helix Universal Server, which supports a single digital compression format that works on all media players, including Windows. (To achieve universality, Real programmers had to reverse-engineer the Windows Media Player.) This was a boon to programmers, who could save themselves—or, more important, their companies—time and money by encoding once for all formats, instead of encoding the same media several times.

> *The way you beat Bobby Fisher is you don't play chess.*
>
> —*Glaser to* Business Week, *September 2001*

Although companies must pay for the Helix Universal Server, the underlying Linux operating system is free, making Helix less costly than the free Windows Media Player because businesses still had to pay for the Windows operating system.

In the months following the debut of the Helix Universal Server, Real released key parts of the source code—referred to as its DNA—under various open-source licenses. Although the Helix initiative itself was built upon nearly a decade of Real's proprietary software development, its future products lay in the Helix community of programmers, which, with the support of industry leaders such as Sun Microsystems, Hewlett-Packard, and Nokia, grew quickly to 10,000.

For Glaser, the Helix initiative was not only the best route to continue head-to-head competition with Microsoft, but also the best way to embrace the growing convergence of digital media. "We've done hundreds of millions of dollars of R&D [research and development] ourselves," he told Tim O'Reilly during a talk at the O'Reilly Open Source Convention in 2002. "But it is clear that the scale of what is going to happen in the future is hundreds of millions of dollars more of commercial activity to make this work. And there are only two possible sources of that: Collectively, us, and what we're doing with Helix and with the community; and Microsoft."

As the Helix initiative gained momentum, Glaser strove to retain his company's standing on more familiar ground. In August 2003, Real recovered from the MusicNet debacle by acquiring the successful Rhapsody music subscription business at Listen.com. Weeks later, Apple debuted its iTunes Music Stores, which became an overnight hit, though Glaser insists the universality of Real's platform will prevail in the end. In the realm of video, RealPlayer helped launch *ABC News Live* in 2003, the first twenty-four-hour online news broadcast. However, at the same time, a deal with major league baseball to broadcast games using RealPlayer on mlb.com fell through, and, by March 2004, had resulted in legal action by RealNetworks. In another lawsuit, in December 2003, RealNetworks brought a $1-billion suit against Microsoft, claiming that the decision in the previous antitrust suit did not address Microsoft's anticompetitive actions in the streaming-media market. Real's case mirrors a larger antitrust case brought against Microsoft by the European Union (EU) in the late 1990s, which cites Microsoft for illegally bundling the Windows Media Player with its operating system. The initial EU decision against Microsoft, released in March 2004, bodes well for Real, as the EU is expected to require Microsoft to remove its media player from the operating system and allow for more competition. At long last, Glaser may be able to meet his fiercest competitor on a level playing field.

## FURTHER READING

### In These Volumes

Related Entries in this Volume: Andreessen, Marc; Dyson, Esther; Gates, Bill

Related Entries in the Chronology Volume: 1992: First Audio Multicast on the Internet Gives Rise to the Multicast-Backbone; 1996: The Browser War Heats Up; 1998: Department of Justice Files Suit Against Microsoft; 1998: Starr Report Is Released Online; 1999: Napster Roils the Music Industry; 2004: Microsoft Is Fined by European Union

Related Entries in the Issues Volume: Internet Broadcasting; Peer-to-peer Networking

### Articles

Greene, Jay. "Rob Glaser in Racing Upstream." *Business Week,* September 3, 2001.

Hattori, James. "Profile of RealNetworks CEO Rob Glaser." *CNN.com,* August 11, 2000, http://www.cnn.com/2000/TECH/computing/10/28/index.glaser/cover.glaser/ (cited September 16, 2004).

Kover, Amy. "Is Rob Glaser For Real?" *Fortune,* September 4, 2000.

Rothenberg, Randall. "Rob Glaser, Moving Target." *Wired* 7.08, August 1999, http://www.wired.com/wired/archive/7.08/glaser.html (cited September 16, 2004).

Silberman, Steve. "Martyred by Microsoft." *Time,* September 7, 1998.

Swisher, Kara. "Move Over, Beethoven." *Wall Street Journal,* April 17, 2000.

Walker, Rob. "Between Rock and a Hard Drive." *New York Times Magazine,* April 23, 2000.

### Websites

The Glaser Progress Foundation. Online home to the Glaser Progress Foundation, founded in 1993, as part of Rob Glaser's continuing dedication to progressive social causes, http://www.progressproject.org/ (cited September 16, 2004).

Helix Community. Online base of the Helix open-source development community, initiated by RealNetworks in 2001 as a new venture in streaming media, http://www.helixcommunity.org/ (cited September 16, 2004).

MusicNet Home. Online home to the MusicNet music subscription service, the first major collaboration between an Internet company, Real, and major recording labels, http://www.musicnet.com (cited September 16, 2004).

Real.com. RealNetworks' commercial outlet for Real's audio, video, and gaming products, as well as Real's premium subscription services, http://www.real.com (cited September 16, 2004).

RealNetworks. The main RealNetworks corporate Website, with links to company information, Real streaming media products, and the Helix development community, http://www.realnetworks.com/ (cited September 16, 2004).

RealNetworks. Online home of RealNetworks' charitable arm, http://www.realfoundation.org/ (cited September 16, 2004).

Rhapsody Digital Music Service. Online home of the Rhapsody subscription music service, a subsidiary of Real, http://www.listen.com/ (cited September 16, 2004).

# Charles Goldfarb (1939–)

## INVENTOR OF SGML

**Charles Goldfarb is known** as the Godfather of Markup for developing Standard General Markup Language (SGML), a set of rules—similar to grammar or syntax—that provides the basis for much of modern-day electronic publishing. Although people unfamiliar with the computer world may not immediately recognize the term *SGML,* most have heard of its progeny, such as HTML, the markup language that revolutionized publishing on the Web in the early 1990s, and XML, which, since 1998, has been hailed as the lingua franca of the Web. Because of his landmark work in the development of SGML in the 1970s, for leading the twelve-year effort to make SGML an international standard in the 1980s, and for being one of the leading authorities on XML, Goldfarb is considered to be one of the primary forces behind the shaping of electronic publishing and the Internet.

## THE ROAD TO IBM

Goldfarb, a Brooklyn native, attended Columbia College in New York City. He graduated with a bachelor's degree in 1960, and then went to Harvard Law School. After graduating with a law degree in 1964, Goldfarb began practicing as an attorney in Boston, Massachusetts.

At the time, Goldfarb admits, he knew nothing about computers. As a practicing lawyer who spent much of his time researching, preparing, and publishing legal documents, however, he was very familiar with the long, arduous process of publishing without the aid of a computer. Most lawyers relied solely on yellow legal pads and pencils. Research consisted of thumbing through books of case law, poring over legal findings, then weaving that information together with new arguments and supporting evidence. Attorneys then dictated their notes to legal secretaries, who typed up documents

on electric typewriters; those documents were sent back to the lawyer for review, then retyped at least one more time. The margin for error was large, and the process was time-consuming and inefficient.

In Goldfarb's spare time, he was a rally master for sports car races run through the streets of Boston. Rally masters wrote the directions for the racecourse, which wound through the city. Rally masters usually gave simple directions like, "Left on Massachusetts Avenue," or "Right onto Boylston Street." Goldfarb, however, offered far more cryptic directions. He took great joy in giving drivers a map of Yugoslavia, with directions like "Head toward Sarajevo," which might have meant "Head east" or "Head west," depending on one's location. Other times, he would write:

26. Left at light onto Jones Rd.
27. (Repeat instructions 20–26, substituting left for right.)
28. Second right.

In writing these directions, Goldfarb divorced the meaning of the directions from the words themselves, causing people to think abstractly. A friend mentioned to Goldfarb that his directions looked like computer programs. Goldfarb had never seen computer code, but his friend's comment planted a seed.

In November 1967, Goldfarb stopped practicing law to work for IBM, in part to get practical experience that could help him gain clients from the burgeoning high tech companies in Boston. Primarily, his job was to design electronic accounting systems for small businesses, using punch-card tabulating machines. After a year or so, Goldfarb began to tire of this work, until he started another project that entailed installing a computerized typesetting system for a Boston-area newspaper.

## IN THE BEGINNING

While installing the newspaper's computer, an IBM 1130, Goldfarb learned the difference between specific coding and generic coding. Typesetters in the publishing industry had formatting rules (style sheets) that regulated the specific format for different aspects of a document, which helped lengthy documents like newspapers and books maintain a consistent look. For example, a particular newspaper might have headlines typeset as "12pt, bold, Times News Roman, centered," but a byline might be "8pt, italicized, Courier, aligned left." Such formatting directions are considered to be specific coding. (Specific coding has existed since typesetting began.) The style sheet might give these formatting directions particular names, such as "headline" and "byline," which is considered generic coding. Using generic coding, an editor could tell a typesetter how to style each page of the paper without explicitly writing out each of the instructions—"headline" existed as a style, as did "byline" and other integral aspects of a typical news article. While this may seem commonplace to anyone who has used any word processing computer program, at the time, it was as advanced as typesetting had become.

Goldfarb often points out that his revelation about typesetting and codes occurred around the same time that William Tunnicliffe, chair of the Composition Committee of the Graphic Communications Association (GCA), first explored the concept of separating the content in a document from the formatting of the document, in a September 1967 presentation for the Canadian Printing Office. During this same period, Stanley Rice, a book designer in New York, proposed to the GCA "editorial structure" codes—which he presciently called tags—that could standardize book publishing. GCA saw potential in both Tunnicliffe's and Rice's ideas and convened the GenCode Committee to investigate their ideas about generic coding. The GenCode Committee would later come to be a key ally in the development and standardization of the markup language that Goldfarb was about to create.

## SPEAKING THE SAME LANGUAGE

In 1969, just as Goldfarb began to consider returning to his legal practice, he was given one final project. He joined a small research project at IBM's Cambridge Scientific Center that was trying to find a way to automate legal research and publishing—an area with which Goldfarb was very familiar.

By the late 1960s, many legal offices had adopted computer programs that could help perform certain time-consuming activities. One program could edit text, another program could retrieve information, and yet another program could be used to compose and print briefs and other documents. However, each of these programs had its own programming code that was not compatible with that of the other programs, and they all ran on separate operating systems. Goldfarb's job was to find a way to make them work together.

Using an IBM CP-67 computer, which allowed several operating systems to run at the same time, Goldfarb and his fellow researchers, Edward Mosher and Ted Peterson, were able to get three separate IBM computer programs—an editing program, an information retrieval program, and a composition program—to communicate with each other. However, a document from one system still had its own unique rules for formatting and describing text—its own markup language, in a sense—that was different from each of the others. The next step was creating a markup language that was compatible with all three.

First, Goldfarb experimented with labeling aspects of legal briefs with what they were (e.g., title or abstract),

> *With a product as complex as a Boeing 747, which needs about 4 million pages of documentation, no single human can understand everything. If you make mistakes, then people die. SGML was the solution for these information management problems.*
>
> —*Goldfarb to Red Herring, March 2000*

instead of what they should look like (e.g., italics or bold). His idea was to treat different aspects of the document as data elements instead of as content. In this way, each legal document was actually a database of all its parts, with formatting code to describe each part of the text. The document-cum-database could then be sorted, searched, and manipulated by its various elements—by date, by case law, by legal findings, by name—then re-arranged and published. This advancement in text processing made updating, compiling, and printing legal briefs incomparably quicker and more efficient.

IBM encouraged the group to go one step further and expand the project beyond legal documents to text processing in general. Ray Lorie, another IBM researcher, joined Goldfarb and Mosher on the project, which had been named Integrated Text Processing. The first prototype was called Integrated Textual Information Management Experiment (InTIME).

Goldfarb, Mosher, and Lorie decided that, instead of writing a standard set of tags—or formatting codes—they would write the rules that would create the tags. Their intent was to create a common language to be shared by all electronic texts—not just legal briefs or other highly specific types of documents—so that any computer could read any document. This would free the user from becoming locked in to any single computer program.

By 1971, the three had developed a generic but very powerful and flexible grammar that could describe the components of any document regardless of the technical roots of the system. Goldfarb coined the term *markup language* to describe what they had done, and dubbed their invention GML, for Generalized Markup Language. Not coincidentally, GML was also the acronym of the last names of the three scientists who developed it. Indeed, Goldfarb told *Web Techniques* in 1998, "The name . . . was my way of labeling the technology so that its origin would be unmistakable."

The first prototype of GML was ready in 1973, in what Goldfarb calls a "relatively primitive" implementation in IBM's Advanced Text Management System (ATMS). A year later, Goldfarb developed a "validating parser," a program called ARCSGML that checked the validity of markup language in a document. The basic elements of what would become SGML—Standard Generalized Markup Language—had fallen into place.

## BECOMING THE STANDARD

In 1975, Goldfarb moved to another position within IBM, in San Jose, California, where a main part of his job was promoting GML-based products. He succeeded in having GML added to IBM's Script product. (By 1980, GML was an integral part of nearly 90 percent of all IBM text-processing products.) The growing success of GML in the 1970s prompted IBM to have Goldfarb submit GML to the American National Standard Institute (ANSI). In 1978, ANSI's Computer Languages for the Processing of Text Committee asked Goldfarb to lead the technical effort in developing a standardized markup language.

For the next twelve years, Goldfarb helped GML become SGML. Collaborating with a worldwide team of users, programmers, and academics, including the GCA's GenCode project members, Goldfarb hammered out the various additions to GML that made SGML the robust language that it is. The first working draft of SGML was released in 1980. The sixth working draft, released in 1983, was endorsed by the GCA and adopted by several government agencies, including the Department of Defense and the Internal Revenue Service. The following year, the International Organization for Standardization (ISO) joined ANSI's efforts. Goldfarb led both the ANSI and ISO efforts. Finally, in 1986, SGML became an international standard—ISO 8876/1986.

> *If you can afford custom tailoring and get a suit made exactly the way you want and the way you look the best, you'll do it. The rest of us go shopping in department stores. The problem was that up till now, SGML was strictly a custom-tailoring thing. . . . XML is for the mass market.*
>
> —*Goldfarb to* Web Techniques Magazine, November 1998

As a grammar, instead of a language, SGML defined how markup language was to be used—including what markup is allowed, what markup is required, and how markup is distinguished from text. It was written in ASCII—one of the most basic computer languages, as well as one of the first to have been standardized. It was both powerful and adaptable. Goldfarb later told *MacWeek*, "SGML doesn't limit you to just one way of doing something . . . What it does is give you a standard way to state which encoding method you are using for an object type, to point to a program that can interpret that coding and to describe the relationships between different object types."

SGML, it turned out, was best suited for large organizations with extensive in-house publishing and information management needs. (Ironically, SGML did not do as well in its original capacity—the legal profession.) In addition to the Department of Defense and the IRS, SGML was also embraced by entire industries, including the book publishing, telecommunications, and aerospace industries. Indeed, SGML was used to publish the 4 million pages necessary to create and maintain a Boeing 747 airplane. Some of the most vital institutions in America—governmental and otherwise—used SGML. (IBM published nearly 90 percent of its material using SGML.) For that reason, in a 2000 interview with *Red Herring*, Goldfarb said, "SGML powers the infrastructure of modern society."

However, SGML was expensive to set up and install, thus only the largest institutions had the means—or the need—to do so. Once people began to publish on the Internet, SGML also proved to be too complex for Web browsers. The specifications for SGML filled more than 500 pages—a cumbersome amount of information. By the mid-1990s, it had become clear that something new was needed.

## ANOTHER KIND OF MARKUP

While SGML was well suited to organizations that needed efficient *internal* exchange of documents, the Internet required a new kind of grammar that could cater to the *external* exchange of documents. In 1989, Tim Berners-Lee, a scientist at the European Laboratory for Particle Physics (CERN), began developing what would become HTML—Hypertext Markup Language. HTML was rooted in SGML, but it was far simpler and

*Charles Goldfarb. (Photo courtesy of Charles Goldfarb)*

easier to use. (The relationship between SGML and HTML meant that SGML documents could be published on the Web often without changing any of the markup language itself.) HTML quickly became the predominant markup language used to publish on the Web, and because, in the beginning, most Web browsers could support it, it spread like wildfire.

As the Internet and Web browsers grew more complex, however, HTML, too, revealed its limitations. In 1994, the World Wide Web Consortium (W3C) was formed to manage standards and specifications for the Web, but by 1996, the evolution of HTML had become inconsistent and insufficient. After all, HTML was designed for the simple hypertext documents that appeared in the early days of the Web. By the late 1990s, the W3C began to explore another markup language that was as extensible and flexible as SGML—HTML's "father"—but not nearly as complex. They called it XML—for eXtensible Markup Language. It, too, was rooted in the original SGML, but while HTML dictated how parts of a document should look, XML dictated what each part meant. (The difference is similar to the difference between specific and generic coding.)

The first working draft of XML was released at the SGML '96 conference, held in Boston. (The name XML was chosen over the alternative, MAGMA.) At just fifty pages, XML's specification was one-tenth the size of SGML's, but it still retained the inherent power and flexibility of Goldfarb's SGML. (Indeed, as editor of the SGML standard, Goldfarb was closely consulted in the development of XML.) XML made the server-browser incompatibility experienced with HTML obsolete, going further to allow for the free-flowing exchange of data between any two XML-based programs. As Goldfarb told *Red Herring* in 1993, "XML is a Web-optimized version of SGML."

In 1996, Microsoft put XML in a product called Channel Definition Format (CDF) and became one of XML's greatest advocates. Some believe this was a challenge to the early success of Java, a popular computer language developed by Sun Microsystems, one of Microsoft's key competitors, which, like XML, allowed for the unfettered exchange of data between different programs. By 1998, the W3C had released XML version 1.0 and reformulated HTML. That same year, Goldfarb's *The XML Handbook* was published by Prentice-Hall.

By the late 1990s, XML was being hailed as a revolution, dubbed the "holy grail" of computing. Goldfarb himself became something of a celebrity in geek circles. In-store book signings of *The XML Handbook* were crowded beyond capacity. Books flew off the shelf. Indeed, the buzz surrounding XML was deafening. Even so, early on, many misunderstood XML as just extensible HTML—HTML with tags that the user could write himself—when really it was the realization of Goldfarb's earliest hopes for SGML. "It frees data from a hostage relationship to a particular software," he told the *South China Morning Post* in 2000.

Through the 1990s and into the 2000s, Goldfarb has been one of the leading authorities on XML. Though books and Websites on the subject abound, the 1998 edition of *The XML Handbook* was rated the number one XML book by Amazon.com, and updated editions were printed in 2000, 2001, and 2002. *The XML Handbook* is part of "The Definitive XML Series from Charles F. Goldfarb," a series Goldfarb edits for Prentice-Hall, along with another series of books, the "Charles F. Goldfarb Series on Open Information Management." In addition, Goldfarb has his own SGML/XML consulting business, Information Management Consulting.

Undoubtedly, Goldfarb is a major figure in the history of text processing, from the early days of GML to the more recent days of XML. In addition to serving as the editor of ISO 8879/1986—the SGML standard, which is still in use—he has also consulted on the development of many SGML and XML products. His efforts have helped the computer world evolve from program-centric computing toward more data-centric computing, which many believe is the path to the future.

## FURTHER READING

### In These Volumes

Related Entries in this Volume: Berners-Lee, Tim; Nelson, Ted

Related Entries in the Chronology Volume: 1990: The World Wide Web Is Invented; 1991: The World Wide Web Is Developed at CERN; 1998: Extensible Markup Language (XML) Is Introduced for the World Wide Web

Related Entries in the Issues Volume: E-books

### Works By Charles Goldfarb

*The SGML Handbook.* New York: Oxford University Press, 1990

*The SGML Buyer's Guide.* Upper Saddle River, NJ: Prentice Hall, 1998

with Paul Prescod. *The XML Handbook.* Upper Saddle River, NJ: Prentice Hall, 1998.

### Articles

Anderson, Tim. "Workshop: XML—The XML Files." *Personal Computer World,* December 1, 2001.

Churbuck, David C. "Document Esperanto." *Forbes,* June 7, 1993.

Floyd, Michael. "Beyond HTML: A Conversation with Charles F. Goldfarb." *Web Techniques Magazine,* November 1998, http://www.sgmlsource.com /press/Floyd1.htm (cited September 16, 2004).

———."Beyond HTML: XML Opportunities Knocking." *Web Techniques Magazine,* December 1998, http://www.sgmlsource.com /press/Floyd2.htm (cited September 16, 2004).

Hibbard, Justin. "Charles Goldfarb on Why XML Matters." *Red Herring,* March 1, 2000.

Michalski, Jerry. "Content in Context: The Future of SGML and HTML." *Release 1.0,* September 27, 1994.

"Players Debate Future Net Rules." *South China Morning Post,* September 12, 2000.

"Tagging Along: How SGML Works." *MacWEEK,* June 21, 1993.

### Websites

Charles F. Goldfarb's SGML Source Home Page. Goldfarb's Website about SGML, including links to several key papers in the development of SGML, http://www.sgmlsource.com/ (cited September 16, 2004).

Overview of SGML. The World Wide Web Consortium's resources on SGML, http://www.w3.org/MarkUp/SGML/ (cited September 16, 2004).

The Roots of SGML—A Personal Recollection (1996). Goldfarb's first-person essay about the development of SGML, http://www.sgmlsource.com/history/roots.htm (cited September 16, 2004).

The XML Handbook. Website for Goldfarb's authoritative book in XML, http://www.xmlhandbook.com/index.htm (cited September 16, 2004).

# James Gosling (1956–)

## INVENTOR OF JAVA

**In the mid-1990s,** James Gosling, a Canadian-born computer programmer, invented Java—a new programming language that revolutionized the Internet. Although he had developed important programs before Java and has since gone on to do more groundbreaking work for Sun Microsystems, Java is Gosling's true legacy. Ironically, Java was never meant to be a programming language. It actually started out as a simple solution to a problem in another of Sun's research projects, the Green Project. However, with the help of Sun luminary Bill Joy, Java quickly emerged on its own and became the first universal programming language, able to run on almost any computer platform. Within a decade of its debut, Java had been used for everything from Web pages to cell phones, and Gosling has become something of a rock star in the computer world.

## PROGRAMMING PRODIGY

Gosling grew up on a farm near Calgary, in Alberta, Canada. His fascination with computers began when he was young. At age 12, he fashioned a primitive machine that played tic-tac-toe out of a television set and discarded switches and relays that he had retrieved from dumpsters behind the local telephone company. Gosling took first prize in a local science fair. Two years later, he began breaking into the computer lab at the University of Calgary. Despite his illegal entries, Gosling quickly impressed his peers in the lab with his prodigious programming abilities and desire to learn. Within a year, the head of the physics department had hired him, part-time, to create a computer program for an early PDP-8 computer that would help analyze data from a Canadian satellite. Gosling's teachers at William Aberhart High School excused his many absences, knowing he was, at just 15 years old, a core member of one of the country's most important research projects.

Gosling earned $2 per hour writing code for the university, in addition to the dime he received in allowance each week. However, money was not the goal; Gosling wanted access to a computer at a time—the early 1970s—when computers were scarce. In 1996 Gosling explained to the *San Francisco Examiner,* "At home, I was always trying to build things with no money. . . . On the computer, I could create things of infinite complexity and I didn't have to buy any parts."

After high school, Gosling enrolled at the University of Calgary as a computer science major, graduating in 1977. In college, as in high school, Gosling spent more time on the computer than in class, and his somewhat spotty academic record made acceptance into graduate school difficult. Finally, Pittsburgh's Carnegie-Mellon University accepted him into its computer science program.

At Carnegie-Mellon, in addition to working on his degree, Gosling helped transfer the school to a UNIX computing platform, a new, powerful, and flexible operating system that had been developed in the early 1970s. He wrote a powerful program that translated the university's existing computer code into UNIX code, a feat—in terms of interoperability—that some consider to be an early predictor of Gosling's later work with Java. In 1981, Gosling also wrote the first real-time text editor for UNIX based on EMACS, a real-time text editor developed in the mid-1970s. Written in the popular programming language C, Gosling's creation became known as GOSMACS, or Gosling EMACS.

While working on his doctorate, Gosling first encountered Bill Joy, then a computer science graduate student attending the University of California–Berkeley. Joy was a UNIX wizard, best known for developing BSD UNIX, one of the most dominant computing platforms of the era. (Indeed, BSD was chosen by the Defense Department as the network-computing environment for ARPANET, the predecessor to the Internet.)

The two computer geniuses traded frequent emails, forging a professional relationship that would last for decades.

When Gosling graduated from Carnegie-Mellon in 1983, Joy attempted to recruit him to Sun Microsystems, the fledgling computer company Joy had cofounded with Andy Bechtolsheim, Scott McNealy, and Vinod Khosla in 1982. Instead, Gosling went to IBM. "IBM research group had some really cool stuff. They had built some prototypes that completely demolished anything Andy and his people were trying to do," Gosling recalled in the *Ottawa Citizen* in 2001. "I was convinced Sun was going to be dead in months." Gosling was wrong. Within a year, he had grown bored with his work at IBM. In 1984, he reconsidered Joy's offer and moved to Silicon Valley to join Sun.

## IT'S NOT EASY BEING GREEN

Gosling spent his first five years at Sun developing technology for Sun's UNIX workstations. The fruit of his labor—dubbed NeWS, for Network Extensible Windowing System—was one of the leading technologies of its time. NeWS was the embodiment of a then-radical idea: it allowed a program running on one computer in a network to be viewed in a window on another computer on the same network. (Some see Gosling's development of NeWS, with its distributed networking model and compact code, as yet another root of Java.) By 1990, however, NeWS had been snubbed by developers in favor of a rival technology—X-Windows—even though many deemed X-Windows inferior.

Gosling was devastated and nearly quit Sun. Instead of leaving, he joined with a handful of other Sun employees, most notably Patrick Naughton and Mike Sheridan, to challenge Sun's leadership to do something new and ambitious. The result of their challenge was the Green Project, initiated in 1991 under a different name, the Stealth Project.

Green Project members were housed well apart from the Sun headquarters, and their research was kept secret. After months of brainstorming, the group, inspired by

> *In my heart of hearts, I'm an engineer, and what makes me happy is building something that works and having someone use it. That's cool.*
>
> —Gosling to CNET News, *January 2002*

the early advances of digital technology in Japanese consumer electronics, set out to develop a prototype for a new kind of distributed computing environment, one in which various consumer devices—VCRs and telephones, for example—could communicate through a wireless network. They refined their ideas into one product: a handheld digital butler, dubbed *7. The name, pronounced "star seven," came from a key code on the Sun telephone system that allowed a person to answer a ringing phone from any other phone on the network. The simplicity and usefulness of *7 code captured the spirit of what the Green Project was attempting.

Gosling's role was to develop the software that would make *7 work. Early on Gosling realized that the existing programming languages were not efficient enough to run on small electronic devices, so he developed his own language. Named Oak, for the tree outside his window, the new language was based loosely on the computer language C. Gosling didn't set out to write an entirely new language, but he felt compelled to do so because of his frustrations with C, particularly with the C compiler, which is used to translate computer code written in C into machine code.

By August 1991, after about five months of nonstop work, Oak was up and running. For *7 to work on various devices, each running possibly different programs, Oak had to be a platform-independent language that could interpret other languages and had to function on small applications, or applets, that could fit on the small computer chips embedded in consumer devices. In addition, it had to be both secure and reliable. Oak was and did all.

Gosling continued to write Oak code to be integrated into the *7 during the next year. In fall 1992, the Green Project debuted *7 to Sun CEO Scott McNealy, who was duly impressed—so impressed that in October 1992 he set up a separate company, First Person, Inc., to sell the prototype. However, the Japanese and European electronics companies that First Person was targeting were not interested in *7. By 1993, First Person had shifted gears from consumer devices to interactive TV. After losing a bid for a contract with Time-

Warner cable in 1994, however, First Person was deemed to be a failure and reincorporated into Sun. Once again, one of Gosling's cutting-edge programs faced imminent extinction.

## NOT YOUR AVERAGE CUP OF JOE

Luckily, Bill Joy stepped in. While First Person had been caught up in the hype over interactive products, Joy saw an opportunity for the underpinnings of *7—Gosling's language—in the burgeoning Web. Joy brought up the issue at a 1994 corporate retreat near Lake Tahoe. "I told them roughly that 'the game's afoot,' and that the game is the Internet," Joy recalled to the *Ottawa Citizen* in 2001. The rest of the Sun executives soon fell into step, but not without some pressure. As Gosling told the *Ottawa Citizen,* Joy "did a lot of serious screaming and yelling" to save Oak from going down with First Person. Then began the arduous process of readying Oak for the Web.

The first step was to rechristen Gosling's language, since a programming language named Oak already existed. Eventually, Sun chose Java; predictably, the media have since liberally sprinkled articles about Gosling and Joy "brewing up" the language, and Java's "caffeinating" effect on the Web. Step two was to transform the newly named Java into an Internet-ready product. Over the next year, Joy and Gosling hashed out Java's basic framework. It was difficult work, marked by more than a few shouting matches. At the center of the debate was the struggle between power and simplicity: Joy argued for a robust implementation; Gosling advocated for clean, small code, adhering to the maxim, "when in doubt, leave it out."

The end product was both simple and powerful. Java was a lean language, written in a C-like syntax, which appealed to the growing number of C programmers, but without the endless amount of code usually involved with C programming. Java also solved many of the memory problems associated with programs written in C, while at the same time featuring built-in security functions that C lacked. Most important, however, was its "write once, run anywhere" capabilities. Indeed, Java developers could write an array of applets that could run on any platform, be it Apple's operating system, Windows, or UNIX. Such interoperability dovetailed perfectly with the rise of the Internet, which, for the first time, allowed an endless number of disparate computers to be connected through the same network. Java also revolutionized the Web page itself. Java applets allowed for real-time interactivity, multimedia capabilities, and animation. Previously, such features were only available on CD-ROM. With Java, anyone with a browser could experience dynamic Web pages, not just static documents published on the Internet. Some liken the way Java transformed the Web to the way color transformed black-and-white television.

Many of Java's key attributes were ideas that had simmered in computing circles for years, and early critics of Java were eager to dispel its revolutionary hype. Still, Gosling was truly the first to bring these ideas together into a single product. No one—not even Gosling himself—could have predicted Java's resounding success.

*Java is most successful when you don't realize it's there.*

*—Gosling to the Edmonton Journal, March 2003*

## BREAKING WINDOWS

Java also capitalized on the growing frustration in the computing community about the dominance of the Microsoft Windows operating system, which at its height held nearly 90 percent of the market. For the first time, developers using Java could write software for the Web, instead of software for Windows. Within months of Java's debut, in May 1995, Microsoft began to see it as a threat. Many at Sun—a longtime Microsoft rival—hoped the same.

Microsoft licensed Java from Sun in 1996 and began circulating its own flavor of Java soon after. Executives from Microsoft's Redmond, Washington, headquarters claimed their Java variant was "optimized" for Windows; Sun's executives saw the variant as Redmond's way of undermining Java's potential. By creating a noninteroperable ("polluted") version of Java and saturating the Windows market with it, Microsoft was trying to protect Windows from becoming obsolete.

In 1997, Sun filed suit against Microsoft over the Java variant. The following year, Java also played a role in the larger antitrust suit brought against Microsoft by the U.S.

government and twenty states. Testifying on behalf of the government, Gosling explained to the court how Java functioned and outlined the various ways in which it could threaten the Windows monopoly. "Because the Java technology allows developers to make software applications that can run on . . . multiple platforms, it holds the promise of giving consumers greater choice in applications, operating systems, and hardware," explained Gosling, in his 34-page prepared written testimony. "This may give new operating systems and hardware platforms a chance to compete in markets previously dominated by a particular vendor." The government further supported Gosling's claims with rather damning evidence, including an email in which Bill Gates wrote that Java "scares the hell out of me" and a Microsoft memo that outlined plans to "kill cross-platform Java by growing the polluted Java market."

Sun settled the initial case over Java with Microsoft in 2001, with Microsoft agreeing to pay Sun $20 million and to stop marketing its Java variant with the "Java compatible" mark. The following year, however, Sun sued Microsoft again, claiming that by not supporting Java programs in Windows, Microsoft was trying to undermine interoperability. When Microsoft settled its antitrust case, in January 2003, Gosling expressed his dismay over the ongoing battle over Java. "In many ways," he told the *Toronto Star,* "we're right back to square one."

## HITTING THE JACKPOT

For much of the late 1990s, Gosling traveled across the globe evangelizing for Java. He was often met by hordes of adoring Java disciples, many of whom asked for his autograph on their T-shirts and laptops. (Indeed, an attendee of one of Gosling's speeches held at his alma mater, the University of Calgary, explained to the *Edmonton Journal* in 2003, "It's a geek cult and he's the leader.") Gosling also took part in the development of new Java-based technologies, such as the Jini networking product, which, in many ways, closely resembled the initial incarnation of *7.

By the end of the decade, however, Gosling's role at Sun began to change. As he told *Javaworld* in 2001,

> *[Java is] just a language, but, my God, language is one of the most important developments of the human race.*
>
> —*Gosling to* Network Week, May 1999

"Folks are pretty converted these days. I hardly give talks at all now." Instead, he embarked on a separate project for Sun, dubbed Jackpot, in mid-2000. With Jackpot, Gosling has focused his energies on making tools, integrated development environments (IDEs), for technology developers. In short, Gosling's developments—most notably NetBeans, which is written in Java but supports other programming languages as well—try to simplify the chore of writing source code.

In November 2003, Gosling's work with Jackpot and NetBeans earned him the title Chief Technology Officer of Sun's Development Tools division. Even though Gosling has moved away from direct development of Java products, he is still commonly referred to as "the Java guy." It is a testament to the way he and Java have revolutionized computing. Nearly a decade since its debut, Java boasts more than 3 million developers, and, in 2003, Sun boldly predicted more than 10 million developers by 2005. Gosling's work in Sun's Development Tools division is the key to Sun's plan to triple Java's consumer base and help it retain its reputation as the most popular programming language for the Web.

## FURTHER READING

### In These Volumes

Related Entries in this Volume: Joy, William
Related Entries in the Chronology Volume: 1995: Sun Microsystems Introduces Java

### Works By James Gosling

*NeWS Book: An Introduction to the Network/Extensible Window System.* Berlin: Springer Verlag, 1990.
*The Java Language Specification.* Reading, MA: Addison-Wesley, 1996.

### Articles

Abate, Tom. "Code Warrior." *San Francisco Examiner,* December 8, 1996.
Bank, David. "The Java Saga." *Wired* 3.12, December 1995, http://www.wired.com/wired/archive /3.12/java.saga.html (cited September 16, 2004).
Lohr, Steve. "Canada's Programmer." *Ottawa Citizen,* November 19, 2001.

————. "A 'Rare Form' of Imagination." *Ottawa Citizen,* November 26, 2001.

Manjoo, Farhad. "Is There Hope for Java?" *Salon.com,* January 21, 2003. http://archive.salon.com/tech/feature/2003/01/21/java/index.html (cited September 16, 2004).

Markoff, John. "Making the PC Come Alive." *New York Times,* September 25, 1995.

### Websites

James Gosling: On The Java Road. Gosling's blog, which features links to his most important developments for Sun—including Java and NetBeans. Topics run the gamut from personal to professional, http://weblogs.java.net/jag/ (cited September 16, 2004).

Java Technology, http://java.sun.com/ (cited September 16, 2004).

NetBeans, http://www.netbeans.org/ (cited September 16, 2004).

NeWS: Network Extensible Window System. Online digest of information on the origins of NeWS compiled by Don Hopkins, a member of the NeWS development team at Sun, http://www.art.net/studios/Hackers/Hopkins/Don/lang/NeWS.html (cited September 16, 2004).

World Wide Web—Beyond The Basics: Java. Chapter 21 of *World Wide Web—Beyond The Basics* (Prentice-Hall, 1998), which covers the history and functionality of Java, http://ei.cs.vt.edu/~wwwbtb/book/chap21/background.html (cited September 16, 2004).

# William Joy (1954–)

## PROGRAMMER; FOUNDER OF SUN MICROSYSTEMS

**For more than three decades,** Bill Joy has quietly but steadily revolutionized the computer industry. In the 1970s, he led the development of Berkeley UNIX (BSD), a brand of the UNIX operating system that became the standard networking protocol in the early years of the Internet. Throughout the 1980s, he created applications, program architectures, and even a microprocessor for the cutting-edge workstations and servers that made Sun Microsystems, a company he co-founded in 1982, one of the most powerful companies in Silicon Valley. In the 1990s, he brought the computer world Java—the first universal, platform-independent software—which transformed what could be done on the World Wide Web. Java paved the way for a generation of Joy's platform-independent programs, including Jini networking software and JXTA peer-to-peer software. By the turn of the 21st century, however, Joy was probably best known for his dark predictions about the future of technology, spelled out in an April 2000 *Wired* cover story, "The Future Doesn't Need Us." Indeed, since leaving Sun in September 2003, Joy has dedicated himself to finding a safe path through the technological future.

## D-I-Y PROGRAMMING

Born and reared in Michigan, Joy was preternaturally smart. A skilled reader by the age of 3 and a math prodigy, he skipped several grades, graduating from high school at age 15. He entered the University of Michigan as a math major, and there discovered computers. After experimenting with numerical supercomputing and computer programming, he changed his major to electrical engineering. Joy graduated from the university in 1975 and was then pursued by two top engineering schools, Caltech and Stanford, for their graduate programs. Joy, however, opted to go to the University of California–Berkeley for a joint electrical

engineering and computer science degree. One of the deciding factors, so it is said, was the sad state of the Berkeley's computer science facilities: Joy believed that he would have more to do at a place where the equipment could use some fixing.

Joy's academic focus was theoretical computing, but in his seven years on the University of California campus, his most important—and now legendary—accomplishments involved developing part of the Berkeley UNIX operating system. UNIX, which had been developed by AT&T in the early 1970s, was then a relatively new operating system. AT&T supplied it free to government institutions and universities. Along with a small group of graduate students in computer science, Joy began exploring UNIX, initially just identifying and fixing bugs. Soon, however, he was developing entire applications. Frustrated by the difficulties of the existing UNIX file-editing programs, Joy wrote an editor from scratch—the *vi* editor, which allows a user to edit the text of a file through a window with a standard set of keystrokes. This efficient, powerful program is still standard on UNIX systems. Joy also developed the C shell; a shell is a command interpreter—it passes on instructions from the user to the operating system. Joy's shell, created for the popular computer language C, helped programmers using C interact more easily with BSD.

In 1978, Berkeley UNIX was awarded a Defense Department Advanced Research Projects Agency (DARPA) contract to develop the networking capabilities of ARPANET, the predecessor to the Internet, for the popular VAX computer. Joy became dissatisfied with the existing TCP/IP (Transmission Control Protocol over Internet Protocol) networking protocol, so he did what he had so often done in the past: he wrote his own. Joy's version of the TCP/IP code, written for Berkeley UNIX, was so technologically sound that the DARPA adopted BSD as the universal computing environment for ARPANET. Indeed, as ARPANET grew,

Joy's program grew, or scaled, with it beautifully, forming a solid foundation for early networking. Many cite this aspect of Berkeley UNIX as one of the most important innovations in the history of the Internet.

In addition to such technological feats, Joy's method of developing Berkeley UNIX helped pave the way for the open-source movement. The university sold Berkeley UNIX for $50, which covered basic costs, such as shipping and publishing the user manual. For that $50 a purchaser also received the operating system's source code; thus the software was "open" for users to see both how it worked and to adapt if they wished. Hackers and other enthusiasts were encouraged to make changes and send them back to Joy, who could then integrate various improvements into the basic system. This process was later adopted by other open-source projects, including Linux.

Although Joy left Berkeley in the early 1980s, Berkeley UNIX continued to evolve into one of the most widely used operating systems. Indeed, almost thirty years later, incarnations of BSD are the basis of Apple Computer's OS X operating system and various other operating systems that compete with Microsoft's Windows.

## THE SUN ALSO RISES

Word of Joy's accomplishments with Berkeley UNIX spread. In the early 1980s, three Stanford graduate students—business school students Vinod Khosla and Scott McNealy, and electrical engineering student Andreas Bechtolsheim—began toying with the idea of a new computer company built around a cheap but powerful UNIX workstation. Bechtolsheim had already developed a prototype; Khosla and McNealy had the business vision and skills. Joy's programming abilities would be indispensable. By the time Khosla, McNealy, and Bechtolsheim approached Joy with the idea, Joy had tired of Berkeley UNIX and was ready for change. Joy left Berkeley in early 1982, his doctorate unfinished, to join the company they called Sun Microsystems (Sun originally stood for Stanford University Network).

*[A]ll successful systems were small systems initially. Great, world-changing things—Java, for instance—always start small. The ideal project is one where people don't have meetings, they have lunch.*

—*Joy to* Wired, *December 2003*

The core of the new company was networking. One of Sun's most successful slogans, coined in 1988, says it all: "The network *is* the computer." Initially, the company had set out to build a server that could compete with one of IBM's mainframes but at a fraction of the cost. Sun Microsystems grew to be much more.

Early on, Joy's primary task was to bring the advances of BSD to Sun. In fact, BSD, including its renowned TCP/IP program, became the core of Sun OS, the company's first operating system, which debuted in 1983. By 1984, Joy had created the Network File System (NFS), a UNIX file-sharing architecture that has since become a networking standard. Two years later, he was able to extend the NFS technology to all PCs. In addition to developing networking capabilities, Joy was also instrumental in developing the RISC microprocessor, Sun's first computer chip, for which he later designed a program architecture called SPARC. Few in the computer world were as well suited to both the software and hardware areas of the process.

In 1987, after the SPARC workstations hit the market, Joy was—for the first, but not only, time—ready to leave the company. However, Sun's continuing success, along with a new contract with AT&T to rework UNIX, convinced him to stay. By 1988, Sun's annual revenues had reached $1 billion, a record for Silicon Valley. Never before had a start-up achieved such financial figures in so little time. Joy continued work on the SPARC architecture, as well as the new UNIX-based operating system called Solaris. But by the late 1980s, he had grown tired of Silicon Valley culture, the traffic in particular. In 1989, he moved to Aspen, Colorado, where he started Sun Aspen Smallworks, a satellite research lab. There, in the peace of the Rocky Mountains, he brewed up his next idea.

## WAKE UP AND SMELL THE JAVA

The first inklings of Java began at Sun's Silicon Valley headquarters, soon after Joy left for Aspen. Back then, the highly secretive project, first code-named Stealth, then Green, and later, Oak, focused on developing interactive technology for computer devices, most notably interactive TV. For many years, Joy remained on the fringes of the project. Indeed, by 1994, he was once again contemplating leaving Sun. Soon, however, it became increasingly clear that the promises of interactive TV were being quickly surpassed by the realities of the World Wide Web, and the Oak project began to founder. Instead of leaving, Joy stepped in and led the creation of Java.

Joy convinced Sun to rework Java, as Oak had come to be called, for the Web, encouraging James Gosling, Sun's lead programmer on the project, to write Java as a stand-alone programming language. In December 1994, Sun posted Java on a secret Web site and invited parts of the programming community to test it out. Word of Java's capabilities—primarily, its previously unheard-of ability to create programs that could run on any computer regardless of the operating system—spread like wildfire. After its official release in January 1995, Java got a seal of approval from Marc Andreessen, the Internet wunderkind whose browser, Mosaic, which later became Netscape Navigator, jump-started the Web's rise. Everyone, it seemed, wanted Java.

Despite the tidal wave of hype surrounding Java at its debut, Joy still had to advocate and negotiate endlessly on its behalf. In May 1995, he scored a coup when Netscape agreed to license Java as part of Netscape Navigator, then the number one Web browser, with more than 75 percent market share. The agreement gave Java a vast, ready-made user population, and brought on a Java programming frenzy. Sun also provided Java for free to noncommercial users. Within months, Web sites that had previously been static came alive with Java-written animations and new levels of interactivity. Web design had undergone a revolution.

Joy's vision had Java transforming computing. It was, at once, a programming language and, in many ways, a simple operating system that was not tied to any specific desktop, particularly not to any Windows desktop. Java was supposed to put an end to frustrating software incompatibilities, the scourge of computing in an environment full of different platforms. Also, Joy hoped

> *My own biggest mistake in the last twenty years was that sometimes I designed solutions for problems that people didn't yet know they had.*
>
> —*Joy to* Fortune, *October 13, 2003*

that Java would release programmers from the stranglehold of writing software strictly for Windows, the dominant computing environment, and from the time-wasting endeavor of writing the same software for each different platform.

Though Java did revolutionize how work was done on the Web, it did not immediately reach what many, Joy included, saw as its full potential. One obstacle arose in 1996. Microsoft, which licensed Java from Sun, began developing a Windows-only variant of Java, thus undermining Java's universality, the central power of the application. Sun's ensuing $35-million lawsuit against Microsoft, begun in October 1997, alleged that Microsoft was using its monopoly power to stamp out innovation and to quash a small, simply written program that, combined with the Internet, threatened Windows, the heart of and profit engine for the Microsoft empire. In January 2001, Sun and Microsoft settled the case for $20 million and the courts decreed that Microsoft could not use the Java-compatible trademark for its "polluted" Java variant. The following year, Sun brought a different case against Microsoft, asking the courts to require Microsoft to carry Sun's version of Java for Windows, which had been excluded from Microsoft's latest operating system, Windows XP. As of 2004, the case is still in the appeals process.

Outside the courtroom, Java inspired and underlay a new generation of platform-independent products developed by Joy for Sun. Around the same time that the first case against Microsoft began, Joy began to conceptualize his next great breakthrough. The idea was first sketched out on a placemat in an Aspen restaurant in 1997. Joy began to envision ways in which Java-enabled technology in home electronics as varied as toasters, computers, thermostats, and stereos could interact through a vast communications network. Joy thought of the network as almost biological—able to evolve and adapt when new elements were introduced into the computing environment.

Two years later, Joy had not just an idea, but a product: Jini. In a sense, Jini was to operating systems what Java had been to programming languages—a universal platform that solved the problems of incompatibility. Jini provided the basic architecture to network various devices in any given environment: allowing a computer to communicate with the thermostat in a home, for instance, or allowing a camera to communicate with a printer. Larger applications could then be built upon the network. Early on, Jini was adopted by the military, the automotive and telecommunications industries, and in high-end computing.

While Jini relied entirely on Java technology, which some developers deemed a drawback, Joy's next major innovation took distributed computing to another level. In 2000, he began a secretive project to develop a standard platform for peer-to-peer (P2P) technology. At the time, the Napster music-sharing service had already captured the nation's imagination, and other P2P programs began to emerge. Joy believed that, if a standard architecture could be built, programmers could develop more complex and effective P2P applications, without having to reinvent the wheel each time with new, often incompatible, connection technology. His answer was JXTA.

JXTA (pronounced "juxta," as in *juxtaposition*) emerged as six basic protocols that let users find and connect with one another, written primarily in XML, a programming standard promoted by the World Wide Web Consortium (W3C). JXTA was released, much like BSD UNIX, as an open-source program, giving users the freedom to enhance and upgrade according to the needs of this emerging type of computing. As Joy explained to the computer industry publication *ZDWire* in 2002, "It's like an agreement to have a standard electrical outlet; it is a standard way to get things to plug together. . . . We are trying to make it as simple as possible."

With JXTA, as with Jini and Java before it, Joy has helped Sun get ahead of the curve in new trends in computing. Java opened the door for platform-independent computing, Jini enabled users to reach new levels of interconnectivity in distributed computing, and JXTA has helped invigorate P2P applications. However, in many ways, Joy's work has also served the future. His simple, secure programs, which are virus-resistant, stand in contrast to the buggy, insecure systems and the ever-increasing complexity of much of today's computer technology. Joy believes that the faults in such technology, coupled with exponentially increasing computer power, could spell doom for humanity.

> *Failing to understand the consequences of our inventions while we are in the rapture of discovery and innovation seems to be a common fault of scientists and technologists; we have long been driven by the overarching desire to know that is the nature of science's quest, not stopping to notice that the progress to newer and more powerful technologies can take on a life of its own.*
>
> —Joy in Wired, April 2000

## IT'S THE END OF THE WORLD AS WE KNOW IT

"Having struggled my entire career to build reliable software systems, it seems to me more than likely that this future will not work out as well as some people may imagine," Joy wrote in the April 2000 cover story for *Wired*, titled "Why the Future Doesn't Need Us." "My personal experience suggests we tend to overestimate our design abilities." In this article, Joy discussed far more than the familiar and almost daily dangers of network instability and insecurity. He argued that three specific new technologies—genetic engineering, nanotechnology, and robotics—could threaten human existence if scientists, and, indeed, humanity itself, did not step back to consider the ramifications of technological "progress." He drew from a vast array of sources, from science fiction to the Unabomber manifesto to the writings of some of the most prescient thinkers and futurists.

The reaction to Joy's article was mixed. Some disregarded his dystopian predictions as paranoid and baseless. For others, Joy's reputation as a visionary in the world of computing made them take heed. One, Melvin Schwartz, recipient of the 1988 Nobel Prize for physics, told *Business Week*, "[Joy is] thinking about the things that should be thought about. What sounds wild today won't be in twenty to thirty years."

Joy followed the *Wired* article with an op-ed piece in the *Washington Post* and various speaking engagements across the country, all of which contained the same message of caution and a call for increased attention to scientific ethics. At times, he has even suggested a Hippocratic oath for scientists and technologists, an idea that quickly found support in the scientific community. In the fall of 2000, a London-based group, the Institute for Social Invention, came up with this suggestion:

> I vow to practice my profession with conscience and dignity; I will strive to apply my skills only with the utmost respect for the well-being of humanity, the Earth, and all its species; I will not permit considerations of nationality, politics, prejudice, or material advancement to intervene between my work and this duty to present and future generations. I make this Oath solemnly, freely, and upon my honor.

However, the question of how to implement such an oath—and what to do with scientists who break it—remains. Even Joy does not claim to have the answers.

As part of Joy's wake-up call, he intended to write a book, but, as he would later state, the terrorist attacks of September 11, 2001, served as wake-up call enough to the perils of this century. Plans to write a book filled more with answers than questions have also stalled. "The problem is," he told *Wired* in December 2003, "I'm not satisfied with the prescriptions I have. You don't get two shots at something like this, so I'm holding off."

After several failed attempts to leave Sun Microsystems, in September 2003, Joy finally succeeded. "There is no ideal time to leave a company," he told *Fortune* the day after his departure. "But I feel now that all projects and strategies at Sun are in good hands." While Scott McNealy, Sun's CEO and the only founder remaining, called Joy's departure amicable, the *Los Angeles Times* declared it "another sign of brain drain" at Sun, which, by 2003, was showing signs of serious struggle.

Since his departure, Joy has focused first on finding a way to make the Internet more secure from viruses and free it from spam, its newest scourge. Admittedly, his plans are still at the thinking stage. "We need an evolutionary step of some sort," he told *Fortune*. "Or, we need to look at the problem in a different way." Luckily, Joy—a man who has been nicknamed The Edison of the Internet—is on the job.

*Bill Joy. (Associated Press)*

## FURTHER READING

### In These Volumes

Related Entries in this Volume: Andreessen, Marc; Gosling, James

Related Entries in the Chronology Volume: 1995: Sun Microsystems Introduces Java

### Works By William Joy

"Why the Future Doesn't Need Us." *Wired,* April 2000, http://www.wired.com/wired/archive/8.04/joy.html (cited September 16, 2004).

### Books

Hall, Mark. *Sunburst: The Ascent of Sun Microsystems.* Chicago: Contemporary Press, 1990.

## Articles

Bank, David. "The Java Saga." *Wired,* December 1995, http://www.wired.com/wired/archive/3.12/java.saga. html (cited September 16, 2004).

Cave, Damien. "Killjoy." *Salon.com,* April 10, 2000, http://www.salon.com/tech/view/2000/04/10/joy (cited September 16, 2004).

Corcoran, Elizabeth. "Bill Joy's Big Idea." *Washington Post,* February 14, 1999.

Hudson, Kris. "Legend Seeks Ideas Away from Sun." *Denver Post,* September 29, 2003.

Kelly, Kevin, and Spencer Reiss. "Joy Shtick." *Wired,* August 1998, http://www.wired.com/wired /archive/6.08/joy.html (cited September 16, 2004).

———. "One Big Computer." *Wired,* August 1998, http://www.wired.com/wired/archive/6.08/jini.html (cited September 16, 2004).

Leonard, Andrew. "BSD Unix: Power to the People, From the Code." *Salon.com,* May 16, 2000, http://www.salon.com/tech/fsp/2000/05/16 /chapter_2_part_one (cited September 16, 2004).

London, Simon. "Bill Joy of Sun." *Financial Times,* September 4, 2002.

Menn, Joseph. "Pied Piper of Technology." *Los Angeles Times,* December 20, 1999.

"The Other Bill." *The Economist,* September 21, 2002.

Reiss, Spencer. "Hope Is a Lousy Defense." *Wired,* December 2003, http://www.wired.com /wired/archive/11.12/billjoy.html (cited September 16, 2004).

Schlender, Brent. "An Ode to Joy." *Fortune,* September 29, 2003.

———. "Joy After Sun." *Fortune,* October 13, 2003.

Vaughn, Susan. "Making It." *Los Angeles Times,* September 24, 2000.

## Websites

The Center for the Study of Technology and Society. Special Focus: Bill Joy's Hi-Tech Warning Web site dedicated to the exploration of the predictions Bill Joy made in his April 2000 *Wired* article, "The Future Doesn't Need Us." Includes many external links to Related Entries, http://www.tecsoc.org /innovate/focusbilljoy.htm (cited May 26, 2004).

Executive Perspectives: Bill Joy. http://www.sun.com /executives/perspectives/joy.html (cited May 27, 2004).

Java.com: The Source For Java Technology. Web home for Java technology and resources, sponsored by Sun Microsystems, http://www.java.com/ (cited May 27, 2004).

Java Lawsuit. http://www.sun.com/lawsuit/ (cited May 27, 2004).

Jini.org. Web home of the Jini programming community, http://www.jini.org/ (cited May 27, 2004).

Project Juxta. Online home for JXTA technology, sponsored by Sun Microsystems, http://www .jxta.org/(cited May 27, 2004).

Sun Microsystems. Web home of Sun Microsystems, with links to information about many of Bill Joy's developments, from NFS to JXTA, as well as some of Joy's own writing; also contains ongoing information about the Java lawsuit against Microsoft, http://www.sun.com (cited May 27, 2004).

# Robert Kahn (1938–)

## INVENTOR OF TCP/IP

**Robert Kahn is considered** to be one of the founding fathers of the Internet for his role in developing TCP/IP (Transmission Control Protocol/Internet Protocol), the set of rules for computer networking that opened the door for today's Internet. TCP/IP is just one of Kahn's many contributions to the Internet. In the 1960s, while working at the computing think tank Bolt, Beranek and Newman, Kahn wrote key technical specifications for the minicomputers that eventually ran the ARPANET, the Internet's predecessor. His tireless work on the early stages of the ARPANET earned him a position at ARPA's Information Processing Techniques Office (IPTO), where he remained for thirteen years—long enough to see TCP/IP implemented as the Internet's networking standard. In 1986, Kahn left to create the Corporation for National Research Initiatives (CNRI), a private sector agency that would perform the research and development that IPTO once did. Early on, CNRI led the government's quest to develop a national information infrastructure. CNRI has since championed various new ideas in Internet-age computing, for example, "knowbots" and digital object architectures, and has provided the home base for such vital organizations as the Internet Society.

## BROOKLYN'S FINEST

Kahn was born and raised in Brooklyn, New York. After high school, he attended City College of New York, where he earned his bachelor's degree in electrical engineering in 1960. That year, on a National Science Foundation fellowship, he entered Princeton University's electrical engineering program. While working toward a master's degree, which he received in 1962, Kahn worked at AT&T Bell Laboratories in Murray Hill, New Jersey, studying communications systems. In 1964, Kahn was awarded a doctorate and headed north to Cambridge, Massachusetts, for postdoctoral work at the Massachusetts Institute of Technology (MIT). There, as an assistant professor of electrical engineering, Kahn began studying information theory as well as communications.

The move to MIT was awkward for Kahn. "I felt sort of like the coach who's never played the sport he's coaching," he told *The Bent of Tau Beta Pi,* a publication of the Tau Beta Pi Engineering Honor Society, in 2002. (Kahn was elected to Tau Beta Pi in 1960.) After two years of teaching, Kahn decided to leave academia to get some real world experience to back up his skills as a theoretician. In 1966, he went to work at a small technical think tank in Cambridge, Bolt, Beranek and Newman (BBN), firmly intending to return to MIT and teaching after a year or two.

BBN had just started to branch out into the field of computing. (J.C.R. Licklider, one of the earliest Internet visionaries, helped turn BBN's focus toward real-time, interactive computing in the late 1950s, when he was vice president of the fledgling firm.) Kahn, BBN's first communications specialist, began working on computer network design and distributed computing techniques, writing memos on topics such as controlling the flow of traffic through networks. Kahn never returned to teaching.

## CALLING ALL IMPS

In the summer of 1968, two years into Kahn's tenure at BBN, the Information Processing Techniques Office (IPTO) of the Pentagon's Advanced Research Projects Agency (ARPA) sent out a Request for Proposals to build a special kind of minicomputer, an interface message processor (IMP), geared toward networking. IMPs would be the cornerstone of IPTO's landmark experiment in computer networking—the ARPANET.

Over the next several months, BBN poured nearly $100,000 into researching and developing its proposal for the IMP. (Meanwhile, IBM and AT&T dismissed outright the idea of a network based on packet switching, which involved transmitting information by dividing it into small data packets.) As a senior scientist and resident networking theorist, Kahn wrote the bulk of the proposal's technical material. The BBN proposal was, by far, the most detailed that IPTO received—more than 200 pages offering initial hardware designs, proposed system architectures, and tables detailing issues such as timing, routing, and transmission delays. The work paid off: that December, IPTO awarded BBN the one-year contract, passing over much larger and better-equipped companies.

On January 1, 1969, Kahn and a small team of BBN scientists led by Frank Heart began work on the IMP, with a deadline of Labor Day. Although the detailed proposal gave them a jumpstart on the actual construction of the IMP, the work was intense and extremely difficult. Several attempts were needed before the software for the IMP actually worked on the powerful Honeywell DDP-516 minicomputer that BBN was using, but by Labor Day weekend of 1969, the IMP was on its way to the University of California–Los Angeles (UCLA), the first of the four initial nodes of the ARPANET. The box bore graffiti that read, "Do it to it, Truett!" an encouraging message intended for BBN engineer Truett Thach, who accompanied the IMP from the airport to UCLA.

Kahn had written the technical specifications for connecting the IMP to the various host computers. By January 1970, the IMPs were in place at all four nodes of the early ARPANET: UCLA, the Stanford Research Institute (SRI) in Menlo Park, California, the University of California–Santa Barbara (UCSB), and the University of Utah in Salt Lake City. Kahn and another BBN engineer, David Walden, flew to UCLA to test the network. At UCLA, Kahn first met Vinton Cerf, then a graduate student working as a chief programmer of the ARPA-funded project in UCLA's computer lab.

Kahn and Cerf worked closely during the two-week trip. Kahn arrived with serious doubts about the performance of the network under the existing protocols.

> *In the beginning it wasn't clear how computer networks were going to be used. . . . A lot of the research community originally thought I had gone off my rocker.*
>
> —*Kahn to the* New York Times, *September 1990*

Together, he and Cerf bombarded the network with more data than it was intended to transmit, finding key points where the network would suffer from congestion, blockage, and resultant lockup. When Kahn returned to BBN, he used his findings to lobby for better protocols and error control. By December, ARPA-funded scientists, including Cerf, had replaced the initial BBN protocols with the network control protocol (NCP) and had begun work on the file transfer protocol (FTP). (Telnet, a remote login protocol, had already been completed.)

As new nodes were added to the ARPANET—more than a dozen by January 1971 and nearly two dozen by April 1972—the pressure to finish the network's programs and protocols grew exponentially. To help push the ARPANET to completion, Larry Roberts, then the head of IPTO, called for a public demonstration of the network at the International Conference on Computer Communications (ICCC), slated for October 1972 in Washington D.C. He appointed Kahn head of the project.

Over the next nine months, Kahn pulled together a group of about fifty scientists who worked nonstop to put the finishing touches on the ARPANET. That October, the ARPANET demonstration, involving nineteen different programs on forty different terminals, drew more than 1,000 attendees from various disciplines and from countries all over the globe. The ICCC was a landmark event in the history of the ARPANET. In that October alone, use of the ARPANET jumped 67 percent.

## THE NETWORK OF ALL NETWORKS

Soon after the ICCC ended, Larry Roberts lured Kahn away from BBN to work on IPTO's new networking technologies. Capitalizing on Kahn's experience in developing the first generation of packet-switching network architecture with the ARPANET, Roberts set Kahn to work on similar systems for use with packet-switching networks involving satellites and mobile radios.

Kahn began work as the principal architect of ARPA's packet-radio network and manager of its packet-satellite network, in addition to working on aspects of the ARPANET. Soon, however, his primary task evolved into finding a way to make each of these networks compatible, in terms of communication, with the other two. Kahn needed to develop a network of networks. Kahn dubbed the task "internetting."

Kahn had actually been thinking about a similar issue while still at BBN. In an internal memo from 1972, written shortly before he arrived at IPTO, Kahn had outlined his thoughts on a new networking idea he called "open architecture." He had first considered open architecture in terms of a communications-oriented set of operating system principles, but, with respect to internetting, the essential thinking was the same: how do you get disparate systems to communicate with one another?

"From an architectural point of view, I knew how I wanted to approach the problem," Kahn explained in a 2003 interview with *Ubiquity*, an information technology trade magazine. "But I did not have a good idea about how to embed it into the machines that would be on the Net." In the spring of 1973, Kahn turned to his colleague from the early days of the ARPANET, Vinton Cerf, then a professor at Stanford. Cerf, Kahn figured, had tackled many of the technical issues in developing a networking protocol with the creation of NCP a couple years earlier. With prodding, Cerf joined the project.

During the summer of 1973, Kahn and Cerf worked relentlessly on "the Internet problem"—developing the ideas now known as Internet protocol (IP) addresses, routers, and various networking protocols. By September, they had a working draft, which they presented at a meeting of the International Networking Group (INWG), in England. Six months later, a revised paper appeared in the May 1974 issue of the *IEEE Transactions on Communications,* a now-historic article titled, "A Protocol for Packet Network Intercommunication."

In the following months, Cerf led a group of graduate students in developing the basic host-to-host protocol—the transmission control protocol (TCP). The first

*People often ask, "When you installed the first node on the ARPANET, who was there from the press?" and my answer always is: "You've got to be kidding."*

*—Kahn to* Ubiquity, *March 2003*

specification, published in December 1974, allowed scientists elsewhere—specifically at BBN and University College in London—to implement TCP on various computing platforms. Between 1974 and 1978, four increasingly more refined versions of the protocol were developed and tested, with the fourth iteration eventually becoming the networking standard, still widely in use thirty years later.

ARPA-funded scientists began using TCP to make the first links between ARPANET and the mobile radio network and, by 1977, the satellite network as well. That summer, scientists working under Cerf (who had by then taken over IPTO's Internet Program) were able to demonstrate the first successful transmission across a three-network Internet: packets sent via the radio network from a van on a California freeway went through the ARPANET to a satellite network on the East Coast, traveled via satellite and ground networks throughout Europe and then back to the van in California via the ARPANET. Kahn's earliest goals for internetting were being achieved.

In 1979, Kahn was named director of IPTO and continued to push forward the Internet effort he had started in 1973. By then, ARPA researchers had split TCP into two components—TCP for host-to-host communication, and IP for internetwork communications. This greatly increased the flexibility and versatility of the network. In 1981, Kahn channeled funding toward a brilliant young computer scientist at Berkeley, Bill Joy, to write the TCP/IP program for Berkeley UNIX, expanding the network further still. (Joy's TCP/IP for Berkeley UNIX became the universal computing environment for the early Internet.) In 1982, when Cerf left IPTO to work for the telecommunications company MCI, Kahn once again took the reins of the Internet Program. With his guidance, TCP/IP was already well on its way to becoming the military's de facto networking standard. On January 1, 1983, the switch to TCP/IP (from NCP) for all networks connected to the ARPANET was made official. The Internet, as we now know it, was born.

Kahn remained at the helm of IPTO until 1985. In addition to guiding the evolution of IPTO's Internet

Program for more than a decade, he set the stage for the government's $1-billion Strategic Computing Program, the largest computer research and development program ever undertaken by the federal government. By the mid-1980s, however, legislation had been passed that threatened to limit government spending. "I thought staying in the government wouldn't help me advance my notion of a national information infrastructure," Kahn told *Ubiquity* in 2003. "Instead I'd be in the mode of just trying to protect what budget we already had." IPTO was already mired in political infighting and bureaucracy. Dismayed, Kahn stepped down from his post after thirteen years.

## MY OWN PRIVATE DARPA

Kahn's solution to shrinking government support of new Internet technologies was to create a nonprofit, private-sector version of IPTO—the Corporation for National Research Initiatives (CNRI). Kahn believed that CNRI could provide the R&D necessary to build what he termed the *national information infrastructure* (NII)—often referred to in the popular media as the "information superhighway." "I thought CNRI could be helpful in nurturing the Internet, helping it to grow and to apply the Internet in related areas with the business community, the banking community, on building knowledge banks, digital libraries and a whole variety of similar infrastructure ideas," he told *Ubiquity* in 2003.

Kahn's first act was to bring Vinton Cerf on board as vice president. Cerf had spent the past several years developing MCI Mail, an early commercial email program. As one of CNRI's first endeavors, he and Kahn sought to connect MCI Mail to the Internet. Work began in 1988, with Cerf lobbying the government for permission to open the Internet up to commercial services. (The Internet had been strictly the domain of research facilities and the government.) By June of 1989, the first experimental relay between MCI Mail and the Internet was complete. CNRI's application opened the door for other commercial email systems to connect with the Internet, and soon companies such as AT&T, Sprint, and CompuServe were flocking to it.

> *The Internet is an architectural philosophy, rather than a technology.*
>
> —*Kahn to the Washington Post, May 2002*

In 1990, the ARPANET was phased out and replaced by a faster network backbone developed by the National Science Foundation (NSF), known as NSFNET. Soon after, the NSF awarded a three-year, $15.8-million contract to CNRI to develop a new high-speed "gigabit" network, which would operate nearly 1,000 times faster than any existing networks. The initiative made a big splash in the media, landing on the front page of the *New York Times* in June 1990. (As head of the project, Kahn was profiled in the Sunday Business Section just a few months later.)

As CNRI organized the gigabit initiative, which involved more than forty different organizations nationwide, Kahn and Cerf were also hard at work on various other projects, including the creation of information retrieval agents called "knowbots" and putting the funding in place for the development of an early point-and-click browser at the University of Illinois.

At the time, the early 1990s, the Internet was just beginning to take off, with interest growing in many different sectors—government, research, and commercial. In 1991, the NSF decided to step back from funding the Internet Engineering Task Force, one of the Internet's primary technical bodies, which CNRI had housed since the late 1980s. In response, in January 1992, Cerf and Kahn founded the Internet Society (ISOC), which would take over full support of the IETF in addition to serving as an international, interdisciplinary body dedicated to the advancement of and education about the Internet.

The following year, CNRI brought together another multidisciplinary group, the Cross-Industry Working Team (XIWT). Initially founded with the help of twenty-eight major U.S. corporations—including AT&T, IBM, Citicorp, and Sun Microsystems—the XIWT was designed to make key technical decisions about the NII, one of the Clinton administration's major technology goals. In 1993, a member of Clinton's National Economic Council told the *Federal Technology Report*, "Before we can realize the full potential of NII there are serious technical issues that need to be addressed, or we'll be in the same situation as someone trying to buy a house from 100 different subcontractors." He added confidently that the XIWT would prove successful, because, "Bob Kahn has had a remarkable record of making things happen." Later, in 1997, CNRI would serve a similar purpose in organizing IOPS.ORG, a group of Internet services providers (ISPs)—including AT&T, BBN, and EarthLink—to help streamline technical coordination and information exchange between ISPs.

By the mid-1990s, after Cerf had left CNRI to return to MCI, Kahn had turned most of his attention to developing a new way to handle the glut of information on the Internet, particularly once the World Wide Web took the network by storm. He began to envision another layer built into the information infrastructure of the Net—one that would be able to describe elements on the network as digital objects, telling computers and users the *what* of a piece of information, not just the *where*. (The URL, for example, simply tells *where* an object is on the Web.) "My work is focused on the re-conceptualization of the Internet to deal with information as a fundamental commodity," Kahn explained to the *The Bent* in 2002. "This is important because almost everything in the real world has some notion of ownership, but the Internet was never created with that in mind. We have packets moving across the network, but there's fundamentally no way to tell what's what and who's who."

Kahn called his new schema the Handle System. Each piece of information, or digital object, would be assigned a permanent "handle," or digital object identifier (DOI), that would have a description of the contents of the digital object, including any copyright or privacy restrictions. The publishing industry was one of the first to champion Kahn's new system, particularly in the realm of e-books. By the late 1990s, CNRI was collaborating with the Association of American Publishers on adapting the DOIs to their needs, and the Library of Congress and the Department of Defense were implementing DOI architectures at the National Digital Library Program and the Defense Virtual Library, respectively. DOIs have come to be known as "bar codes for intellectual property."

By the turn of the 21st century, some of CNRI's programs and projects had lost their initial impetus; nevertheless, Kahn remains one of the primary figures in the history of the Internet—and one of the most decorated. He has been awarded the National Medal of Technology, the Charles Stark Draper Prize, and the IEEE's Alexander Graham Bell Medal. From his early

work on the IMPs to his continuing developing of DOIs, Kahn has, for more than forty years, used his strengths as a theoretician to develop the key technological architectures that make the Internet what it is today.

## FURTHER READING

### In These Volumes
Related Entries in this Volume: Cerf, Vinton; Goldfarb, Charles; Joy, William; Postel, Jonathan; Roberts, Lawrence; Taylor, Robert

Related Entries in the Chronology Volume: 1967: Plans for ARPANET Are Unveiled; 1969: The ARPANET Is Born; 1972: ARPANET's Public Debut; 1974: "A Protocol for Packet Network Intercommunication" Is Published; 1979: DARPA Establishes the Internet Configuration Control Board; 1983: Internet Is Defined Officially as Networks Using TCP/IP

### Works By Robert Kahn
"Putting It All Together with Robert Kahn." *Ubiquity*, March 11–17, 2003, http://www.acm.org /ubiquity/interviews/r_kahn_1.html (cited September 16, 2004).

with Vinton G. Cerf. "A Protocol for Packet Network Interconnection." *IEEE Transactions on Communications*. COM-22, no. 5 (May 1974).

### Books
Abbate, Janet. *Inventing the Internet*. Cambridge, MA: MIT Press, 2000.

Hafner, Katie, and Matthew Lyon, *Where Wizards Stay Up Late: The Origins of the Internet*. New York: Simon & Schuster, 1996.

### Articles
Anthes, Gary. "Reinventing the Internet." *ComputerWorld*, August 27, 2001.

Bert, Raymond. "Founding Father." *The Bent of Tau Beta Pi*, Winter 2002, http://www.tbp .org/pages/publications/BENTFeatures/BertW02 .pdf (cited June 15, 2004).

Henry, Shannon. "Visions of a Wild and Wireless Future." *Washington Post*, May 23, 2002, http://www.washingtonpost.com/ac2/wp-dyn/A60488-2002May22 (cited September 16, 2004).

Markoff, John. "Creating a Giant Computer Highway." *New York Times*, September 2, 1990.

Peterson, Shane. "The Return of Kahn." *California Computer News,* May 2001, http://www.ccnmag .com/index.php?sec=mag&id=12 (cited September 16, 2004).

Spira, Jonathan B. "20 Years—One Standard: The Story of TCP/IP." *Basex: TechWatch,* January 1, 2003.

Zaret, Elliot. "Internet Pioneer Urges Overhaul." *MSNBC.com,* November 21, 2000, http://www .doi.org/news/msnbc_rkahn_interview.pdf (cited September 16, 2004).

### Websites

The Corporation for National Research Initiatives (CNRI). http://www.cnri.reston.va.us (cited September 16, 2004).

Digital Object Identifier System, http://www.doi.org (cited September 16, 2004).

The Handle System, http://www.handle.net/ (cited September 16, 2004).

IOPS.ORG, http://www.iops.org (cited September 16, 2004).

Online home of CNRI, founded by Kahn in 1986. Includes Kahn's biography, links to publications, as well as links to several of CNRI's programs, tools and organizations, including: The Cross-Industry Working Team, http://www.xiwt.org (cited September 16, 2004).

# Lawrence Lessig (1961–)

## CYBERLAWYER

**Few were familiar with** the name Lawrence Lessig when, in 1997, the Harvard law professor was consulted on various technical and legal matters involved in the *United States v. Microsoft,* one of the most closely watched antitrust trials in American history. Lessig is a constitutional law professor with programming skills—he has the rare ability to explain technology to lawyers and explain law to technologists. Indeed, Lessig was one of the earliest pioneers in the rocky terrain where the Internet ran up against the Constitution. Following his short stint on the Microsoft case, Lessig has become one of the most strident advocates against American copyright law. His books, particularly *Code and Other Laws of Cyberspace,* published in 1999, form the basis of what we now consider cyberlaw.

## PORTRAIT OF THE LAWYER AS A YOUNG MAN

Born in South Dakota, Lessig spent most of his youth in Williamsport, Pennsylvania, where his father ran a steel-fabricating business. Even as a child, Lessig had an astonishingly sharp mind. By high school, he had added to his various intellectual pursuits a passion for politics, particularly the right-wing Republican politics of his father. By 1980, when he graduated as valedictorian of his high school class, Lessig was been a devoted member of the National Teen Age Republicans and had played the governor of Pennsylvania in a mock-government exercise. After graduation he joined the campaign of a Republican candidate for the Senate. That summer, he was the youngest delegate to the Republican Convention.

In the fall of 1980, Lessig entered the University of Pennsylvania, where two generations of Lessigs had gone before him. He earned undergraduate degrees in both economics and management, then headed to Trinity College in Cambridge, England, for what was supposed to be one year of postgraduate study. He stayed for three years and, in 1986, earned a master's degree in philosophy. At Cambridge Lessig began to rethink his right-wing views and became more liberal, even libertarian.

Returning to America, Lessig entered law school at the University of Chicago. After just one year, he transferred to the law school at Yale University, in Connecticut, where he focused on constitutional law. After graduating in 1989, he became a law clerk for Judge Richard Posner, a well-respected federal judge in the Seventh Circuit Court of Appeals in Chicago. (Within ten years both men would be summoned to assist the government in the federal antitrust case against Microsoft.)

In 1990, Lessig secured a clerkship with Justice Antonin Scalia of the Supreme Court. Justice Scalia, like Judge Posner, is on the conservative end of the judicial spectrum. "We had a lot of battles at times," Lessig told the *Boston Globe* in 2000. "But . . . it was a lot of fun working with someone who took seriously trying to get to the right answer." Though he worked for Scalia for just a year, Lessig still left a legacy—technically, if not legally. Lessig convinced the Court to finally leave its archaic computer system behind and adopt a new desktop publishing system. (In 1998 a fellow Supreme Court clerk told the *Chicago Tribune* that the change came after Lessig showed Justice Scalia the thesaurus function on his own computer, saying the judge called the function "the greatest thing he had ever seen.") To help ease the transition from old systems to new, Lessig, who had learned to program in college, himself wrote various short programs for the system.

In 1991, after a brief sabbatical in Costa Rica, Lessig joined the ranks of his former professors at the University of Chicago Law School. His appointment was at the behest of Judge Posner, who often joked that his own life's ambition was to be "Larry Lessig's law clerk." Indeed, Lessig came so highly recommended that, for one of the only times, ever, the University of Chicago hired a law professor so recently out of law school. Lessig was barely 30 years old.

## NOW ENTERING CYBERSPACE

For the next six years, Lessig taught constitutional law and enjoyed growing renown in academic circles as one of the most incisive scholars in his field. By the mid-1990s, just as the Internet began to reach the mainstream, Lessig branched out into new territory—cyberspace.

As a visiting professor at Yale University in 1995, he taught a course, Law and Cyberspace. The course—one of the first of its kind—was inspired by a 1993 *Village Voice* article, "A Rape in Cyberspace," written by Julian Dibell (who has since taught cyberlaw courses alongside Lessig at Stanford University's Law School). The article described a series of virtual sexual assaults that occurred in the LambdaMOO online community, a MUD (multiuser dungeon). In the description of the LambdaMOO community's struggle to deal with the virtual rapes, Lessig found odd echoes of the arguments set forth in *Only Words* (1993), a book by antipornography activist Catherine Mackinnon. Usually the *Village Voice* and Mackinnon would lie on opposite ends of the ideological spectrum, particularly on the issue of the First Amendment, but in the context of "A Rape in Cyberspace," they saw eye to eye—that some words and images, pornography in Mackinnon's case or the virtual rape in Dibell's case—should not be considered protected speech. In 1998, Lessig told the *Pennsylvania Gazette,* "I realized that if cyberspace could get McKinnon and the *Voice* to agree on anything, it had an amazing power to knock people off their usual political bearings, to force them to look at old issues anew."

In addition to teaching at Yale, Lessig began writing articles—first, for legal journals and, later, for mainstream publications like *Wired*—and slowly began to gather his thoughts about law, regulation, and the Internet for a future book. An article he had written for a law journal in the summer of 1996, "Reading the Con-

> *I'm not in favor of copyright theft, I'm just opposed to shutting down all technologies merely because copyright theft may occur on them.*
>
> —Lessig to Reason, June 2002

stitution in Cyberspace," found its way into a 1997 Supreme Court opinion by Justice Sandra Day O'Connor against the 1996 Communications Decency Act. She cited from it repeatedly.

Cyberspace was not the only new terrain that drew Lessig's attention. As a professor, he also paid close attention to the political evolution of Eastern Europe, a region that had captured his imagination when he hitchhiked through it in his Trinity College days. In 1995, Lessig was invited by the former Soviet Republic of Georgia to help write the country's new constitution. The new Georgian president, Eduard Shevardnadze, thanked him personally.

To the average citizen and, indeed, to most lawyers, the Republic of Georgia's new constitution would seem to have little in common with the technological terrain of cyberspace. Lessig, however, found the two profoundly related. "Constitutional law is about trying to set up structures that embed certain values within a political system," he explained to *Reason* magazine in 2002. "It was a tiny step to see that that's exactly what the architecture of cyberspace does: it's a set of structures embedding a set of values. . . . The architecture is analogous to the Constitution."

## THE MASTER

By the summer of 1997, Lessig had left the University of Chicago to join Harvard's new Berkman Center for Internet and Society, a $5.4-million endeavor to incorporate cyberlaw into Harvard's law curriculum. In addition to a full load of constitutional law and cyberlaw courses, Lessig planned to finish his first book while at Harvard. However, near the end of November, only a few months into his new position, he received a call that would make the name Lawrence Lessig as well known in mainstream media as he was in legal circles.

The call came from Judge Thomas Penfield Jackson, the U.S. District Court judge presiding over *United States v. Microsoft* in Washington, D.C. Judge Jackson wanted help in sorting out the thorny technical issues involved in the *Microsoft* case, someone as equally versed in the language of software as in the language of the law. After reading a profile of Lessig, Judge Jackson believed he had found his man.

Lessig's formal appointment to the court came on December 11, 1997. He was to be a Special Master, a court-appointed official who carries out a specific set of tasks required by the court. In Judge Jackson's words, Lessig was appointed " . . . in the interest of justice to resolve as expeditiously as possible the complex issues of cybertechnology and contract interpretation connected therewith."

As special master, Lessig was imbued with many of the same powers of a federal judge: the power to issue subpoenas, gather testimony, examine witnesses, evaluate evidence, and even find parties to the case in contempt of court. His ensuing report would provide the basis for Judge Jackson's later rulings on the two different aspects of the *Microsoft* case—the findings of fact and the conclusions of law.

While Lessig logged eleven-hour days working on the case, Microsoft's lawyers went to work trying to get Lessig removed. On January 5, 1998, Microsoft's attorneys faxed Lessig and asked him to disqualify himself because they believed him to be biased—an assertion based primarily on a 1997 email Lessig sent to a friend, in which Lessig joked that, by installing Internet Explorer, he had "sold his soul" to Microsoft "and nothing happened." (The fact that Lessig used a Mac, not a PC, was also brought up as an issue.) The following day, in a conference call, Lessig told lawyers for both the Department of Justice and Microsoft that he would not stand down. A week later, Judge Jackson ruled against Microsoft's attempt to disqualify Lessig, calling Microsoft's lawyers' remarks to the press, and to Lessig himself, "defamatory" and "not made in good faith."

Microsoft pushed ahead and appealed Judge Jackson's ruling. On February 3, 1998, the federal appeals court ruled in Microsoft's favor, and Lessig was dropped from the case after only fifty-four days. Lessig was crushed. He told *Wired* in 2002, "You know, the

> **Never in our history have fewer people controlled more of the evolution of our culture. Never.**
>
> —Keynote Address, 2002 O'Reilly Open Source Software Convention

Microsoft case was such a gift, and the problem was so interesting and fun. . . . Not getting a chance to finish was extraordinarily frustrating. And not getting a chance to finish it in the context where lots of people thought I was kicked off because I was biased was doubly frustrating."

As the case dragged on, Lessig remained on the sidelines until, in early 2000, two years after his dismissal, Judge Jackson asked Lessig to write a friend-of-the-court brief, one of several briefs Jackson requested to aid him in deciding the conclusions of law in the case. (Judge Jackson had already found, in the findings of fact, Microsoft to be a monopolist.) On February 20, 2000, Lessig turned in forty-five pages of legal analysis. Lessig found the proposed breakup of Microsoft "illogical"—an unexpected boon to the defendant—and while he agreed, in general, with the Justice Department's case, he advised Judge Jackson not to rule against an earlier appellate court decision in favor of Microsoft because such a ruling could harm the government's case on appeal. Legal analysts and other experts who watched the trial closely claim Lessig's brief was indispensable. In 2000, a George Washington University law professor following the case told the *Boston Globe,* "For somebody who is not a witness, for someone who is truly an outsider, those 45 pages represent a big contribution."

## LIFE AFTER MICROSOFT

Once the Microsoft case put Lessig in the spotlight, he did not leave. Shortly after his dismissal, Lessig began writing a column for the *Industry Standard,* one of the leading publications to chronicle the Internet revolution. He tackled such issues as patents, spam, broadband Internet connections, and privacy. His first book, *Code and Other Laws of Cyberspace,* was very well received in mainstream and academic circles and provided the foundation of the growing field of cyberlaw.

Lessig also weighed in on various landmark cases involving cyberlaw and the Internet, testifying before Congress on the Child Online Protection Act and before the Justice Department on the merger of AT&T

*Lawrence Lessig. (Photo courtesy of Lawrence Lessig)*

like the *Boston Globe,* the *Wall Street Journal,* the *Los Angeles Times,* and the *New York Times.* Lessig's efforts culminated in 2002 in a constitutional challenge to the 1998 Sonny Bono Copyright Term Extension Act, a case known as *Eldred v. Ashcroft.*

## FREE MICKEY

On October 27, 1998, President Bill Clinton signed into law the Sonny Bono Copyright Term Extension Act (CTEA), named in memory of the singer, musician, and former Republican Congressman who had died in January of that year. (Bono's widow, Mary Bono, said her husband believed "copyright should be forever.") The CTEA extended copyright protection—which, at the time, was about to expire for works created in the 1920s—by an additional twenty years. It was the eleventh time in forty years that Congress had extended the term of copyright protection.

One of those works was a short cartoon, "Steamboat Willie" (1928), the first incarnation of Walt Disney's Mickey Mouse character. To protect its claim on Mickey Mouse, the Disney Corporation lobbied endlessly on behalf of the CTEA, giving money to three-quarters of the twenty-five sponsors of the bill. Disney's efforts earned the CTEA the nickname the Mickey Mouse Preservation Act.

To Lessig, the CTEA was yet another blow to intellectual freedom in the age of Internet, and he sought to strike down the law as unconstitutional. Article 1, sec. 8 of the Constitution not only secured copyright "for limited times . . . the exclusive right to . . . writings and discoveries," but also declared that copyright existed "to promote the progress of science and useful arts." Lessig believed the CTEA actually inhibited progress, particularly with respect to the Internet.

That Mickey Mouse had become the symbol of the CTEA was a great advantage to Lessig. In pleading his case—both in court and out—Lessig liked to remind his audience that Disney itself stole "Steamboat Willie" from Buster Keaton's "Steamboat Bill." Indeed, had modern copyright protections existed in 1928, the entire Disney empire would have never come into being. In his 2002 keynote address to the O'Reilly Open Source Software Convention, Lessig explained, "the Disney Corporation could do this because that culture lived in a commons, an intellectual commons, a cultural

and MediaOne. He filed friend-of-the-court briefs in the case of *Microsystem Software, Inc. v. Scandinavia Online,* a case involving code that cracked the Cyberpatrol filtering software used by parents to restrict their children's access to online material. In 2001, he testified on behalf of the groundbreaking peer-to-peer file-sharing network Napster in *A&M Records v. Napster, Inc.,* and submitted a brief on behalf of the defendants in the Motion Picture Association of America's case against DeCSS, a computer program that cracked the anticopy security code used on DVDs. Outside court, Lessig railed against how American copyright law was being used to quash new Internet technologies in publications commons, where people could freely take and build. It was a lawyer-free zone." The CTEA, Lessig believed, denied that commons.

In January 1999, just months after the CTEA was passed, Lessig filed suit. The lead plaintiff, Eric Eldred, a retired UNIX administrator from New Hampshire, had, since 1995, been building an online library of free, annotated versions of classic literature, such as Nathaniel Hawthorne's *Scarlet Letter*. His Website, www.eldritchpress.org, received more than 3,000 hits each day, mostly from students, and had been lauded by the National Endowment for the Humanities. Eldred was eager to publish poems by Robert Frost, which were slated to come into the public domain in 1998, until the CTEA made them unavailable until 2019. Enraged, Eldred published Frost's poems, along with other works from 1923 and 1924, until he was arrested for violating the 1998 Digital Millennium Copyright Act and the 1998 No Electronic Theft Act, which made publishing of copyrighted works on the Web a felony.

Lessig recruited Eldred as lead plaintiff in his already planned strike against the CTEA. (Other plaintiffs included Laura Bjorklund, who published genealogies and obscure local histories on the Web.) The suit rested on two claims: that copyright extensions violated the "limited times" language of the Constitution, and that twenty-year extensions violated the First Amendment.

*Eldred v. Ashcroft* moved through the lower courts with little success. The first judge found in favor of the government without even hearing oral arguments. When Lessig appealed to the U.S. Circuit Court of Appeals in Washington D.C., the judges heard extensive arguments, but upheld the CTEA. However, Lessig garnered a dissent from the most conservative judge, Judge David Sentelle, who agreed that the CTEA violated the "limited times" clause of the Constitution. On a second appeal, Lessig earned an additional dissent from Judge David Tatel, one of the most liberal judges on the D.C. circuit court. Bolstered by support from both ends of the political

> *[T]he ultimate picture here is leaving the market . . . open to evolution . . . and not permitting existing dinosaurs to control evolution to make sure we never develop into mammals.*
>
> —Lessig to the San Francisco Business Times, *November 2001*

spectrum, Lessig filed an appeal with the Supreme Court.

In February 2002, after postponing the decision three times, the Supreme Court agreed to hear *Eldred v. Ashcroft* that October. For the first time the practice of extending copyright was being challenged on constitutional grounds. Lessig spent the summer preparing for his first time presenting a case before the Court he had clerked for nearly a decade earlier. He amassed a wide variety of friend-of-the-court briefs ranging from the cyber-libertarian Electronic Freedom Foundation (of which Lessig is a board member), to the conservative antifeminist activist Phyllis Schlafly, to a half-dozen Nobel Prize–winning economists.

That October, Lessig argued the case. Three months later, on January 15, 2003, the verdict was returned. Seven justices upheld the CTEA, with Justices Breyer and Stevens dissenting. (Breyer's dissent claimed that copyright terms had become so long that they were almost forever.) Once again, in the realm of copyright, Lessig had lost.

## IN THE COMMONS

*Eldred v. Ashcroft* was the latest in a long line of legal decisions—including those in the Napster and DeCSS cases—that seemed to validate the accuracy of Lessig's pessimistic vision of the future of the Internet that he had set forth in his first two books, *Code and Other Laws of Cyberspace* and *The Future of Ideas* (2001). As he predicted, the dinosaurs of the pre-Internet age—embodied by the Recording Industry Association of America and the Motion Picture Association of America—were using copyright to protect themselves against Internet-age innovations like peer-to-peer file-sharing (P2P) and hacked code. Lessig's third book, *Free Culture* (2004), recounted what went wrong in the Eldred case.

In a gesture toward the cultural commons he holds dear, in March 2004, Lessig released *Free Culture* on the Internet free, under a new type of copyright—a Creative Commons Attribution/Noncommercial license. Lessig had cofounded Creative Commons in 2001 as a way for individual creators of music, graphics, text, and other original work to decide their own version of copyright, adding a bevy of work to the public domain, to be copied, reused, or altered by others in the commons.

Creative Commons was just one of Lessig's efforts to put his intellectual convictions into action. As director of Stanford's Center for Internet and Society, founded in April 2000, Lessig is both professor and activist. He has used the cyberlaw clinic, where Stanford law students can work with practicing attorneys on current cases and legislation, to tackle some of the most pertinent issues affecting the Internet. An ongoing project is to teach law students how to get a piece of legislation — The Public Domain Enhancement Act, which would limit copyright to fifty years—through Congress. (It is also known, informally, as The Eldred Act.)

Lessig's influence in the field of cyberlaw is impossible to overstate. Lessig has done some of the most cogent and forward-thinking analyses of the intersection between the 21st-century world of the Internet and the 18th-century world of the United States Constitution. Lessig's status as self-taught programmer, lawyer, scholar, and cyberactivist gives him a unique perspective and grants him cachet in the many worlds that collide on the Web.

## FURTHER READING

### In These Volumes

Related Entries in this Volume: Fanning, Shawn; Gates, Bill

Related Entries in the Chronology Volume: 1998: Department of Justice Files Suit Against Microsoft; 1999: Napster Roils the Music Industry; 1999: DeCSS Creates Storm of Controversy

Related Entries in the Issues Volume: Copyright; Peer-to-peer Networks

### Works By Lawrence Lessig

*Code and Other Laws of Cyberspace.* New York: Basic Books, 1999.

Also available online at: http://www.code-is-law.org/ (cited September 16, 2004).

*The Future of Ideas: The Fate of the Commons in a Connected World,* New York: Random House, 2001.

Also available online: at http://the-future-of-ideas .com/index.shtml (cited September 16, 2004).

"Free Culture: Lawrence Lessig Keynote from OSCON 2002," *O'Reilly Network,* August 15, 2002, http://www.oreillynet.com/pub/a/policy/2002/08/1 5/lessig.html (cited September 16, 2004).

*Free Culture: How Big Media Uses Technology and the Law to Lock Down Culture and Control Creativity.* New York: Penguin Press, 2004.

Also available online at: http://www.free-culture.cc/ (cited September 16, 2004).

"How I Lost the Big One." *Legal Affairs,* March/April 2004, http://www.legalaffairs.org/issues/March-April-2004/story_lessig_marapr04.html (cited September 16, 2004).

### Articles

Bendavid, Naftali. "Lawyers in Microsoft Case Cut Teeth at U. of C." *Chicago Tribune,* January 26, 1998.

Boynton, Robert S. "The Tyranny of Copyright." *New York Times Magazine,* January 25, 2004.

"Copy Fight." *Chicago Tribune,* March 28, 2004.

Leggiere, Phil. "Constitutionalist in Cyberspace." *Pennsylvania Gazette,* November/December 1998, http://www.upenn.edu/gazette/1198/leggiere.html (cited September 16, 2004).

Levine, Daniel S. "One on One with Lawrence Lessig." *San Francisco Business Journal,* November 30, 2001, http://sanfrancisco.bizjournals.com/sanfrancisco /stories/2001/12/03/newscolumn10.html (cited September 16, 2004).

Levy, Steven. "Lawrence Lessig's Supreme Showdown." *Wired,* October 2002, http://www.wired.com /wired/archive/10.10/lessig.html (cited September 16, 2004).

Streitfeld, David. "The Cultural Anarchist vs. the Hollywood Police State." *Los Angeles Times,* September 22, 2002.

Walker, Jesse. "Cyberspace's Legal Visionary." *Reason,* June 2002, http://reason.com/0206/fe.jw .cyberspaces.shtml (cited September 16, 2004).

Zitner, Aaron. "The Net at What Price?" *Boston Globe,* March 19, 2000.

### Websites

Creative Commons. Online home of Creative Commons, cofounded by Lessig in 2001. Site contains links to public domain works for artists, writers, musicians, photographers, filmmakers, and other creative people, all protected by various Creative Commons copyrights, http:// creativecommons.org/ (cited September 16, 2004).

Eldritch Press. The Website of Eric Eldred, who was the lead plaintiff in *Eldred v. Ashcroft,* http://www .eldritchpress.org/ (cited September 16, 2004).

The Eric Eldred Act. Online home of the Public Domain Enhancement Act, which seeks to redress the failure of *Eldred v. Ashcroft*. Contains links to various writings on the Eldred case by Lessig, http://eldred.cc (cited September 16, 2004).

Lawrence Lessig. Lessig's professional site, with links to much of his written work, including columns for the *Industry Standard* and *Wired* and various congressional and legal testimony. Also includes external links to his three books, http://www.lessig.org/ (cited September 16, 2004).

The Microsoft Case. Online home of Lessig's Harvard seminar on the Microsoft case, http://cyber.law.harvard.edu/msdoj/ (cited September 16, 2004).

Stanford Center for Internet and Society. Online home of Stanford's Center for Internet and Society, founded by Lessig in 2000. The site contains links to academic activities as well as ongoing legal cases represented by the Center's Cyberlaw Clinic, http://cyberlaw.stanford.edu/ (cited September 16, 2004).

# J.C.R. Licklider (1915–1990)

## INTERNET VISIONARY

**J.C.R. (Joseph Carl Robnett) Licklider,** a psychologist-cum-computer scientist, sowed the seeds of the Internet in the 1960s. As a one-time director within the Defense Department's Advanced Research Projects Agency (ARPA), Licklider articulated the vision that became the ARPANET, the precursor to the Internet. At a time when computers filled entire rooms and were programmed by punch cards and paper tape, Licklider envisioned seemingly far-fetched computer developments that have become today's reality—desktop computers, graphical user interfaces, digital libraries, and intelligent agents. He was also instrumental in finding the funding for the various research institutions where those inventions were first conceived and developed. Because his visions were often carried out by others, Licklider is sometimes relegated to the shadows of Internet history. But, in fact, his careful planting of ideas into the next generation of ARPA scientists and researchers has earned him a reputation as the "Johnny Appleseed" of computing.

## SON OF A PREACHER MAN

Licklider was born in St. Louis, Missouri, the son of a Baptist minister and his wife. As a young boy, he showed a great aptitude for mechanics and technology, building and rebuilding model planes and cars to see how they worked. After high school, he attended college at St. Louis's Washington State University, where, in 1937, he earned three bachelor's degrees—in physics, math, and psychology. In 1938, Licklider earned a master's degree in psychology from Washington State University.

Licklider specialized in the field of psychoacoustics—the psychological study of how the human mind turns vibrations in the air into intelligible sound. After earning his master's degree, he pursued a doctorate at the University of Rochester in New York. There,

he mapped parts of the auditory cortex and studied pitch perception. For his doctoral thesis, Licklider investigated the neural activity associated with different types and volumes of sound in the brains of cats.

In 1942, having earned his doctorate, Licklider joined Harvard's Psychoacoustics Laboratory as a research associate. The U.S. Air Force had turned to the lab at the beginning of America's involvement in World War II to tackle the problem of radio signal distortion and noise in warplanes; pilots were having extreme difficulty hearing radio communications amid the clamor in the cockpit. Licklider devised an ingenious way to distort radio transmissions so that consonants were transmitted more prominently than vowels, which helped pilots pick out words. While at Harvard, Licklider also performed related research in "clipped speech," investigating the intelligibility of speech made with limited sound waves.

Once the wartime work ended, Licklider continued to do research and lecture in Harvard's psychology and statistics departments. By the spring of 1948, he joined a weekly gathering in Cambridge led by Norbert Weiner, the brilliant MIT mathematician who coined the term *cybernetics,* the study of complex systems in man and machine. At these weekly meetings, Licklider met with some leading scientists and mathematicians—many of them stationed just down the road from Harvard at the Massachusetts Institute of Technology (MIT)—to discuss ways in which systems of control and communication found in human physiology and psychology could be relevant to machines. The discussions drew Licklider closer to what would become a new religion for him—computing.

## REAPING THE WHIRLWIND

In 1950, Licklider left Harvard for an associate professorship at MIT. Although he was initially hired to develop a psychology program linked to MIT's electrical engineering department, soon his time was split between developing a psychology program, building an acoustic lab, and working on various federally funded military projects through MIT's Lincoln Laboratory.

One of the most noteworthy Lincoln Lab projects was SAGE (Semi-Automatic Ground Environment). SAGE was a computer-based air-defense system planned for use against Soviet warplanes and bombers in domestic airspace. SAGE was a significant achievement because it linked hundreds of radar stations in a large-scale network, operating in real time. The ideas used in SAGE had grown out of a World War II–era navy project (Project Whirlwind), begun in 1944.

The Whirlwind computer was the first real-time computer. Computers of the 1940s and '50s operated primarily as batch processors, in which mathematical programs were punched on cards and then fed in batches into a computer. Results could sometimes take days to emerge. The Whirlwind, however, computed almost instantaneously, displaying results on a video console. By the time the Whirlwind went online, in April 1951, the navy had dropped the project, but the technological advances were picked up by the air force, which sought to use Whirlwind's real-time technology to create a way to analyze quickly information from radar and other defense systems.

Along with scientists at IBM, scientists at MIT working on SAGE developed a computer based on the Whirlwind. It came to be known as the IBM AN/FSQ-7. (SAGE was a huge military effort, with the software for the SAGE computer being developed across the country in California.) Licklider, the sole psychologist working with a team of mathematicians and electrical engineers, was the human-factors expert for the project and was responsible for analyzing human capabilities and limitations in relation to the computer.

*In not too many years, human brains and computing machines will be coupled together very tightly, and the resulting partnership will think as no human being has ever thought and process data in a way not approached by the information-handling machines we know today.*

—Licklider in "Man-Computer Symbiosis," 1960

The IBM AN/FSQ-7 opened Licklider's eyes to new ways of conceiving the relationship between computers and humans. He began to contemplate how computers could better serve the needs of humans. By 1957, Licklider was eager to work outside the university. When he left MIT for Bolt, Beranek and Newman (BBN), Licklider brought with him a belief in several key concepts that would soon revolutionize computing.

## MAN AND MACHINE

BBN, an acoustical consulting firm founded in 1948 by professors from MIT, welcomed Licklider as its new vice president of psychoacoustics, engineering, psychology, and information systems. This move permitted Licklider to continue his work in psychoacoustics, as well as to develop his interests in digital computing.

At Licklider's urging, BBN purchased an LGP-30 analog computer in 1958, and then, in the spring of 1960, a PDP-1, one of the first commercial real-time, interactive computers with a built-in display screen, developed by the then-obscure Digital Equipment Corporation. Unlike analog batch processors, the PDP ran on paper tape, permitting programs to be stopped, altered, and restarted in the midst of processing. Compared with IBM's AN/FSQ-7, which weighed nearly 300 tons and took up 20,000 square feet of floor space, the PDP was small, barely the size of a two refrigerators. (A later version of the PDP computer became the first minicomputer.)

Licklider's research team programmed the PDP for experiments with educational software, including a language program that Licklider had written, as well as experimental data retrieval and search programs. The PDP so transformed work at BBN that, by the end of 1960, the consulting firm had branched out of acoustical sciences into computing, and, by 1962, had developed the first time-sharing program for the PDP. (BBN would later win the contract to provide computers for the ARPANET.)

The PDP had a profound impact on Licklider as well. (In an interview for the book, *Tools for Thought* (1985), Licklider would call using the PDP-1 "a kind of religious conversion.") While still at MIT, Licklider had begun to realize that the mathematical models he used to analyze pitch perception were becoming almost too complex to compute, even with analog computers. During the spring and summer of 1957, he began to record the amount of time he spent doing various tasks involved in his job—which centered on collecting, sorting, and analyzing data. He quickly figured out that nearly 85 percent of his time was spent "getting into a position to think" and that only a fraction of the remaining time was spent actually thinking about the problem at hand.

Licklider discussed his experience in a March 1960 paper, "Man-Computer Symbiosis," published in the *IRE Transactions of Human Factor in Electronics*. He wrote, "[The experiment] showed that almost all my time was spent on algorithmic things that were no fun, but they were all necessary for the few heuristic things that seemed important. I had this little picture in my mind how we were going to get people and computers really thinking together." Licklider called for computers to play a greater role in people's lives—to function not just as grand calculators but as tools to help humans solve a variety of problems. The goal, he asserted, was "to enable men and computers to cooperate in making decisions and controlling complex situations without inflexible dependence on predetermined programs." He even called for various advances in computing that would not be developed for another forty years, for example, the pen and tablet computing of a PalmPilot and speech recognition programs.

"Man-Computer Symbiosis" has become one of Licklider's hallmark works. With it, he inspired a generation of budding computing scientists, mathematicians, and engineers, many barely in their 20s, to pursue his pioneering vision of computing.

> *In a few years, men will be able to communicate more effectively through a machine than face to face.*
>
> —*Licklider in "The Computer As Communications Device," 1968*

## SOWING THE SEEDS

By the time Licklider's paper had been published, the United States was already two years into its efforts to surpass what the American government feared to be the Soviet Union's technological superiority. Back in February 1958, four months after the Soviet Union launched the *Sputnik* satellite into orbit, President Dwight D. Eisenhower created the Department of Defense's Advanced Research Projects Agency (ARPA), with a mandate to quickly and effectively advance American defense technologies.

From the start, ARPA gathered an interdisciplinary collection of engineers, physicists, mathematicians, and a whole array of other scientists under its wing, granting them the freedom and considerable funding necessary to undertake bold new projects in the name of national defense. In that same spirit, in October 1962, then-director Jack Ruina, impressed by "Man-Computer Symbiosis," hired Licklider to head two ARPA departments—Behavioral Sciences and Command and Control—and charged him with finding new uses for ARPA's behemoth Q-32 computer. Ruina would later say that hiring Licklider was the most significant thing he accomplished at ARPA.

With Ruina's support, Licklider set up a new department within ARPA, the Information Processing Techniques Office (IPTO), which was dedicated to bringing the power of computing into all levels of day-to-day ARPA activities—working toward the broader goal of information processing instead of the narrow military goal of air defense. Before Licklider, computing was primarily relegated to the Command and Control department, where scientists developed computer-simulated war games. From the start, Licklider geared his new department's work to projects that would develop time-sharing programs, real-time and interactive computer displays, early computer graphics, and networking.

With a budget of nearly $10 million, Licklider spent the first several months criss-crossing the country to find the scientists to carry out his vision. Eventually, he cobbled together a loose-knit group of universities and tech-minded think tanks scattered across the country: MIT in Cambridge; three University of California campuses (Berkeley, Santa Barbara, and Los Angeles); Carnegie-Mellon University in Pittsburgh; the University of Utah; System Development Corporation and Rand Corporation in Los Angeles; and Stanford Research Institute (SRI) in Menlo Park. Through IPTO, Licklider allotted these institutions huge budgets, 30 to 40 times the average, to undertake long-term research that would help achieve his lofty goal of man-computer symbiosis.

It is impossible to overstate the effect that Licklider's vision, and the funding he put behind it, has had on computing. Licklider funded Project MAC (Multi-Access Computer) at MIT—a large-scale computer time-sharing project that also drew from MIT's Artificial Intelligence department—and several other time-sharing projects at SRI and UC–Berkeley, which opened the door for the personal computing revolution. (Project MAC eventually developed Multics, a time-sharing operating system that was a progenitor of the UNIX operating system.) Licklider also funneled monies to Douglas Engelbart, a then-unknown scientist at SRI, whose ideas, set forth in "The Augmentation of Human Intellect," were neatly in line with what had become known as the "Licklider vision" at ARPA. The funding helped Engelbart and fellow researchers develop such vital advances in computing as the mouse, onscreen window technology, and one of the first working hypertext programs, which Engelbart called NLS (oN-Line System). In addition, Licklider and IPTO funding provided the foundation for the fledgling academic field of computer science. In a 2000 interview in *MIT Technology Review*, a former chair of Carnegie-Mellon's computer science department explained, "Lots of very smart people made a career decision to go into a field [computer science] that didn't exist yet, simply because ARPA was pouring money into it."

## THE INTERGALACTIC NETWORK

The success of ARPA's early time-sharing experiments, particularly Project MAC, introduced Licklider to an entirely new type of computing. With time-sharing, many different users could interact with a computer simultaneously. Eventually an informal community of users would grow up around each multiaccess computer. Licklider began to envision various interconnected time-sharing networks, in which users could share ideas, information, and even computer programs, regardless of proximity. In Licklider's mind, the network could reach across the galaxy itself.

In a memo sent in April 1963, Licklider jokingly addressed his group of university labs and think tanks as "Members and Affiliates of the Intergalactic Computer Network"—but his message was serious. He urged them to contemplate standardization in order to help computers operate in a then-hypothetical integrated network. He imagined a whirlwind of computers, large and small, being used simultaneously across the nation, and hypothesized about software that would exist only on the network itself—not unlike modern-day Java applets. "It will possibly turn out that only on rare occasions do most or all of the computers in the overall system operate together in an integrated network," Licklider wrote in the memo. "It seems to me important, nevertheless, to develop a capability for integrated network operation."

Many consider Licklider's memo the first written description of what has become the Internet. However, Licklider left ARPA before his dream could be achieved. After a standard two-year term, in 1964 Licklider passed the IPTO torch to Ivan Sutherland, a brilliant twenty-six-year-old computer scientist from MIT's Lincoln Laboratory. Sutherland then hired Robert Taylor, who soon succeeded Sutherland as director of IPTO. In February 1966, Taylor started the process that would make Licklider's Intergalactic Network a reality with the advent of the ARPANET Project.

## AFTER ARPA

In his two years as director of IPTO, Licklider laid the very foundation for the following decades of computing.

As Sutherland, Taylor, and others at ARPA and its various research institutions carried out the plans Licklider set in place, Licklider went to work for IBM's Thomas J. Watson Research Center in New York. At the time, IBM was still rooted in batch-processors and mainframes—computing dinosaurs in Licklider's mind. Frustrated after just two years, Licklider left and returned to MIT as director of Project MAC in 1968. Though he had founded the lab with IPTO funding, his time there was also quite frustrating. Licklider's distaste for bureaucracy and paperwork made him an indifferent administrator; two years into his directorship, a faction of Project MAC broke off to form MIT's first artificial intelligence lab (which later became the Laboratory for Computer Science). His struggles as a manager nearly ruined his contemporary reputation as one of the leading figures in computer research; by 1971, he had stepped down and returned to lecturing at MIT.

Although he struggled managerially, Licklider continued to put forth his ideas on the future of computing. In his book, *Libraries of the Future* (1965), Licklider described what has come to be known as electronic publishing and predicted the coming of digital libraries by the turn of the 21st century. In his seminal paper, "The Computer as Communications Device," written with former IPTO Director Robert Taylor and published in the April 1968 issue of *Science and Technology,* Licklider envisioned the growth of online communities, bringing up issues of security and the possibility of a digital divide. He also coined the term *Netizen,* a word most people would not even understand for nearly thirty years.

By the time Licklider returned to IPTO for a second tour in 1974, the federal government was not as taken with his vision of the future. Licklider came back to a culture very different from the one that had existed a decade earlier; whereas the early 1960s were marked by the freedom to explore new ground in computing, the mid-1970s were bureaucratic and driven by a need for results more than ideas. Licklider, out of sync, did not thrive. In 1975, he returned to lecture at MIT, where he stayed until he retired in 1986.

In the late 1980s, just as many of his computing predictions were breaking into the mainstream, Licklider's health began to fail. He suffered from Parkinson's disease and prostate cancer as well as severe asthma. Shortly after receiving the Common Wealth Award for Distinguished Service—one of the few distinctions he earned for his landmark work in computing—Licklider died from complications related to an asthma attack on June 26, 1990.

A decade after his death, and more than forty years since he first walked through the agency's doors, DARPA (renamed in 1972) is still pursuing Licklider's vision. Indeed, Licklider was so forward-thinking that, in 2003, a DARPA manager working on a recent cognitive computing initiative, told *T&D,* a trade magazine,

"If DARPA succeeds in this strategic thrust just now getting under way, then in another 10 to 20 years much of Licklider's vision may finally be realized." Nevertheless, a great deal of Licklider's vision has been achieved. As Robert Taylor told M. Mitchell Waldrop, Licklider's biographer, "Most of the significant advances in computer technology were simply extrapolations of Lick's vision. He was really the father of it all."

## FURTHER READING

### In These Volumes

Related Entries in this Volume: Taylor, Robert

Related Entries in the Chronology Volume: 1953: The SAGE System; 1958: The Advanced Research Projects Agency Begins Operation; 1963: Memo to the Intergalactic Computer Network

Related Entries in the Issues Volume: Digital Libraries

### Works By J.C.R. Licklider

"Man-Computer Symbiosis." *IRE Transactions of Human Factor in Electronics,* March 1960.

*Libraries of the Future.* Cambridge, MA: MIT Press, 1965.

"The Computer as Communications Device." *Science and Technology,* April 1968.

### Books

Hafner, Katie. *Where Wizards Stay Up Late: The Origins of the Internet.* New York: Simon & Schuster, 1996.

Hauben, Michael, and Ronda Hauben. *Netizens: On the History and Impact of Usenet and the Internet.* Los Alamitos, CA: IEEE Computer Society Press, 1997.

Also available online at: http://www.columbia.edu /~rh120/ (cited September 16, 2004).

Norberg, Arthur, and Judy O'Neill. *Transforming Computer Technology: Information Processing for the Pentagon, 1962–1986.* Baltimore: Johns Hopkins University Press, 1996.

Rheingold, Howard. *Tools for Thought.* New York: Simon & Schuster, 1985.

Also available online at: http://www.rheingold .com/texts/tft/7.html (cited September 16, 2004).

Waldrop, M. Mitchell. *The Dream Machine: J.C.R. Licklider and the Revolution That Made Computing Personal.* New York: Viking, 2001.

### Articles

Buchan, James. "Nice Guy Behind the Net Dreamt of Happy Digital Age." *New York Observer,* September 17, 2001.

Hauben, Ronda. "Licklider's Legacy: The Vision, The Program, The Achievements: Is Licklider's Vision Still Viable?," http://www.columbia.edu/~rh120 /other/misc/lick001.txt (cited September 16, 2004).

Waldrop, M. Mitchell. "Computing's Johnny Appleseed." *MIT's Technology Review,* January 1, 2000.

———. "No, This Man Invented The Internet." *Forbes ASAP,* November 27, 2000.

### Websites

KurzweilAI.net: Big Thinkers—J.C.R. Licklider. Short bio, plus links to Licklider's most seminal works— "Man-Computer Symbiosis," "The Computer As Communications Device," and his "Intergalactic" memo, http://www.kurzweilai.net/bios/frame .html?main=/bios/bio0145.html (cited September 16, 2004).

# Pattie Maes (1961–)

## PIONEER OF SOFTWARE AGENTS

**Pattie Maes broke new** ground in the field of intelligent agents, an aspect of technological research and development that focuses on building practical, proactive, and adaptive software that operates independently on behalf of the user—think e-secretary or e-butler. Since joining the faculty of the Massachusetts Institute of Technology (MIT) in the early 1990s, Maes has spearheaded several breakthrough projects, including the first collaborative filtering applications and the first prototypes of simple, autonomous agents. She channeled some of her work into two successful commercial ventures: Firefly Networks and Open Ratings, each of which have helped transform e-commerce and online culture itself. In addition to her work as a cutting-edge researcher and entrepreneur, Maes is also one of the most fervent evangelists for her field, called upon to explain and defend the role of agents in the technological future.

## FROM BRUSSELS TO BOSTON

Maes (pronounced "mahs") got her start in technology at Vrije Universiteit Brussel, in her native country, Belgium. Although her initial academic interests ranged from architecture and business to technology and art, she narrowed her focus to the field of computer science, in which she earned both a bachelor's degree in 1983 and a Ph.D. in 1987. For much of her academic career, she focused on artificial intelligence (AI). Indeed, her doctoral thesis, "Computational Reflection," covered how the human concepts of self-reflection and reasoning can be programmed.

During an academic conference in 1987, Maes came to the attention of Rodney Brooks, director of the Artificial Intelligence Laboratory at MIT. As Brooks told *Wired* in 1997, "[Maes] touched on a whole range of issues that I had been struggling with in behavior-based AI—having it be a more dynamic, fluid system. . . . She

had a way to make the system learn. It was very exciting. It was very new." Impressed, Brooks invited Maes to come to MIT's AI lab and work alongside him and Marvin Minsky, another visionary in the field and cofounder, in the 1950s, of the AI lab.

Maes could hardly refuse. Brooks was leading a new wave of thinking in the field of AI. While much of AI research centered on creating intelligence through the writing of fundamental rules and programming for every possible scenario, Brooks championed what some called the "bottom-up" approach. His teams developed simple programs and learning algorithms—strings of mathematical equations used to find and compare patterns—designed to let intelligence evolve as the programs interacted. Although the programs could not be said to actually think, they did react and learn from their environment. Early critics decried his work as lowbrow, but Brooks's robots could maneuver within and manipulate their physical surroundings more quickly and more easily than their "top-down" peers. The pragmatism of Brooks's work drew Maes to his lab.

When Maes arrived at MIT, in 1989, she began working in the Mobile Robots Group, known familiarly as the "Mobots Lab." There, she programmed Ghengis, the lab's first robot with legs, to teach itself to walk. (Genghis now resides at the Smithsonian Air and Space Museum.) Such cutting-edge work earned her an invitation to stay at MIT, first as a research scientist, then as a tenure-track professor. As she told the *Boston Globe,* "I nearly fell over . . . I was 27. I thought, 'My God, I'm going to be a professor at MIT.'" (She officially left her position as senior research scientist at the Belgian National Science Foundation in 1991.)

In 1990 Maes also began working with MIT's interdisciplinary Media Lab, where, under the direction of the lab's founder, Nicholas Negroponte, she conducted research in the realm of human-computer interaction. In 1991, she was named an assistant professor at the

Media Lab. Initially, she developed Silas T. Dog, a two-dimensional, virtual pet that responded and learned through digital interaction with humans. Soon, however, she concentrated on intelligence agents. She founded the lab's Software Agents Group in 1992.

## TALK TO MY AGENT

The concept of computerized agents has been an aspect of AI research since the late 1950s. AI researchers John McCarthy and Oliver Selfridge coined the term *agent* with the meaning of an autonomous program that can carry out tasks without direct human supervision. In the 1970s, Negroponte popularized the idea of agents as "alter-egos" and "butlers." (Indeed, the Media Lab's first symposium on agents, held in 1991, was emceed by an actor playing Jeeves, the English butler.)

By the time Maes had begun to explore the topic as a doctoral candidate in the mid-1980s, intelligence agents had been a long-talked-about but little-seen phenomenon. Most research scientists had pursued the development of knowledge-based agents—programs based on copious amounts of preprogrammed information. At MIT, Maes took a different tack. "I thought something must be wrong about the way in which we approached the problems," she said to the technology journal *Mass High Tech* in 1992. "So I started building learning agents instead of knowledge-based agents." She described her work as IA, for intelligence augmentation, instead of AI.

Like Brooks's robots, Maes's early agents were based on simple algorithmic programs that performed basic tasks. Her research rested on the idea that pattern recognition is one of the simplest forms of intelligence. If a computer program could recognize patterns in a user's actions, Maes believed, it could learn to automate those actions. More developed agents could shed old knowledge and acquire new knowledge; they could, in other words, evolve.

Maes's first prototype was a scheduling agent. The program observed when Maes did and did not accept appointments and by detecting patterns in Maes's behavior over time, it had learned, for instance, not to

*I got into this field for purely selfish reasons. I'm trying to simplify my life. It's way, way too complex.*

—Maes to the Boston Globe, 1997

schedule any meetings before 9 A.M. Her second prototype, Maxims, was a filtering agent that learned to sort, prioritize, copy, save, forward, and delete email messages according to Maes's emailing habits, which helped her manage the large number of emails she received daily. NewT, a third prototype, applied similar learning and filtering concepts to Usenet newsgroups; it scoured Usenet for relevant information based on Maes's past reading habits, prioritizing some articles and compiling recommended reading lists every half hour.

In 1992, Maes unveiled her agents at the Media Lab's first Interface Agents Symposium to an audience of more than 400 industry leaders, researchers, and engineers from companies such as Apple, IBM, AT&T, Sony, and Lotus. Her prototypes were some of the first practical—and potentially commercial—applications of AI-based agent technology that many of these companies had ever seen.

## THE POWER OF WORD OF MOUTH

The following year, Maes and a team of research assistants began to develop a new type of agent that functioned as "electronic word of mouth." The prototype, known as HOMR (Helpful Online Music Recommendations), and, later, as Ringo, was a music-recommendation program. (Like most of her projects, personal necessity was the mother of invention—since coming to MIT, Maes could not find music she liked on Boston's commercial radio stations.) The program would ask its users to rate various albums. It then compared that information to the ratings of like-minded users and made recommendations based on its analysis. As each user rated more titles, the agent would become more tailored to his or her tastes; the more users, the more accurate the recommendations, overall. The "intelligence" of the system lay in the aggregate wisdom of its users.

Like Maes's other prototypes, Ringo worked by matching patterns, not by understanding anything about music or musical tastes. To manage the enormous quantities of disparate information, Maes pioneered the practice of collaborative filtering. Collaborative filtering agents fill the gaps in their own knowledge by learning from other agents. Learning and communicating with other agents brought agent technology to a new level of intelligence.

Ringo began in July 1994 with twenty users and 575 albums. Over the next week, the prototype attracted nearly 20,000 visitors. "We weren't trying to create a company at all," Maes told the *Boston Globe* in 1997. In less than year, however, they had.

The founding of Firefly Networks, Inc., was somewhat of a fluke. Nicholas Grouf, a Harvard Business School graduate, met David Waxman, one of Maes's Media Lab research assistants, on a plane en route from Boston to San Francisco. The two men chatted about the value of email agents, not unlike Maes's Maxims prototype. When Grouf wondered aloud about the possibilities of software that could prioritize all online data, Waxman told him Maes had software that could do just that.

In March 1995, Grouf, Maes, Waxman, and three other MIT researchers—Max Metral, Upendra Shardanand, and Yezdi Lashkari—founded Agents, Inc., a private company that would market collaborative filtering software and personalization technology. That same year, Maes was promoted to associate professor. After eighteen months, with offices in Cambridge, Massachusetts, New York and San Francisco, more than 100 employees and nearly $20 million in backing, the company was renamed Firefly Networks, Inc., and began licensing its agents to online businesses.

Like Ringo, Firefly examined a user's preferences and found, in the jargon of Firefly software, a user's "nearest neighbors." The word choice was apt, as Firefly software functioned not just as a word-of-mouth tool but also as a community-building tool. Users found their "taste-mates" online. Firefly software was not limited to just music, however. Clients like Barnes & Noble used the software to recommend books, Reuters for news stories, and Yahoo! for Websites.

Firefly software (and its imitators) quickly became ubiquitous, to be found on music sites like CDNow.com, film sites like MovieCritic.com, and even

> *Just as machines extend our control of the physical world, so agents will give us greater control over the world of information, software and networks—the world of the 21st century.*
>
> —Maes to Business Wire, 1996

product-oriented sites like Deja.com. Agent-based technology is something most Internet users have come to expect on e-commerce sites, though most recognize the agent not by name, but by links that read: "People who like this also bought X."

Microsoft acquired Firefly Networks, in 1998, for an undisclosed sum (estimated at $30 million). Firefly.com disappeared on August 18, 1998, subsumed by Microsoft's Passport shopping site.

Just as Firefly.com disappeared, another of Maes's Media Lab–based projects, Frictionless, came online. While Firefly worked well for items that consumers bought on a regular basis—such as books, movies, music—Frictionless software could recommend products that people bought infrequently, expensive items like camcorders, computers, or cars. The Frictionless ValueShopper agent rated products found online according to various aspects of the product, like price, warranty, special features, or delivery time, removing the "friction" from extensive online comparison-shopping. (The idea came after Maes spent weeks researching the purchase of a new car.) After three years of research and fine-tuning at the Media Lab, Frictionless, the private company, quickly grew from its customer-based ValueShopper software to business-to-business sourcing agents. Clients for Frictionless's intelligent purchasing technology were as varied as Lycos and the U.S. Army.

## FIREFLIES IN THE OINTMENT

The success of collaborative filtering stoked the flames of an age-old Internet debate: privacy. To address the issue, Firefly pledged to keep its user profiles confidential, hiring Coopers & Lybrand, an accounting firm, to certify its privacy practices with audits every six months. In addition, Maes led the movement for the Open Profiling Standard (OPS).

OPS is a technical specification, developed first by Firefly and later championed by both Netscape and Microsoft, for handling user profiles and the exchange of personal information on the Internet. The specification provides for "trusted communication (mediated by computer software and communication systems): 1) between people and services, 2) between services mediated by people, and 3) between people." In short, OPS allowed e-consumers to control what kinds of information—particularly personal information—would be given out on the Internet, while allowing the collection of marketing information by e-commerce sites. As of 2003, OPS is just one of several suggested protocols under consideration for the World Wide Web Consortium's Platform for Privacy Preferences (P3P) industry standard.

Maes's agents also drew criticism from those who questioned the validity of automated computer programs making "decisions" about a user's tastes and interests. One notable critic was Jaron Lanier, a computer scientist-cum-Internet philosopher, who argued that intelligence agents might actually make users dumber, not smarter, by making users believe that a computer algorithm was a more accurate predictor of what music or films they would like than their own tastes and whims. In a 1996 online debate with Maes on *Hotwired.com*, Lanier quipped, "Imagine living your life so that a moron could understand your interests."

Other critics claimed that agents provided only a narrow set of suggestions. Many pointed to the way an agent-based newspaper—*The Daily Me,* as some theorists called it—would select news stories based only on a user's past reading habits, perhaps missing major news stories simply because the user had never previously taken an interest in, for example, stock market crashes or international terrorism. Maes countered such arguments by asserting that users still had the power to make their own choices and seek out albums, films, or news stories that the agent had missed. In this sense, agents were just a way to augment a user's ability to seek out and find things that interested them, not to replace it.

## THE DOWNLOAD DIVA

In addition to commercial success, by the late 1990s, Maes was also receiving worldwide acclaim. Within her field, she chaired the first International Conference on Autonomous Agents (Agents '97), which featured cutting-edge agent-based technology from various institu-

tions, including NASA, Stanford University, and Mitsubishi. The World Economic Forum listed Maes as one of its "100 Global Leaders of Tomorrow." The Association for Computing Machinery named her one of the industry's top fifteen visionaries. Her renown had seeped into the mainstream as well. In 1997, *Time Digital* named her one of its 50 "Cyber Elite," and *Newsweek* dubbed her one of the "100 People to Watch in 2000." Even *People* magazine chimed in, including Maes in its 50 Most Beautiful People issue (May 1997).

In many ways, *People* magazine brought Maes the most public attention, but not all of it was positive. The magazine called Maes, a part-time model during college in Belgium, a "download diva" and quoted her saying that MIT was "almost a wasteland in terms of beauty. It's not hard to be the prettiest woman if there aren't any other women. There's an expression that goes, 'In the land of the blind, the one-eyed man is king.'" Students and industry professionals were incensed. Maes claimed she had been misquoted and issued an apology. Maes had initially agreed to be part of *People's* 50 Most Beautiful issue as a way to combat the stereotype of female computer geeks. "There's this image of the female computer scientist not looking very attractive," she told the *Boston Globe* in 1997. "I wanted to show that that wasn't necessarily accurate. People have this natural tendency to stereotype other people. I hope that my appearance in *People* magazine in a way shook some people's preconceptions up. I hope it made some of them think, 'Hey, scientists do look very different from the stereotype belief I held so far.'"

Indeed, Maes felt that her position as a woman in male-dominated labs was somewhat of an asset. "If you're a woman in a field where there aren't as many women, you get more attention, rather than less. Or more attention than men at the same level," she told *Wired.* "So I haven't experienced it to be a negative thing at all. I have never made a professional distinction between men and women. I've never compared myself with other women. I have always compared myself with everybody else."

## BACK TO BUSINESS

In the late 1990s, as e-commerce evolved, Maes turned her efforts to the realm of business-to-business (B2B) applications. In 1999, she founded the e-markets special interest group at the Media Lab, which helped MIT engineers collaborate with members of the Sloan Business School to create new methods of online transactions and to develop a deeper understanding of the new social and economic orders created by online commerce.

Maes initially poured this knowledge into Frictionless, helping the company's shopping-agent technology evolve to fit the purchasing and sourcing needs of B2B companies—the company's current mainstay. In May 1999, she founded another commercial project, Open Ratings, Inc., which combined new e-market knowledge with the ever-evolving collaborative-filtering technology. (The prototype for the company was a semifinalist in the MIT $50,000 business plan competition.) In March 2000, Open Ratings was one of only a dozen start-ups to debut at the PC Forum, one of the industry's most prestigious conferences.

Open Ratings provides neutral, third-party ratings for e-commerce businesses. It was not the first company to do so. Sites like Rateitall.com and Bizrate.com polled volunteer users, while Bizrate.com used questionnaires to create ratings. Open Ratings added statistical data, for example, credit histories and past contract performance, to the standard word-of-mouth ratings and weighted recommendations according to the reputations of both the raters and the rated. Other learning-based agents in the software countered the effects of ballot-stuffing and other fraudulent behavior that skew results in other, less sophisticated ratings systems. Through these mechanisms, Open Ratings can more reliably assess the reputation of each business in its network, which, in turn, can help streamline decisions that affect the supply chain and the general efficiency of any given company. The result, Maes explained to *Business Wire,* is increased trust between online buyers and sellers, which translates into more online transactions at higher dollar amounts. Current clients include such heavy hitters as Xerox and United Airlines.

Through her research, prototypes, commercial applications, and fervent support of intelligence-based agent technology, Maes has forever transformed the relationship between individuals and the Internet. Her pioneering work has become commonplace technology on major e-commerce sites, key tools in both consumer and business-to-business transactions. Indeed, her early prototypes of scheduling assistants and newsgroup filters helped cement the public's ideas of what intelli-

gence agents could do and how they worked. Perhaps her greatest gift to the computing world, however, was to bring a level of pragmatism to the field. "In all of my research I try to strike a difficult balance between doing basic research, coming up with results that are general and can be used by the research community, and building prototypes that are usable, that can inspire others to deploy them in commercial applications," Maes explained in a 1997 interview with *Internet Computing Online.* "I think it's important to try to do both of these things, and never to lose contact with the real world."

## FURTHER READING

### In These Volumes

Related Entries in this Volume: Sherman, Aliza
Related Entries in the Issues Volume: E-commerce; Online Communities; Privacy

### Works By Pattie Maes

ed. *Designing Autonomous Agents: Theory and Practice from Biology to Engineering and Back.* Cambridge, MA: MIT Press, 1991.
and Rodney Brooks, eds. *Artificial Life IV: Proceedings of the Fourth International Workshop on the Synthesis and Simulation of Living Systems.* Cambridge, MA: MIT Press, 1994.
with Maja J. Mataric, Jean-Arcady Meyer, Jordan Pollack, and Stewart W. Wilson, eds. *From Animals to Animats IV: Proceedings of the 4th International Conference on Simulation of Adaptive Behavior.* Cambridge, MA: MIT Press, 1996.

### Articles

Brockman, John. "Intelligence Augmentation: A Talk with Pattie Maes," http://www.edge.org /3rd_culture/maes/ (cited September 16, 2004).
Cobb, Nathan. "Face of the Future: From Software to Stereotypes, Pattie Maes of MIT's Media Lab Is an Agent for Change." *Boston Globe,* September 2, 1997.
Germain, Ellen. "Software's Special Agents." *New Scientist,* April 9, 1994.
Holloway, Marguerite. "Pattie." *Wired* 5.12, December 1997, http://www.wired.com/wired/archive/5.12 /maes.html (cited September 16, 2004).
"Humanizing the Global Computer." *Internet Computing Online,* July–August 1997, http://www .computer.org/internet/v1n4/maes.htm (cited September 16, 2004).
Johnson, Steven. "The Soul Encoded." *Harper's,* September 1998.

Lawrence, Andy. "Agents of the Net." *New Scientist,* July 15, 1995.

Maes, Pattie, and Jaron Lanier. "Brain Tennis: Intelligent Agents=Stupid Humans?" *Hotwired,* July 15–24, 1996, http://hotwired.wired.com /braintennis/96/29/index0a.html (cited September 16, 2004).

McMahon, Timothy E. "Just What Is an Intelligent Agent?" *Information Outlook,* July 1, 2000.

Patton, Phil. "Buy Here, and We'll Tell You What You Like." *New York Times,* September 22, 1999.

Porter, Patrick L. "Pattie Maes Introduces Learning Agents." *Mass High Tech,* November 2, 1992.

Williams, Mark. "Firefly's Pattie Maes Is Still an Agent of Change." *Red Herring,* March 1, 2000.

## Websites

Frictionless Commerce, Inc. Online home of Frictionless Commerce, which Maes founded in 1998, http://www.frictionless.com/ (cited September 16, 2004).

Intelligent Software Agents. Carnegie-Mellon University's Intelligent Software Agent Lab, http://www-2.cs.cmu.edu/~softagents/ (cited September 16, 2004).

MIT: Software Agents Group. Online home of the Media Lab's Software Agents Group, with links to past and present projects, http://agents.media .mit.edu/index.html (cited September 16, 2004).

Open Ratings, Inc. Online home of Open Ratings, Inc., which Maes founded in 1999, http://www .openratings.com (cited September 16, 2004).

Pattie Maes Home Page. Maes's Media Lab home page, with biography, extensive list of publications (with several links), and project information, http://web.media.mit.edu/~pattie/ (cited September 16, 2004).

UMBC Agent Web. The University of Maryland Baltimore County's clearinghouse of information and news about intelligence agents, http://agents.umbc.edu/ (cited September 16, 2004).

# Marshall McLuhan (1911–1980)

## MEDIA THEORIST

**Although Marshall McLuhan,** one of Canada's most influential academics, died before the personal computer revolution began, his prescient theories from the 1960s have come to define the Internet Age. McLuhan's great insight was that all media—from print to the telephone to the black-and-white television of his day—indelibly shape the people who use it. He is best known for succinct sayings, including "the global village," and "the medium is the message," both of which refer to the way technology changes the nature of social relations and society itself. A polarizing figure at the best of times, McLuhan and many of his ideas were dismissed by academics in the 1980s, but a decade later they were again embraced by those trying to explain the cultural changes brought on by the newly wired world. McLuhan was resurrected as the patron saint of the Internet age.

## MIND THE GAP

McLuhan was born in Edmonton, Alberta, in 1911. In 1915, his family moved to Winnipeg, where the young McLuhan spent his elementary and high school years. It was a semirural area, and the McLuhans lived modestly. (McLuhan would later describe himself as a "prairie boy" and "Winni-pigeon.") After being discharged from the army after World War I, McLuhan's father sold insurance and real estate. His mother was an actress and elocutionist. From her, Marshall learned to love wordplay, puns, and language.

As a child, McLuhan was technically inclined and fascinated with radios, so when he had to choose a major at the University of Manitoba, he enrolled in an engineering program. A year later, his love of reading prompted him to leave engineering for a multidisciplinary program with emphases in literature, history, and philosophy. When asked what he wanted to do with his life, McLuhan quoted writer Samuel Johnson, "Outside of being a great and influential man, sir, I have not the faintest idea."

After graduating with a bachelor's degree in 1933, with a University Gold Medal in Arts and Science, McLuhan stayed on at the University of Manitoba another year, earning a master's degree in English. In 1934, upon winning an IODE War Memorial Scholarship, he left Winnipeg for Cambridge, England, to begin postgraduate work. There, McLuhan met professors and theorists who would put him on track to become the leading figure in media studies.

One such was Professor Ivor Armstrong Richards, a psychologist-cum-literary critic who taught a course at Cambridge on the philosophy of rhetoric; another was literary critic F.R. Leavis. Both men—founders of an intellectual movement dubbed the New Criticism—inspired McLuhan to use the tools of literary criticism, such as analyses of imagery, metaphor, rhythm, and language, on popular cultural artifacts, including advertisements, radio shows, and comic strips.

McLuhan brought these tools and insights with him to the United States in 1936, when he began teaching at the University of Wisconsin. McLuhan continued to work on his doctoral thesis, an analysis of the life and work of Thomas Nashe, a little-known 16th-century writer. Teaching a course in Elizabethan rhetoric, McLuhan, at age 25, began to feel a vast cultural gap between him and his students, even though most of them were only five or six years younger than he. "I was confronted with young Americans I was incapable of understanding," McLuhan said in a now-famous 1965 interview with *Playboy* magazine. "I felt an urgent need to study their popular culture in order to get through."

Thus, McLuhan became one of the first academics —and, by the 1960s, one of the most celebrated—to so closely study the effects of American popular culture. He continued to teach, moving to St. Louis University, a Catholic college, in 1937—the same year he converted to Catholicism. Biographers and critics have made much of the influence of Catholicism in McLuhan's life and work. By the mid-1940s, McLuhan, now married, moved back to Canada, first to teach in Windsor, Ontario, and then to settle in Toronto, where he would live the rest of his life. He joined the faculty of the University of Toronto in 1946. There, he met Harold Innis, a Canadian political economist, who would broaden McLuhan's theories about popular culture and mass media to include their effects on social organization and society.

## POPULAR MCLUHANISM

McLuhan's formative experiences at the University of Wisconsin–Madison and with Innis at the University of Toronto laid the groundwork for his first book, *The Mechanical Bride* (1951). Using New Criticism techniques of literary analysis, McLuhan discussed a series of advertisements and their social and psychological effects on society. (The "mechanical bride" of the title refers to the marriage of sex and technology in advertising.)

At the time, McLuhan was still an unknown, obscure forty-year-old professor of literature from Toronto. Although the book was little noticed by the mainstream, it did earn McLuhan a $40,000 grant and the opportunity to lead a Ford Foundation seminar on culture and communications. He used part of the grant to create a periodical, *Explorations,* published irregularly from 1953 through 1957, which laid out some of the major tenets of what quickly became known as McLuhanism. At the heart of McLuhanism was the idea that all media act as extensions or amplifications of the human nervous system (printed matter, an amplification of the eye; radio, an amplification of the ear), that these amplifications alter human consciousness, and that by altering consciousness on a grand scale, they transform society at large.

By the late 1950s, McLuhan's work had come to the attention of the U.S. government, and he was named di-

*Today, after more than a century of electric technology, we have extended our central nervous system itself in a global embrace, abolishing both space and time as far as our planet is concerned.*

—McLuhan in
Understanding Media,
1964

rector of the Media Project of the National Association of Education Broadcasters in 1959. The research he conducted during this one-year appointment would later form the basis another book, *Understanding Media* (1964). But first, he would return to Toronto and write *The Gutenberg Galaxy* (1962).

In *The Gutenberg Galaxy,* McLuhan formalized his theories about the changes in consciousness and society brought about by the advent of print culture. It was, as McLuhan himself put it, "a footnote to the work of Harold Innis." Indeed, McLuhan drew heavily on Innis's books from the early 1950s, *Empire and Communications* (1950) and *The Bias of Communication* (1951), but added a psychological dimension. According to McLuhan, the "invention of movable type" by Johannes Gutenberg in the mid-15th century, "forced man to comprehend in a linear, uniform, connected, continuous fashion"—a way of thinking that had not previously existed, when manuscript culture and the oral tradition that preceded it defined communications. McLuhan argued that these changes in consciousness, particularly the effect of isolation that reading created, opened the door for powerful cultural forces such as nationalism.

McLuhan also discussed in *The Gutenberg Galaxy* his concept of *the global village,* a term he had originally coined in 1959. The global village, McLuhan posited, was the vast collective space that emerged as electronic media broke down the normal physical and temporal barriers associated with print and oral cultures. Thus, with the advent of television, for instance, the nationalism of print media would be supplanted by the globalism of electronic media.

*The Gutenberg Galaxy* won the Canadian Governor-General's Award for nonfiction in 1964 and brought McLuhan some public recognition. By then, he had founded the University of Toronto's Center for Culture and Technology—a lofty name for what was, for many years, nothing more than his office and printed stationery. Still, his ideas were attracting notice, particularly once his next book, *Understanding Media: The Extensions of Man* (1964), was published.

*Understanding Media* began where *The Gutenberg Galaxy* left off. In it, McLuhan applied his understanding of the societal shifts caused by the advent of print culture to the age of television. He developed the notion of media temperature—whether a particular medium was "hot" or "cool." "Hot" media were high definition, featuring a lot of data—like a photograph, which offers a great deal of visual detail. In contrast, "cool" media were low-definition, requiring the audience to fill in information—like a cartoon, which offers minimal visual detail. He also introduced his now-famous aphorism, "the medium is the message." He wrote, "The message of any medium or technology is the change of scale or pace or pattern that it introduces into human affairs." Thus, the "message" of television was not contained in whatever show one might watch, but in the way television transformed social relations.

As a soundbite, "the medium is the message," made such an impact that, three years later, McLuhan published another book, titled *The Medium Is the Massage* (1967)—relying on his audience to understand the wordplay. *The Medium Is the Massage* (alternately glossed as *The Medium Is the Mass-age*) was a thin volume featuring innovative, almost psychedelic, graphic design. (Thirty years later, *Wired* magazine and various Websites mimicked the look of the book.) Where *Understanding Media* was dense and lengthy, *The Medium Is the Massage* was hip and pithy. It became his first—and only—best-seller.

Taken together, *The Gutenberg Galaxy*, *Understanding Media*, and *The Medium Is the Massage* formed the core of McLuhanism, which had, by the 1960s, become the intellectual movement du jour. By the time *The Medium Is the Massage* was published, McLuhan was being courted by corporations like General Electric and IBM as a sort of cutting-edge media consultant—informing IBM that it was not in the business of manufacturing equipment, but of processing information, and GE that it was not in the business of making light-bulbs, but of moving information. In a 1965 *New York Magazine* profile, Tom Wolfe, the celebrated journalist and author, called McLuhan, "the Freud of our times, the omniscient philosophe, the unshakable dialectician," asking readers, "What if he is *right*? Suppose he *is* what he sounds like—the most important thinker since Newton, Darwin, Freud, Einstein and Pavlov?" *Newsweek* featured McLuhan on its cover in 1967; in 1969, *Playboy* magazine dedicated a full twenty-five pages to a McLuhan interview, dubbing him the "High Priest of Popcult and Metaphysician of Media."

## HOT, COLD, LUKEWARM

By the 1970s, McLuhanism had been distilled into a handful of pithy sayings. Indeed, his ideas had become such a part of the culture that even those McLuhanites who could not endure his books (of which there were many, as his writing was often difficult and abstruse even by academic standards) could regularly employ his notions of "the global village" or "hot" and "cold" media. Still, many got him wrong; even more did not get him at all.

A decade after his meteoric rise from obscure literature professor to pop culture sensation, McLuhan experienced somewhat of a backlash. Fellow academics began to question his work. One, the German writer Hans Magnus Enzensberger, dismissed McLuhan as a "charlatan" who wrote nothing more than "provocative idiocy." Ideas that were celebrated as revolutionary a decade earlier were, by the 1970s, decried as obscure nonsense. In certain circles, McLuhanism got a new name—McLunacy. Other critics wrote him off as a celebrity sellout, citing his cameos on the television show *Laugh In* and in Woody Allen's film, *Annie Hall* (1977), as shameless self-promotion. (Perhaps fittingly, McLuhan, playing himself in the movie, scolds a faux-McLuhanite, saying, "You know nothing of my work.") As television culture lost its novelty and became taken for granted, McLuhan's musings on the electronic age began to seem more and more outdated.

> *Societies have always been shaped more by the nature of the media by which men communicate than by the content of the communication.*
>
> —McLuhan in The Medium Is the Massage, 1967

In 1979, McLuhan suffered a massive stroke. (In 1967, he had had a brain tumor removed.) The stroke left him aphasic, unable to read, write, or even carry on extended conversation. According to some accounts, although he could occasionally muster a laugh or scream, most often he was silent. Forced to leave teaching because of his health, he retired to his home in Toronto, and died on December 31, 1980.

## RESURRECTION

Following his death, McLuhan slipped into relative obscurity. His books went out of print. Lewis Lapham, the editor of *Harper's* magazine, wrote in 1994 that McLuhan was essentially "sent to the attic with the rest of the sensibility . . . that embodied the failed hopes of a discredited decade [the 1960s]." Then came the Internet.

Although the Internet existed as a governmental research project during McLuhan's lifetime, it did not become a mass medium until the mid-1990s, when the World Wide Web and browsers made it widely accessible. McLuhan himself had been only marginally concerned with the effects of computers; his interest had lain in television. Even the personal computing revolution had barely begun at the time of McLuhan's death.

The advent of the first truly new medium since television cast a new light on McLuhan's decades-old assertions about media effects and societal change. McLuhanisms that had seemed obscure and irrelevant in the twilight of the television age suddenly seemed fresh and new with the Internet. Students of McLuhan (by the 1980s, mostly just journalism students) found new, immediate ways to engage with his scholarship. Cultural theorists, working in a context highly informed by pop culture and postmodernism, eagerly re-embraced his work. In an interview for *Wired* magazine's debut issue in 1993, Camille Paglia, a well-known cultural critic, wrote "We all thought, 'This is one of the great prophets of our time.' What's happened to him?"

Indeed, *Wired* itself was a sort of homage to McLuhan. The magazine named McLuhan its patron saint and opened its debut issue with McLuhan's words,

> *We now live in a global village . . . a simultaneous happening.*
>
> —*McLuhan in* The Medium Is the Massage, *1967*

"The medium, or process, of our time—electric technology—is reshaping and restructuring patterns of social interdependence." Kevin Kelly, the magazine's editor in chief, said "Everyone thought that McLuhan was talking about TV, but what he was really talking about was the Internet—two decades before it appeared." Even the look of *Wired* drew from the graphic design of *The Medium Is the Massage*. In 1994, MIT Press issued a thirtieth anniversary edition of McLuhan's *Understanding Media*, with a glowing introduction penned by Lewis Lapham, who said what many were already thinking—that McLuhan made more sense now than he did when he was alive.

Those trying to describe the changes brought about by the Internet returned time and again to McLuhan's "global village." As a metaphor, it explained how Internet technologies were linking the corners of the globe, collapsing time and space even more intensely than television. They spoke of the "retribalization of man," a McLuhanism from the heyday of the hippie movement, which made all the more sense as amorphous groups of like-minded but geographically dispersed individuals gathered on electronic bulletin boards and Websites. McLuhan's idea of *acoustic* space—a term he coined in 1954 to describe pre-Gutenberg man—was reborn as *cyberspace* in the 1990s. The electronically saturated world of the Internet, twenty-four-hour cable, cell phones, and digitization transformed McLuhanism from out-of-date to ahead of its time. As Tom Wolfe wrote, in a 2004 essay for the *Wilson Quarterly,* "The Internet lit McLuhanism up all over again."

The rebirth of McLuhanism in the 1990s was accompanied by a slew of new books on his life and theories, a few of which were posthumously coauthored by McLuhan himself. McLuhanites have made movies and written plays in his honor; admirers in Toronto have worked hard to name buildings and streets in his honor. Though McLuhan is still, nearly twenty-five years after his death, both praised as a genius and derided as an intellectual fraud, his contribution to the scholarship of media effects and popular culture is indisputable. He gave us the language and concepts to adequately describe the cultural changes brought about by the Internet.

*Marshall McLuhan. (Bettmann/Corbis)*

## FURTHER READING

### In These Volumes

Related Entries in the Chronology Volume: 1966 (sidebar): The Medium is the Message; 1976: Richard Dawkins Publishes *The Selfish Gene*

Related Entries in the Issues Volume: Internet Broadcasting; Languages and Cultures; Online Communities

### Works by Marshall McLuhan

*The Mechanical Bride: Folklore of Industrial Man.* New York: Vanguard Press, 1951.

*The Gutenberg Galaxy: The Making of Typographic Man.* Toronto: University of Toronto Press, 1962.

*Understanding Media: The Extensions of Man.* New York: McGraw Hill, 1964.

*Counterblast.* New York: Harcourt Brace, 1969.

*The Interior Landscape: Selected Literary Criticism of Marshall McLuhan, 1943–1962.* Edited by E. McNamara. New York: McGraw Hill, 1969.

*Culture Is Our Business.* New York: McGraw Hill, 1970.

ed., *Selected Poetry of Tennyson.* New York: Rhinehart, 1956.

and Edmund Carpenter, eds. *Explorations in Communications.* Boston: Beacon Press, 1960.

and R. J. Schoek, eds. *Voices of Literature.* 4 vols, New York: Holt, 1964–1970.

with Quentin Fiore. *The Medium Is the Massage: An Inventory of Effects.* New York: Bantam, 1967.

———. *War and Peace in the Global Village: An Inventory of Some of the Current Spastic Situations That Could Be Eliminated by More Feedforeward.* New York: McGraw Hill, 1968.

with Harley Parker. *Through the Vanishing Point: Space in Poetry and Painting.* New York: Harper, 1968.

with Eric McLuhan. *Laws of Media: The New Sciences.* Toronto: University of Toronto Press, 1988.

with Eric McLuhan and Kathy Hutchon. *The City As Classroom.* Agincourt, Ont.: Book Society of Canada, 1977.

with Barrington Nevitt. *Take Today: The Executive As Dropout.* New York: Harcourt Brace, 1972.

with Bruce R. Powers. *The Global Village: Transformations in World Life and Media in the 21st Century.* New York: Oxford University Press, 1989.

with Wilfred Watson. *From Cliché to Archetype.* New York: Viking Press, 1970.

### Books

Crosby, Harry H., and George R. Bond, eds. *The McLuhan Explosion.* New York: American Book Company, 1968.

Day, Barry. *The Message of Marshall McLuhan.* London: Lintas, 1967.

Duffy, Dennis. *Marshall McLuhan.* Toronto: McClelland & Stewart, 1969.

Finkelstein, Sidney Walter. *Sense and Nonsense of McLuhan.* New York: International Publishers, 1968.

Kroker, Arthur. *Technology and the Canadian Mind: Innis/McLuhan/Grant.* New York: St. Martin's Press, 1985.

Levinson, Paul. *Digital McLuhan: A Guide to the Information Millennium.* New York: Routledge, 1999.

Marchand, P. *Marshall McLuhan: The Medium and the Messenger.* New York: Ticknor & Fields, 1989.

McLuhan, Eric, and Frank Zingrone, eds. *Essential McLuhan.* New York: Basic Books, 1995.

Miller, Jonathan. *Marshall McLuhan.* London: Fontana; New York: Viking, 1971.

Rosenthal, Raymond, ed. *McLuhan: Pro and Con.* Funk & Wagnalls, 1968.

Sanderson, F., and F. Macdonald. *Marshall McLuhan: The Man and His Message.* Golden, CO: Fulcrum, 1989.

Stearns, Gerald Emanuel, ed. *McLuhan: Hot and Cool.* New York: Dial Press, 1967.

Theall, Donald F. *The Medium Is the Rear View Mirror: Understanding McLuhan.* Montreal: McGill-Queens University Press, 1971.

Willmott, Glenn. *McLuhan, or Modernism in Reverse.* Toronto: University of Toronto Press, 1996.

## Articles

Adams, James. "Massaging the Medium Man." *Globe and Mail,* October 16, 2003.

Boyle, Harry J. "McLuhan," *Globe and Mail,* January 5, 1981.

Everett-Green, Robert. "Resurrecting the Media Messiah." *Globe and Mail,* July 22, 1995.

Goodheart, Eugene. "Marshall McLuhan Revisited." *Partisan Review,* January 1, 2000.

Kane, Pat. "Saint of the Superhighway." *Independent* (U.K.), April 28, 1997.

McLuhan, Marshall. "The Playboy Interview: Marshall McLuhan." By Eric Norden. *Playboy Magazine* (March 1969), http://www.digitallantern.net/mcluhan/mcluhanplayboy.htm (cited September 16, 2004).

Skinner, David. "McLuhan's World—and Ours." *Public Interest,* January 1, 2000.

Stille, Alexander. "Marshall McLuhan Is Back From the Dustbin of History." *New York Times,* October 14, 2000.

Wolf, Gary. "Channeling McLuhan," *Wired* 4.01, January 1996, http://www.wired.com/wired/archive/4.01/channeling.html (cited September 16, 2004).

———. "The Wisdom of Saint Marshall, the Holy Fool." *Wired* 4.01, January 1996, http://www.wired.com/wired/archive/4.01/saint.marshal.html (cited September 16, 2004).

Wolfe, Tom. "What If He Is Right?" *New York Magazine,* November 1965.

———. "McLuhan's New World." *Wilson Quarterly,* April 1, 2004.

## Websites

The Marshall McLuhan Global Research Network. Online home of a nonprofit clearinghouse and think tank dedicated to McLuhanism, http://www.mcluhan.ca (cited September 16, 2004).

The McLuhan Program in Culture and Technology. Online home of the media studies program founded by McLuhan at the University of Toronto. Site contains links to program information, as well as biographical information on McLuhan, http://www.mcluhan.utoronto.ca (cited September 16, 2004).

The Official Site of Marshall McLuhan. McLuhan's official Website, maintained by the estate of Marshall McLuhan. Includes biographical information, quotes, list of publications, and related links, http://www.marshallmcluhan.com/ (cited September 16, 2004).

# Robert Metcalfe (1946–)

## INVENTOR OF ETHERNET

**Since the days he was a** graduate student working on the ARPANET, the predecessor to the Internet, Robert Metcalfe has worn a number of different hats: engineer, entrepreneur, executive, teacher, journalist, pundit, and, most recently, venture capitalist. In each capacity, Metcalfe has made a name for himself. He is best known for inventing Ethernet, a local area network (LAN) system for connecting computers, while a scientist at Xerox PARC in the 1970s. As founder of 3Com, Metcalfe went on to establish Ethernet as an open networking standard in the 1980s. In the 1990s, Metcalfe left the business world to become a leading information technology columnist for *InfoWorld,* where he famously and mistakenly predicted the collapse of the Internet in 1996. After nearly a decade as one of the most outspoken technology writers in the industry, Metcalfe put aside his other endeavors to explore the world of venture capital. As Metcalfe made his way through the varied worlds of academia, research, business, and journalism, his brainchild, Ethernet, has evolved into the ubiquitous physical networking systems in use today.

## INTERNET FOR DUMMIES

Metcalfe was born in Brooklyn, New York, and grew up in various working-class neighborhoods throughout New York City and Long Island. His father was a technician working in the aerospace industry; his mother worked as a riveter during World War II. Although neither had gone to college, one of their primary goals was to have Robert attend. "I never considered the possibility that I wouldn't go to college," he said in a 1997 interview with the *San Jose Mercury News.*

Metcalfe excelled at school, showing an aptitude for technology early on. By age 10, he claimed in a book report that he wanted to study electrical engineering at the Massachusetts Institute of Technology (MIT)—a dream he achieved less than eight years later. In eighth grade,

for a science project, he built a rudimentary computer of numbered lights, earning him his science teacher's praise and an A. But Metcalfe's first real experience with a computer came during his senior year in high school, when he enrolled in Columbia University's Science Honors Program. Each Saturday for several months, he learned how to write simple programs for an IBM 7894.

In 1964, Metcalfe graduated second in his high school class and then enrolled at MIT. While an undergraduate in the electrical engineering and business programs, he worked various jobs. In his sophomore year, he worked the graveyard shift at Raytheon, the defense technology company, programming on a Univac military computer the size of a refrigerator.

In his senior year, Metcalfe earned the somewhat dubious distinction of being the first person on record to have had a computer stolen from him. It was a $30,000 computer that he had convinced Digital Equipment Corporation (DEC) to lend him. Instead of fining Metcalfe, DEC used the theft for publicity, proclaiming that DEC had the first computer in history small enough to be stolen. (The computer was discovered a year later in the basement of a fraternity at MIT.)

In 1969, Metcalfe graduated with a degree in electrical engineering and another degree from MIT's Sloan School of Management; he then enrolled in the master's program in applied mathematics at Harvard University. When he finished his degree the following year, Metcalfe stayed on to pursue a doctorate in computer science. He still spent much of his time at MIT, however, as a researcher on Project MAC, an early computer time-sharing experiment at MIT's Lincoln Laboratory, a project funded by the Information Processing Techniques Office (IPTO) of the Defense Department's Advanced Research Projects Agency (ARPA). While with Project MAC, Metcalfe first worked on the ARPANET—the government's computer networking experiment—and met the men who would help transform the ARPANET into the Internet.

One of those was Robert Kahn, then a scientist at Bolt, Beranek and Newman, a computer think tank in Cambridge, Massachusetts. In 1972, Kahn was charged with the daunting task of organizing the first public demonstration of the ARPANET, which was to be held at the International Conference on Computer Communication in Washington, D.C., in October. Kahn invited Metcalfe to write the introductory pamphlet for the conference.

Metcalfe's tract, titled "Scenarios for the ARPANET," gave basic instructions for using the ARPANET, outlined nineteen different ways to use the ARPANET, and listed additional resources. (Metcalfe jokingly refers to it as "the first *Internet for Dummies*.") Kahn was so impressed with the pamphlet that he chose Metcalfe to take ten AT&T officials on a virtual tour of the ARPANET at the conference. In the middle of Metcalfe's demonstration, the system crashed. (It was the only crash of the entire conference.) The AT&T officials, who arrived already skeptical about the packet-switching technology behind computer networking, seemed almost pleased that the ARPANET had failed—it confirmed their belief in the circuit-switching technology that AT&T had always used. "This was my life's work, my crusade," Metcalfe recalled to *Wired* in 1998. "And these guys were happy that it didn't work." Nevertheless, the conference was deemed a success, and the population of ARPANET users grew sharply in the following months.

Back at Harvard, Metcalfe was nearly finished with his doctoral studies. He had already accepted a position at Xerox PARC, the research facility in Palo Alto, California, founded by Robert Taylor, himself an alumnus of the ARPANET project. But, at the last moment, Metcalfe's dissertation was not accepted, his professors finding it "not theoretical enough." Irate, Metcalfe called Taylor to break the news that he would not be able to take the job. Taylor suggested that Metcalfe come to California anyway and finish the dissertation at PARC.

## INTO THE ETHER

From the beginning, Metcalfe's dissertation had focused on the ALOHAnet, a radio-based data communications system used in Hawaii—a project that had also been funded by IPTO. Like the ARPANET, ALOHAnet used packet-switching technology to route data through the network—in this case, through radio waves connecting four stations throughout the Hawaiian Islands.

Metcalfe had learned about the system when he read a 1970 computer conference paper written by Norman Abramson, a computer scientist at the University of Hawaii who first developed the network. In the paper, Abramson pointed out that the ALOHAnet was rather inefficient. Messages could be sent at any time, but if packets from different messages collided on a particular radio frequency, neither message would go through and each sender would have to retry. (The missing packets were said to be "lost in the ether.") With low usage, the simplicity of the ALOHAnet worked fine, but when pushed to a mere 17 percent of its operating capacity, the network faltered. Metcalfe's dissertation had centered on the idea that, by using queuing theory, a form of advanced mathematics, the percentage could be increased to 90.

In 1973, Metcalfe submitted his revised dissertation, titled "Packet Communication," and received his doctorate in computer science from Harvard. Meanwhile, at Xerox PARC, he had been given the task of networking the lab's new Altos, the first personal computer. Metcalfe was to develop one of the first office network environments, connecting the Altos with one another to enable file sharing, with the ARPANET to enable data communications, and with another vital PARC development, the then-cutting-edge laser printer. "I was lucky enough to be presented with the problem: How do you network a building full of personal computers?" he told *Computer Reseller News* in 2003. "In 1973, there were no buildings full of personal computers . . . I was the first person in the history of the world to be given that problem."

Drawing from the knowledge he had gained from working on the ALOHAnet, Metcalfe believed the answer lay in a simple, elegant network that used coaxial cables instead of radio waves. He initially called it the Alto Aloha Network.

Work on the network began simply. David Boggs, a Stanford graduate student who eventually worked in

> **Ethernet is the on-ramp for the Internet.**
>
> —*Metcalfe to* CIO Magazine, *December 1999/January 2000*

tandem with Metcalfe on the project, told *Wired* in 1998 about the first time he saw Metcalfe, in the basement of PARC. "I was assembling a Data General Nova," Boggs explained, "and this guy shows up with an armload of coaxial cable, sits down with a soldering iron, and starts sending signals down the coax to see what would happen." Soon, the two men were working side by side on the project, night and day.

On May 22, 1973, Metcalfe sent a memo to the lab about his and Boggs's work, entitled "Alto Ethernet." (Ethernet referred to the 19th-century idea of "luminiferous ether," a mythical substance once believed to carry light and electromagnetic waves through space.) Six months later, on November 11, 1973, the system was up and running.

Metcalfe's Ethernet ran on a kilometer of coaxial cable that connected the entire network. Metcalfe designed a basic set of communication protocols to enable data to flow through the network quickly and easily. The protocol was known as CSMA/CD (Carrier Sense Multiple Access with Collision Detection). *Carrier sense* referred to rules that made computers wait for pauses in the communications stream before sending their own messages, not unlike polite guests at a dinner party. *Multiple access* indicated that all computers in the network were equal, with no one computer taking priority. *Collision detection* was a complex algorithm Metcalfe had developed that would allow computers to detect message collisions, back off, and resend the message at a random interval to avoid further collision.

The original Ethernet had a bandwidth of 2.94 Mbps—roughly 300,000 characters per second. "We thought that was overkill," Metcalfe said in a 2003 interview with *InfoWorld*. Three years later, however, Metcalfe and company had built a 100Mbps Ethernet hub. Over the years, Ethernet has continued to morph and evolve—from a network built on coaxial cables to ones composed of copper wires, optical fiber, and, nearly thirty years later, to wireless networks, and from 10Mb to 100Mb to 1Gb to 10Gb.

*PARC was heaven on earth. . . . There was a lot of esprit de corps, and we were well supported by Xerox, so we had no worries. And my good luck was leading to the invention of Ethernet.*

—*Metcalfe,* Computer Reseller News, *December 2003*

After working for several years on refining Ethernet, in 1976 Metcalfe moved to Xerox's System Development Division, where he helped pave the way for the Xerox Star workstation, the first commercial PC with a graphical user interface and networking capabilities. (The Star helped inspire Apple Computer's Lisa and Macintosh computers.) However, Metcalfe left before Star's debut. In 1977, U.S. patent no. 4063220 was issued for Ethernet, and by the late 1970s, Metcalfe had begun to see an opportunity to turn Ethernet into a business.

## THE THREE CS

In 1979, Metcalfe came up with the idea to make Ethernet an open networking standard. Almost immediately, Gordon Bell, an executive at Digital Equipment Corporation, the minicomputer maker, joined the effort. Then Intel, a computer-chip maker came on board, as did Metcalfe's former colleagues at Xerox. "None of us planned to make money selling the network," Metcalfe explained to *CIO Magazine* in 1999, "so making it an open standard was not a big risk." IBM, however, felt otherwise, and declined Metcalfe's proposition.

Metcalfe, however, believed the open standard would create a new market for Ethernet products; so he left Xerox and founded 3Com, naming the company for the three things he believed to be vital to the coming age of personal computing: computers, communication, and compatibility. "I wasn't the first one to come up with that concept," Metcalfe explained in a 1993 *Red Herring* interview. "But I was the first one to peddle the three ideas: one, that LANs were a good idea; two, they should be standard; and three, the standard should be Ethernet. A triple sell, which was pretty hard at the time." Indeed, Metcalfe has often said that founding 3Com was something he did "against all advice."

One reason for the early difficulties at 3Com—the personal computing revolution had not begun. In the early 1980s, other Silicon Valley companies had utilized Ethernet for various applications, but the PC market did not yet exist. "That made us look bad for the first couple of years," Metcalfe explained to *Computer Reseller News*. Indeed, by 3Com's second year, it seemed that the company might go under.

Fortunes reversed when IBM introduced its PC in 1981, however. By September of 1982, 3Com had its first commercial product, EtherLink, an Ethernet network adapter card. As PCs grew in popularity, Ethernet really began to take off. (In 1982, the Institute of Electrical and Electronics Engineers appointed the Ethernet an official technology standard.) By 1984, 3Com had gone public.

Still, the company faced serious competition, most notably from IBM itself. After declining Metcalfe's 1980 proposal that IBM help establish Ethernet as a networking standard, IBM chose to adopt and promote a different networking technology, Token Ring. "IBM was the Microsoft of its day then, so Ethernet had to go up against that," Metcalfe explained in a 1999 *CIO Magazine* interview. Although Token Ring technology was mathematically superior, it emerged five years after Ethernet had entered the market. "It was the first big battle [IBM] lost," said Metcalfe.

During the early years of 3Com Metcalfe formulated what technology journalist George Gilder later dubbed "Metcalfe's Law"—roughly, that the value of a network grows as the square of the number of users. In a telephone network, for example, one phone is useless. Two phones increase the value of the network slightly, but thousands of phones make the telephone network very useful.

In his early promotions of the Ethernet, Metcalfe brought with him a black-and-white slide with a graph on it. One line on the graph represented the cost of adding individuals to the network. It grew steadily at a 45-degree angle. The other line represented the value of the network as individuals were added. It grew in a quadratic curve. Once the two lines crossed, he told potential customers, the value of the network grew exponentially compared to the cost. Metcalfe also argued that with a technological standard like Ethernet, the network would quickly grow to the point where it would be irresistible.

Metcalfe's tenure at 3Com was turbulent. In 1982, the company's board of directors, which Metcalfe himself had recruited, asked Metcalfe to step down as chief executive officer, demoting him to vice president of sales and marketing. "The hardest thing I did was not to throw a blind fit and slam the door and storm out in anger and rage over their stupidity," Metcalfe explained in a 1997 interview with the *San Jose Mercury News*. "I stayed because . . . I had chosen them for this job. These were the smartest people around, and I had been very competent in recruiting these people, and now they were giving me exactly the advice that I had chosen them to give. How could I ignore them?"

The board's decision proved to be on target. Over the next two years, Metcalfe brought 3Com's sales revenue from zero to $1 million per month. He cites these gains as one of his proudest accomplishments. "There's a lot of young engineers that I meet who think that I made my fortune because I invented Ethernet," he told the *Mercury News*. "That's a slight missing of the point. I made my fortune by *selling* Ethernet, which is different."

In his decade at 3Com, Metcalfe went on to hold a variety of positions, from chief executive officer to chief technical officer, and from vice president of sales and marketing to vice president of engineering. In 1990, however, Metcalfe was once again passed over for the position of president of the company he had founded. (Eric Benhamou, formerly vice president of product operation, was appointed president.) Metcalfe retired from 3Com on June 4, 1990.

## THE MAN OF LETTERS

After leaving 3Com, Metcalfe spent a year as a visiting fellow at Wolfson College, Oxford University, in England. (Metcalfe's teaching experience included eight years as an associate professor of electrical engineering at Stanford University, where he taught a course on distributed computing part-time while working at PARC.) Uninspired by academia, in March 1992 he returned to the United States and became the publisher and chief

*The Internet is to business what North America was to the Europeans in 1492. It's virgin territory, and it's lush.*

—Metcalfe to Inc. Magazine, December 1996

executive officer of the *InfoWorld Publishing Company*, part of International Data Group (IDG). (IDG publishes *ComputerWorld, The Industry Standard*, and more than 300 other magazines, in addition to producing the "For Dummies" series.) "I'd been trying to get a column in *InfoWorld*, and then-editor Stewart Alsop wouldn't give me one," Metcalfe said in a 1998 *Wired* interview. "But they did need a publisher, so I did that for two and a half years, with great success. . . . I got to write a publisher's column. So being publisher was just my way of getting a column in *InfoWorld*."

Each week for eight years, Metcalfe produced a 685-word column for *InfoWorld*, dubbed "From the Ether." In it, he debunked various up-and-coming technologies and lauded others, and often made passionate predictions. Most notably, in December 1995, he wrote what he calls a "largely tongue-in-cheek" column about the impending, supernova-like collapse of the Internet. In the following months, he honed his argument, citing thirteen- and nineteen-hour outages at Netcom and America Online, respectively, in August 1996. "This is just the beginning," he warned, claiming that a gigalapse—an Internet blackout spanning a billion person-hours—was on its way. His claims made international news, with newspapers proclaiming the Internet's imminent death.

The gigalapse never came, and Metcalfe later ate his words—literally and figuratively—at the WWW6 conference in April 1997, putting his column into a blender with some water, then eating the pulp with a spoon in front of 1,000 people. Metcalfe's longtime friend Michael Dertouzos, director of MIT's Laboratory for Computer Science, called the stunt one of Metcalfe's "marvelous shenanigans." At the WWW8 conference in 1999, in response to further heckling from the audience, Metcalfe good-naturedly ate another bite of one of his columns, as a "down payment" for the following year.

At the end of his eight-year stint as a publisher-cum-pundit, Metcalfe had garnered nearly 650,000 readers. In his final column, published on September 25, 2000, he recounted his history of computer crusades, from time-shared computing in the 1960s, to Ethernet and personal computing in the 1980s, to the most recent turn-of-the-21st-century ideas to have caught his eye, such as the pay-as-you-go Internet and the concept of "anticiparallelism," programming computers to antici-

pate various tasks before they are actually needed. And with that, he bid readers adieu.

## A NEW VENTURE

By the time his last column had hit the newsstands, Metcalfe had relocated from Silicon Valley to a 146-acre hilltop farm in Lincolnville, Maine, where he, his wife, and two children run a nonprofit farm dedicated to saving nearly extinct breeds of domesticated animals, including pigmy goats and spotted pigs. Metcalfe traveled between their rural quarters (dubbed Kelmscott Farm, after a 19th-century British farm) and the Boston suburb of Waltham, Massachusetts, where, in 2001, he joined Polaris Venture Partners, a venture capital firm investing in early-stage technology companies. Metcalfe's focus has been Boston-area information technology start-ups, like Ember, which makes wireless networking products. Since joining Polaris, he has tried to curb his tendency to make bold predictions. In a 2001 interview with *Venture Capital Journal*, Metcalfe joked, "The last thing in the world we need is another know-it-all venture capitalist, and I don't plan to be one of those."

As Metcalfe settled into his new role as a venture capitalist, he witnessed Ethernet's thirtieth birthday, which was marked by numerous celebrations and media coverage. His invention had become the most widely used LAN in the world. In 2003 alone, 184 million new Ethernet connections were shipped, costing nearly $12.5 billion. Most attributed Ethernet's success to its simplicity, interoperability, and widespread adoption. Although the technology itself has changed, many aspects have stayed the same. In fact, the current wireless networking standard draws from Metcalfe's early CSMA/CD standards. As Metcalfe told the *Wall Street Journal* in 2003, "We've come all the way back."

## FURTHER READING

### In These Volumes

Related Entries in this Volume: Kahn, Robert
Related Entries in the Chronology Volume: 1970: ALOHAnet; 1970: Xerox Palo Alto Research Center; 1972: ARPANET's Public Debut; 1973: Robert Metcalfe Outlines Ethernet Specifications; 1990 (sidebar): Battle of the Networked Stars

## Works By Robert Metcalfe

*Packet Communication.* San Jose, CA: Peer-to-peer Communications, 1996.

"Bob Metcalfe on What's Wrong with the Internet: It's the Economy, Stupid." *Internet Computing Online,* March-April 1997, http://www.computer.org/internet/v1/metcalfe9702.htm (cited September 16, 2004).

*Internet Collapses and Other InfoWorld Punditry.* Foster City, CA: IDG Books Worldwide, 2000.

with Peter J. Denning. *Beyond Calculation: The Next Fifty Years of Computing.* New York: Copernicus, 1997.

## Books

Burg, Urs von. *The Triumph of the Ethernet: Technological Communities and the Battle for the LAN Standard.* Stanford, CA: Stanford University Press, 2001.

## Articles

Blom, Eric. "Network Inventor Still Shakes Computer World from His Lincolnville Farm." *Portland Press Herald,* June 22, 1997.

Ferriss, Lloyd. "Seventeen Breeds Apart." *Portland Press Herald,* September 8, 1996.

Fest, Paul. "30 Years of Ethernet Gain." *CNET News.com,* May 21, 2003, http://news.com.com/2008-1082-1008450.html (cited September 16, 2004).

Gilder, George. "Metcalfe's Law and Legacy." *Forbes,* September 13, 1993.

Kirsner, Scott. "The Legend of Bob Metcalfe." *Wired,* November 1998, http://www.wired.com/wired/archive/6.11/metcalfe.html (cited September 16, 2004).

McCreary, Lew. "Going with the Flow." *CIO Magazine,* November 15, 2000, http://www.cio.com/archive/111500_flow.html (cited September 16, 2004).

Perkins, Anthony B. "Networking with Bob Metcalfe." *Red Herring,* November 1, 1994, http://www.redherring.com/Article.aspx?a=8759# (cited September 16, 2004).

Wells, Edward O. "Nothing But Net." *Inc. Magazine,* December 1996, http://www.inc.com/incmagazine/archives/12960421.html (cited September 16, 2004).

## Websites

Charles Spurgeon's Ethernet Web Site. An extensive Website dedicated to the Ethernet, includes an image of Metcalfe's original sketch of the Ethernet from 1973, http://www.ethermanage.com/ethernet/ethernet.html (cited September 16, 2004).

"From the Ether." A collection of Metcalfe's *InfoWorld* columns, including the infamous December 4, 1995, piece in which he predicts the Internet's collapse, http://www.infoworld.com/cgi-bin/displayNew.pl?/metcalfe/bmlist.htm (cited September 16, 2004).

Kelmscott Rare Breeds Foundation. Online home of the Kelmscott Rare Breed Foundation, founded by Metcalfe and his wife and housed on their Maine farm, http://www.kelmscott.org/ (cited September 16, 2004).

Monticello Memoirs: "The Internet After the Fad." Transcript of Metcalfe's May 30, 1996, speech on his role in the digital revolution. The Monticello Memoirs series is part of the Computerworld Smithsonian Program, http://americanhistory.si.edu/csr/comphist/montic/metcalfe.htm (cited September 16, 2004).

Polaris Ventures. Online home of Polaris Ventures, where Metcalfe has been a general partner since 2001, http://www.polarisventures.com/ (cited September 16, 2004).

The Revolutionaries: Bob Metcalfe. Transcript of a 1997 interview with Metcalfe, produced in collaboration with the *San Jose Mercury News,* http://www.thetech.org/revolutionaries/metcalfe/ (cited September 16, 2004).

3Com. Online home of 3Com, the company Metcalfe founded in 1979 and left in 1990, http://www.3com.com (cited September 16, 2004).

# Kevin Mitnick (1963–)

## HACKER

In the 1990s, Kevin Mitnick was America's most wanted cyberspace fugitive. Over the course of his career, Mitnick hacked into many prominent companies, including Digital Equipment, Novell, Nokia, and Sun Microsystems, as well as into national telephone systems, the California Department of Motor Vehicles, and, allegedly, several credit-reporting agencies. The ensuing manhunt transformed Mitnick into a national media sensation. Where law enforcement officials saw a threat, however, many hackers saw a martyr, arguing that Mitnick was a scapegoat for mainstream paranoia about computer security. After serving nearly five years in prison, Mitnick joined the wave of notorious hackers who reinvented themselves as security professionals. His company, Defensive Thinking, founded in 2003, works to help companies prevent the very hacking Mitnick once engaged in.

## PART OF THE GANG

Kevin Mitnick was just fifteen years old—a sophomore at Monroe High School—when he joined the Roscoe Gang, a loosely knit group of teenaged "phone phreaks" in suburban Los Angeles in 1978. Phreaking, or hacking into the telephone system's switchboards, had been an underground activity for nearly a decade. By the time Mitnick learned the ropes, budding computer hobbyists had begun to use personal computers and modems to do far more than make free long-distance calls—raising phreaking to the level of computer hacking. The Roscoe Gang mainly played phone pranks, like changing someone's home phone service to payphone status or talking over directory assistance. But, over time, the stakes grew higher, from eavesdropping to computer break-ins to stealing. The group met at a pizza parlor in Hollywood and communicated on computer bulletin boards. Mitnick fit right in.

Mitnick was a bright but troubled kid, the only child of Shelly Jaffe and Alan Mitnick, who divorced when Kevin was a toddler. He grew up with Jaffe, a waitress, in a lower-middle class San Fernando Valley suburb, channeling his prodigious mental energies into hobbies like magic or ham radios. By high school, he was displaying signs of the mischievousness and bravado shared by many hackers. In one instance he took control of the drive-thru intercom of a fast food restaurant and berated a driver for eating like a slob. When he first met the leader of the Roscoe Gang, Mitnick was harassing another ham radio operator on air.

By 1980 Mitnick had successfully broken into U.S. Leasing's computer network. Without a computer of his own, he borrowed time on any available terminal, from the one at the Radio Shack where he worked to student computer labs at local colleges. In May 1980, Mitnick was caught hacking from a UCLA computer lab, though he was not charged.

In a far bolder move, in late May of 1981, Kevin and two Roscoe Gang members talked their way past security and into the inner rooms of a Pacific Bell telephone center in downtown Los Angeles. Inside, they stole operating manuals for COSMOS, Pacific Bell's main database system, as well as passwords and lock combinations. As a final touch, in a Pacific Bell manager's Rolodex they planted their pseudonyms and a phone number that rang in a coffee shop in Van Nuys. (Mitnick usually went by "Condor," after *Three Days of the Condor*, a book and film about an ex-CIA agent who outfoxes his foes by manipulating the telephone system.) Such audacity eventually led to their arrest.

Police had already been tipped off by the Pacific Bell manager who found the pseudonyms when, later that year, one of the few female members of the Roscoe Gang turned the boys in. Mitnick, just seventeen, was sentenced to ninety days in a juvenile detention center and one year probation. It was his first arrest, but far from his last.

## CATCH ME IF YOU CAN

Over the next decade, Mitnick continued to evade authorities. In 1982 he was caught at a University of Southern California student computer center, allegedly trying to hack into ARPANET, the predecessor of the Internet; some accounts claim that he was breaking into Pentagon computers, which Mitnick denies. Although the university dropped the charges, the Los Angeles police department pursued the case and Mitnick spent six months in the California Youth Authority's Karl Holton Training School, a juvenile prison in Stockton, for breaking probation.

Stories of Mitnick's hacking continued to spread. According to various accounts, Mitnick's parole officer had her phone service cut off and her records erased from the telephone company database. Allegedly, a judge who presided over Mitnick's case found her credit report inexplicably altered. In 1984 authorities suspected Mitnick of hacking into credit-reporting agencies from his workplace. The subsequent warrant for his arrest then disappeared from police records. By the mid-1980s, he was suspected of having illegally logged in to the National Security Administration's public computer system, Dockmaster, and to have hacked into the Jet Propulsion Lab in Los Angeles. In 1987 police finally arrested Mitnick for hacking into Santa Cruz Operation, a software company. He pled guilty to a misdemeanor and was sentenced that December to three years probation.

According to fellow hacker Lenny DiCicco, throughout the 1980s he and Mitnick compromised more than fifty telephone company switches throughout the United States, from California to New York, at one point controlling all switches providing phone service to Manhattan. By compromising switches, they were able to hack into networks, slipping in and out almost imperceptibly; authorities were usually unable to trace them. Indeed, DiCicco and Mitnick had been hacking

*"Hacker" is a term of honor and respect.*

—*Mitnick to* The Guardian, *February 2000*

into Digital Equipment's network for nearly two years before Mitnick was arrested in 1988.

Mitnick was undone in the Digital Equipment case by his lack of discipline, not sloppy hacking. Neither the hardware nor the software went wrong, but the wetware—computer slang for humans—imploded. Mitnick and DiCicco worked together to hack into Digital Equipment's Easynet network by night; by day Mitnick, posing as a government agent, harassed DiCicco by calling DiCicco's boss and claiming that DiCicco was in trouble with the Internal Revenue Service. DiCicco cracked under the pressure and confessed his hacking activities to his boss, who alerted the FBI. When authorities seized Mitnick in the parking lot of the Calabasas offices from which he and DiCicco had launched their cyberattacks, Mitnick asked DiCicco why he turned him in. DiCicco replied, "Because you're a menace to society."

Mitnick pled guilty to one count of computer fraud and one count of possessing illegal long-distance access codes, agreeing to one year in prison and six months of counseling for "computer addiction" at the Beit T'Shuvah treatment center in Los Angeles. Fear of Mitnick's hacking abilities was so great that the federal judge kept him in solitary confinement for eight months and ruled he would not be allowed access to a telephone. (Mitnick claims that authorities mistakenly believed he could start World War III by manipulating the phone, as portrayed in the 1983 movie, *War Games*.)

Upon his release, Mitnick put vanity plates on his Nissan that read XHACKER and moved to Las Vegas, where he lived out much of his three years of probation. He worked as a low-level computer programmer and seemingly disappeared until 1992. That year, Mitnick returned to the San Fernando Valley after his half-brother died of a drug overdose. He went to work for the Tel Tec Detective Agency. By September of that year, the FBI had issued a warrant for Mitnick's arrest because he had violated probation by illegally hacking into computer database systems from his new job. Just months later, authorities issued another warrant for Mitnick's arrest for defrauding the California Department of Motor Vehicles. When police closed in on him at a Kinko's in

Los Angeles in late 1993, Mitnick fled, slipping through their fingers as he had on so many occasions.

## TAKEDOWN

On Christmas day, 1994, the fans of three computers hummed quietly in Tsutomu Shimomura's beach cottage, just outside of San Diego, California. Shimomura, a computational physicist and respected computer security expert at the San Diego Supercomputer Center, was in San Francisco, preparing to leave for a long skiing vacation in the Sierra Nevadas. Although no sound came from the house, at 2:09 P.M., an attack had begun. Within minutes, a hacker had slipped into Shimomura's system, masquerading as a friendly computer, and began stealing emails, security programs, and cellular phone software from the Supercomputer Center network.

Programmers at the Supercomputer Center were able to block the attack by 6:00 P.M. the next day, but the damage had already been done. The *New York Times* called the attack "sophisticated," reporting that such a technique had never been used before and strongly suggesting that no network was now safe. Shimomura quickly returned to San Diego to piece together what happened and found, in addition to the break-in, taunting voicemail messages in a computer-altered voice. "My technique is the best," said the voice. "Don't you know who I am? Me and my friends, we'll get you. My style is the best. Your technique will be defeated. Your technique is no good." In subsequent interviews, Shimomura called the perpetrators "ankle-biters" and said, "Somebody should teach them some manners." Shimomura decided that that someone was going to be him.

The first clues in the case came from a hidden program, unnoticed by the intruder, that had copied the computer's activities to a file on another computer. From those files, Shimomura could recreate the keystrokes used to invade his computer system:

```
14:09:32 toad.com# finger -l @target
14:10:21 toad.com# finger -l @server
14:10:50 toad.com# finger -l root@server
14:11:07 toad.com# finger -l @x-terminal
14:11:38 toad.com# showmount -e x-terminal
```

```
14:11:49 toad.com# rpcinfo -p x-terminal
14:12:05 toad.com# finger -l root@x-terminal
```

Shimomura used the information to patch the holes in his network, but he still could not trace the attack's origin.

For nearly a month, the case languished. Then, in late January 1995, administrators at the WELL, an on-line community based in the Bay area, notified Bruce Koball, head of a group called Computers, Freedom and Privacy, that he was millions of bytes over his storage limit. It was an odd notice, because Koball rarely used his WELL account. When Koball checked, he found hundreds of files, none of them his. On January 28, Koball read a news article about the break-in on Shimomura's network and immediately contacted authorities—he had recognized Shimomura's name from the files in his account.

By early February, the WELL and the FBI had begun monitoring the WELL to detect any activity by the perpetrator. Shimomura enlisted the help of two colleagues, Julia Menapace and Andrew Gross, all armed with laptops, and headed north to the WELL's headquarters in Sausalito. Gross discovered that, along with the pilfered computer programs, the hacked WELL account contained 20,000 credit card numbers from Netcom, a large Internet provider. Shimomura then headed south to Netcom's San Jose headquarters.

By tracing Netcom log-ins, the various investigation teams were able to narrow the perpetrator's location to Colorado, Minneapolis, or Raleigh, North Carolina. They quickly discovered that the hacker was using a cell phone and a modem to connect to Netcom, and that he had altered a phone company switch between GTE and Sprint in an attempt to conceal his tracks. It became increasingly clear to all involved that the hacker they were looking for was probably Kevin Mitnick. Once the assistant attorney general of San Francisco subpoenaed the phone company records, all indicators pointed to Raleigh.

Shimomura joined investigators in North Carolina. He had fixed the location in an area near the Raleigh-Durham International Airport. Using a cell-phone signal locator, he and the FBI closed in on the Players

> *The weakest link in any security chain is always human.*
>
> —*Mitnick to the Financial Times, October 2002*

Club, a twelve-unit apartment complex in Duraleigh Hill, just a few miles from the airport. After a twenty-four-hour stakeout, authorities finally secured a warrant. At 2:00 A.M. on Wednesday, February 15, the FBI knocked on the door of Room 202, which Mitnick had rented under the name Glenn Thomas Case two weeks earlier. After two years of eluding the FBI, crisscrossing the country using fake names and working menial jobs, Mitnick had finally been caught.

Mitnick and Shimomura did not actually face each other until the prearraignment hearing the next day. Mitnick turned to Shimomura and said, "Tsutomu, I respect your skills." Shimomura simply nodded.

## FREE KEVIN

After his arrest in February 1995, Mitnick pled guilty to one count of illegal possession of long-distance access codes and was sentenced to eight months in prison, in addition to fourteen months for violating probation in 1992. Authorities moved him to the Los Angeles Metropolitan Detention Center in August 1995, where he awaited trial on an additional twenty-five federal counts.

Over the next four-and-a-half years, Mitnick, who was held without bail, languished in prison. Trial hearings were repeatedly rescheduled. Mitnick was prohibited from touching anything more complex than a calculator, although other detainees were able to use computers to research their cases in the prison library. Meanwhile, Shimomura, along with *New York Times* writer John Markoff, secured a six-figure book deal and film rights to the story of Mitnick's capture.

The hacking community was outraged by what they considered to be a violation of Mitnick's civil rights, the government persecution of a fellow hacker, and punishment that far outweighed the crimes. Many agreed with Mitnick's claim that he was set up as a "cyberbogeyman." As Apple Computer cofounder Steve Wozniak would later say, "[Kevin] didn't do anything disastrous. He didn't destroy files or steal any money. He was punished so severely, so unusually, for things he hadn't done."

The burgeoning group of Free Kevin activists and hacktivists distributed information about Mitnick's case, collected donations for a defense fund, and sponsored protests and demonstrations throughout the United States and as far away as Moscow. Friends of the

*Kevin Mitnick. (Mark Powell/Corbis)*

Free Kevin movement hacked into high-profile Websites. In December 1997, a group calling themselves the PANTS/HAGIS Alliance hacked Yahoo!, promising to set off a computer virus in the name of Kevin Mitnick. The message read, "On Christmas Day 1998, the logic bomb part of this virus will become active, wreaking havoc upon the entire planet's networks." The antidote would only be released when Kevin was. The message was a hoax. In September 1998, another group hacked the *New York Times* Website, shutting it down for an entire day. Part of one message read, "Hi John Markoff, this one is for you." Shimomura and Markoff were often targets of the movement, because they were perceived as profiting from Mitnick's story. Mitnick supporters even picketed Miramax Studios over the film version of Shimomura and Markoff's story, *Takedown,* which they claimed was erronous.

On March 16, 1999, in a plea bargain agreement, Mitnick pled guilty to seven of the twenty-five counts of wire and computer fraud and was sentenced to forty-six months. He was released from prison on January 21, 2000—eight months early for good behavior plus time served. While on probation for three additional years, he was prohibited from using a computer modem. On January 21, 2003—his first day of online freedom—the first Website he logged onto was that of his girlfriend, Darci Wood.

## THERE IS NO PATCH FOR STUPIDITY

After regaining his freedom, Mitnick kept busy remaking himself as a bad-hacker-gone-good. He testified before Congress and hopped onto the lecture circuit at computer security conferences all over the world. Until December 2001, he hosted a weekly Los Angeles radio show on KFI AM 640, *The Dark Side of the Internet.* (Still prohibited from touching computers connected with a modem, he watched the Web from behind his show's producers.) He had a literary agent by September 2001. His book, *The Art of Deception,* came out in 2002. It focused on the human element of computer security, in particular Mitnick's personal specialty—social engineering, the art of manipulating people in order to gain access to computer networks.

In 2003, Mitnick followed in the footsteps of famed hackers like John Draper and Kevin Poulsen, and launched his own security-consulting firm, Defensive Thinking. The company's Website features links to his social engineering class at www.intenseschool.com, his world premier security conference, the 360 degree Security Summit, and another high-end conference, Access Denied 2004.

In his speeches, Mitnick continues to focus on the importance of the human element in computer security. As he likes to say, "There is no patch for stupidity." He also likes to point out that everyone is vulnerable—even Mitnick himself. Indeed, his site is popular among hackers. On January 20, 2003, a hacker known as Bug-Bear hacked into defensivethinking.com, leaving the message, "Welcome back to freedom Mr. Mitnick . . . it was fun and easy to break into your box." Weeks later, another hacker from Texas broke into the site and asked Mitnick to hire him as defensivethinking.com's new se-

curity officer. Four separate hackers defaced the site in late February. Mitnick took the hacks in stride, claiming in subsequent interviews that no harm had been brought to the business. (The site had been set up before the end of Mitnick's probation. Once he was allowed to use a networked computer, Mitnick fixed the vulnerabilities.)

Although some remain skeptical, Mitnick seems to have left his hacking days behind. Indeed, a decade since Mitnick was known as the most wanted hacker in America, Mitnick has become a hacking expert of another stripe, called upon by the likes of CNN, CNBC, and Fox News for his comments of the latest security breach, virus, or worm. For instance, in August 2003, calls poured into Mitnick's home for his thoughts on the arrest of Jeffrey Lee Parsons, the high school senior from Minnesota who had unleashed a strain of the Blaster worm, a virulent computer virus that had—in its various strains—infected more than a half million PCs worldwide. When Fox News asked Mitnick how his crimes related to those of Parsons, Mitnick replied, "Well, see, my case and his case is [sic] like apples and oranges." Indeed, the saga of the high school student and his virus software could hardly compare with Mitnick's twenty-five-year journey from adolescent phone phreak to America's most wanted cyberfugitive to budding computer security professional.

## FURTHER READING

### In These Volumes
Related Entries in this Volume: Draper, John T.
Related Entries in the Chronology Volume: 1989: Hacker Kevin Mitnick Is Convicted of Computer Fraud; 1995: Hacker Kevin Mitnick Is Arrested
Related Entries in the Issues Volume: Crime and the Internet; Hackers; Security

### Works by Kevin Mitnick
and William L. Simon. *The Art of Deception: Controlling the Human Element of Security.* Sebastopol, CA: John Wiley & Sons, 2002.

### Books
Goodell, Jeff. *The Cyberthief and the Samurai: The True Story of Kevin Mitnick and the Man Who Hunted Him Down.* New York: Dell Publishing, 1996.
Littman, Jonathan. *The Fugitive Game: Online with Kevin Mitnick.* Boston: Little, Brown, 1996.

Markoff, John, and Katie Hafner. *Cyberpunk: Outlaws and Hackers on the Computer Frontier.* New York: Simon & Schuster, 1991.

Shimomura, Tsutomu, and John Markoff. *Takedown: The Pursuit and Capture of Kevin Mitnick, America's Most Wanted Computer Outlaw—By the Man Who Did It.* New York: Hyperion, 1996.

## Film

*Freedom Downtime* (2002)
Documentary film about the Free Kevin movement.

*Takedown* (2000) (bootleg copies titled, "Hackers 2: Takedown")
Fictional account of Shimomura's cat-and-mouse game with Mitnick.

## Articles

Flower, Joe. "Cracking the System." *New Scientist,* September 1, 1995.

Johnson, John. "A Cyberspace Dragnet Snared Fugitive Hacker." *Los Angeles Times,* February 19, 1995.

"Questions + Answers: Kevin Mitnick." Financial Times, October 31, 2002.

Shimomura, Tsutomu. "Catching Kevin." *Wired* 4.02, April 1996. Also available online at: http://www .wired.com/wired/archive/4.02/catching.html (cited September 16, 2004).

"They Call Me a Criminal." *The Guardian,* February 22, 2000.

## Websites

Defensive Thinking. Mitnick's cyber-security company, http://www.defensivethinking.com/ (cited September 16, 2004). The site contains links to: Kevin Mitnick Social Engineering Workshop, http://www.intenseschool .com (cited September 16, 2004); 360 Security Summit, http://360securitysummit.com /360/index.html (cited September 16, 2004); and Access Denied Conference, http:// 360securitysummit.com/ad04/index.html (cited September 16, 2004).

Free Kevin. Home base of the "Free Kevin" movement, http://www.freekevin.com/ (cited September 16, 2004).

Takedown. Companion Website to the book, *Takedown: The Pursuit and Capture of Kevin Mitnick, America's Most Wanted Computer Outlaw— By the Man Who Did It,* by Tsutomu Shimomura and *New York Times* writer John Markoff, http://www.takedown.com/ (cited September 16, 2004).

# Ted Nelson (1937–)

## INVENTOR OF HYPERTEXT

**In 1997 Ted Nelson** (Theodor Holm Nelson) was described by *Forbes* as "one of the most influential contrarians in the history of the information age." Although best known for inventing hypertext in the mid-1960s, Nelson had earlier made a name for himself by questioning the standards and practices of publishing and word processing. Nelson's lifelong work on developing a global hypertext publishing system opened the door—in a theoretical more than a practical sense—for the development of the World Wide Web. While the Web grew explosively, Nelson's own vision of hypertext on the Internet, dubbed Xanadu, was for more than thirty years one of the most infamous unshipped software products in computer history.

## ANYTHING AND EVERYTHING

Nelson, the estranged son of film director Ralph Nelson and actress Celeste Holm, was raised by his grandparents in Greenwich Village, New York City. A precocious child with attention deficit disorder (ADD), Nelson was, even in his youth, an eccentric outsider. As early as the seventh grade, he developed the four maxims that, according to *Wired,* still govern his life: most people are fools, most authority is malignant, God does not exist, and everything is wrong.

An inveterate polymath, Nelson has spent much of his life reconciling his quest for thorough knowledge with his ADD, a condition that often prevents him from fully finishing one thought before leaping to another. (Nelson prefers to call his condition "hummingbird mind.") As a young adult, Nelson poured his thoughts into various forms of new media. By the time he graduated from Swarthmore College in 1959, with a degree in philosophy, he had produced a full-length record album, written a book, published both a magazine and a newsletter, and created what he calls "the world's first rock musical," aptly titled "Anything and Everything." Some of these endeavors were truly genre-breaking. The magazine, for example, was shaped like a kite and had to be rotated to be read.

After one year of graduate study at the University of Chicago, in 1960 Nelson began a master's program in sociology at Harvard. During his second year at Harvard, he took a course in computers. "[The computer course] was like lightning striking," he recalled in a 2001 interview with *BBC News.* "The heavens rolled apart. This was it—it was obvious, the human race would spend the rest of its career at computer screens."

For Nelson, computers held a unique promise—freedom from the constraints of the linearity that dominated publishing and the written word. In some ways, computers could potentially work more like Nelson's own hummingbird mind. In a 1996 radio interview, he explained, "I particularly minded having to take thoughts which were not intrinsically sequential and somehow put them in a row because print as it appears on the paper, or in handwriting, is sequential. There was always something wrong with that because you were trying to take these thoughts which had a structure, shall we say, a spatial structure all their own, and put them into linear form."

In 1961, for a term project, Nelson toyed with the idea of a text-processing system that allowed a user to revise, undo, and compare documents—years before word-processing programs, like Microsoft Word, existed. He was particularly interested in the ability to link different versions of the same document on the same page, with marks revealing the various iterations of a single idea. The project consumed him.

## DREAMING IN HYPERTEXT

Nelson continued to work on his project long after the computer course was over and well after receiving his master's degree in 1963. Slowly, his vision evolved from a mere text-processing program into something more comprehensive and complicated, something that closely resembled an idea put forth two decades prior by Vannevar Bush, a forward-thinking electrical engineer.

In 1945, Bush, often referred to as the Godfather of the Internet, published an article in the *Atlantic Monthly* entitled, "As We May Think." In the article, Bush proposed a machine called a Memex that could store and retrieve associative links between different aspects of documents using microfilm. Bush called the system of links "associative trails." Nelson—ever ready to coin a new phrase—dubbed his version *hypertext.* To Nelson, hypertext aptly described the system of sideways links that subverted the hierarchical and linear structure of the printed word by allowing the writer or reader to maneuver through a document nonsequentially.

Though the idea of hypertext had simmered in Nelson's mind since the early 1960s, the first recorded instance of the word *hypertext* in print only came on February 3, 1965. *Miscellany News,* a student newspaper at Vassar, published an article on Nelson, who was teaching sociology at the college. In August of that year, Nelson presented his concept of hypertext to the computing industry in a paper given at the Association for Computing Machinery national conference. The paper, "Complex Information Processing: A File Structure for the Complex, the Changing and the Indeterminate," mapped Nelson's ideas of nonsequential writing and information retrieval, as well as another invention that Nelson called "zippered lists." In this theoretical structure, any item, from a word to a paragraph to an entire document, could be linked—or "zipped" together—with related material in another document.

Although some expressed sincere (though fleeting) interest in Nelson's idea, hypertext did not immediately revolutionize publishing, as he had hoped. In 1967, Nelson briefly collaborated with the IBM-funded researchers at Brown University who created the first, very simple implementation of his idea, the hypertext editing system (HES). For his part, Nelson secured the rights to use Vladimir Nabokov's 1962 novel, *Pale Fire*—itself a sort of literary hypertext with its elaborate annotations—to demonstrate the program but was unable to

*Any nitwit can understand computers, and many do.*

*—Nelson in* Computer Lib, *1974*

convince his fellow researchers to use it. He left the collaboration shortly thereafter, working alone for much of the next several years.

With the development of the HES, Nelson's idea was finally beginning to take shape in the real world. Indeed, the idea itself had become more substantial. Nelson had worked out detailed thoughts about a hypertext publishing system that could revolutionize literature and build a universal electronic library. By the late 1960s, he had given it a name—Xanadu, from Samuel T. Coleridge's poem, "Kubla Kahn."

## THE ROAD TO XANADU

The first to believe in Xanadu, aside from Nelson, was a group of teenage hackers from Princeton, New Jersey, who called themselves the R.E.S.I.S.T.O.R.S (Radically Emphatic Students Interested in Science, Technology and Other Research Studies). Soon, with the help of investors, Nelson was also able to recruit adult programmers. By 1972, Nelson, working with two other programmers, had developed the first demonstration of Xanadu software, including a program that moved large portions of text in and out of the computer's memory, which Nelson dubbed the "enfilade." That year, Nelson announced to the International Conference on Online Interactive Computing, "Much of what Bush predicted is possible now. . . . The Memex is here. The trails he spoke of—suitably generalized, and now called hypertexts—may, and should, become the principal publishing form of the future." Nelson may have spoken too soon—a habit that would plague Xanadu. Soon after the program was finished but before he could show it to possible investors, Nelson was forced to return the computer to its owner.

Frustrated by Xanadu's slowed development, Nelson took a hiatus from the project and accepted a teaching position at the University of Illinois–Chicago in 1973. Once there, he began to write. The result, published in 1974 under two titles—*Computer Lib(eration)* and *Dream Machines*—was a compilation of the various topics that consumed Nelson over the previous year, covering everything from art to programming advice to Watergate. The book itself was reminiscent of Nelson's kite-shaped magazine. Read in one direction, it was *Dream Machines,* a tome on computers and the arts that also contained a brief description of Xanadu, which Nelson proclaimed would be released to the public in 1976. Flipped over, it was *Computer Lib,* which called for a personal computing revolution; the cover declared, "You can and must understand computers NOW." Among its radical suggestions was Nelson's argument that individuals should actually own their own computers—an outrageous idea in 1974, when mainframe computers cost millions and occupied entire rooms. *Computer Lib* has since been hailed as a veritable bible for early hackers and computer enthusiasts. The book brought Nelson and Xanadu a new generation of disciples.

The cult success of *Computer Lib* helped revive Nelson's active interest in Xanadu, as did the development of locally networked computers, which struck Nelson as the perfect way to implement a global hypertext program. In the summer of 1979, Nelson, now teaching at Swarthmore College, gathered together a half-dozen programmers, led by one of Xanadu's most staunch supporters, Roger Gregory, to bring the mythic software program to life. The group spent much of this era—dubbed "Swarthmore Summer"—waxing philosophic on the underpinnings of the Xanadu system and developing the cornerstone data processing algorithms of the software, which they called the General Enfilade Theory.

Though the group basically disbanded as the summer of 1979 waned, Gregory and others, including Mark Miller, a young programmer from Yale, continued the long, arduous process of developing computer code that could be used to deliver the promises of Xanadu— a universal electronic library, a global information index, and xanalogical storage, a copyright-cum-royalty system for use of electronic text. (*Xanalogical storage,* which later became known as *transclusion,* was just one of the many terms Nelson coined in the process of developing Xanadu; *docuverse, poswards, negwards, humbers,* and *intertwingularity* are a few of the others.) In the early 1980s, Gregory and Miller developed a way to give each element of a document—words, pictures, sounds—a unique address that included vital information about the author and the creation of the element,

> **I'm mad as hell, and I'm trying to make things right.**
>
> —*Nelson to* The Economist, *December 2000*

using an arcane form of calculus. Miller then went to work for the advanced research department at Datapoint, a computer company in Texas. Nelson soon joined him. The Xanadu team was spread across the United States, somewhat held together by the Xanadu Operating Company (XOC), which Gregory founded in 1983.

By the mid-1980s, work on Xanadu had nearly stalled and Nelson was on the brink of suicide. In 1987, he published a revised version of his 1981 book *Literary Machines,* this time subtitled *The Report On, And Of, Project Xanadu Concerning Word Processing, Electronic Publishing, Hypertext, Thinkertoys, Tomorrow's Intellectual Revolution, and Certain Other Topics Including Knowledge, Education and Freedom.* In the book, Nelson promised Xanadu's release in 1988.

Though Xanadu did not arrive as promised in 1988, the project received a great boost from industry star John Walker, founder of software company Autodesk, who agreed to take Xanadu and the XOC under his wing. Walker stated in a press release, "In 1964, Xanadu was a dream in a single mind. In 1980, it was the shared goal of a small group of brilliant technologists. By 1989, it will be a product. And by 1995, it will begin to change the world."

Autodesk sponsored the development of Xanadu from 1988 to 1992, pouring an estimated $5 million in money and resources into the project. Walker's company owned 80 percent of XOC, while Nelson, who was given the title autodesk fellow, retained the rights to the Xanadu name. This arrangement, dubbed the Silver Agreement, allowed Autodesk to develop the product, while Nelson could build his vision of a global hypertext system using Autodesk technology.

Much of the group from the Swarthmore Summer came back together at Autodesk's Palo Alto offices, a decade after their first attempt at designing and coding the building blocks of Xanadu; Nelson remained in Sausalito, California, at the Autodesk headquarters. The basic structure of the software had not changed, but the resources available to the programmers had. Soon, however, fissures appeared. New programmers on the project rejected the old Xanadu code and chose to start from scratch. As each year came and went without delivery of the promised product, tension grew. In August 1992, Autodesk released Xanadu back to its programmers.

## WRESTLING WITH THE WEB

In the mid-1990s, as Xanadu sought to find yet another home, it faced an unforeseen roadblock—the World Wide Web. In the early 1990s, Tim Berners-Lee, a scientist working at a lab in Switzerland, had developed hypertext markup language (HTML), the basis of the connected hypertext documents that formed the Web. While it, too, foundered in its first years, by 1993, a young computer scientist at the Center for Supercomputing Applications, Mark Andreessen, had developed Mosaic, an easy-to-use graphical Web browser, and the Web began to grow exponentially.

Nelson, watching hypertext catch fire in the form of the Web, was mortified. Indeed, even as much of the world came to embrace the Web, he quickly became its most vehement detractor. The list of his critiques, biting remarks, and bold put-downs is impressive—both in its scope and its intensity. To *BBC News,* Nelson quipped, "[The Web is] massively successful. It is trivially simple. Massively successful like karaoke—anybody can do it." In a radio interview, he called HTML, "FTP with lipstick" and bitingly referred to the Web as "a brilliant simplification."

Nelson's critiques are not without substance. On the Web, users often face broken links, no form of copyright management, and no reliable procedures for reuse of various materials. Many in Nelson's camp believe that Xanadu, had it been realized in time to compete with the Web, would have been a far more comprehensive and complex system.

*Software as we know it has become a nightmare. Something happened on the way to computer liberation.*

*—Nelson to* MIT Technology Review, *September 1998*

## BIG IN JAPAN

By the mid-1990s, few who were familiar with the thirty-year saga of Xanadu still had the patience for Nelson's predictions about its final delivery. In response, Nelson moved to Japan. As he explained in a 1995 *Wired* interview, "[The Japanese have] done their homework. . . . They understand and they listen. They care about the ideas." Supported by Hitachi and Fujitsu, two of the largest electronics companies in Japan, Nelson

founded a twelve-person software lab, dubbed the Hyperlab. In 1996, Nelson added to his already lengthy list of university appointments a professorship in environmental information at Keio University. Finally, Nelson seemed to be receiving the recognition he so greatly desired. Shortly after joining Keio University, Nelson told the *Wall Street Journal,* "I feel appreciated here. . . . You get Brownie points here for persistence."

Nelson, with Xanadu outposts and colleagues in Japan and Australia, began developing aspects of Xanadu for the World Wide Web environment. He also began to redefine Xanadu by its discrete parts, the most commercially viable of which addressed the problem of reuse and copyright that had been left unaddressed by the Web. Nelson and his team developed programs for transcopyright and text transpublishing, which formed the basic structure for reuse of materials, as well as transpayment, an electronic copyright payment plan. Each of these was an original aspect of Xanadu.

In April 1998, Nelson announced his first-ever non-virtual software product, the ZigZag operating system. The following year, two versions of Xanadu's operating code—Udanax Green and Udanax Gold—were released into the public domain as open-source software. (Udanax Gold was still unfinished.) Programmers and computer hobbyists were encouraged to see what they could do with the software that was once supposed to save the world.

Although the promises of Xanadu have fallen short of predictions, Nelson—visionary or madman—and his contributions to modern computing remain invaluable. Nelson's early vision for computers and hypertext paved the way for groundbreaking technology, such as Apple Computer's HyperCard, Lotus Notes, and for the World Wide Web itself. Though the epic saga of Xanadu *the software* ended with a whimper, the power behind Xanadu *the idea* energized several generations of hackers and programmers. In some ways, the promise of Xanadu's dynamic, global hypertext publishing system still exists each time the Internet world grapples with issues of copyright and royalties, and many still yearn for the ideal of a universal library of interlinked information. Nelson, who has continued, since 1998, to market and develop Xanadu-based products, such as Cosmic-book, which offers hypertext with visible connections, appears to be much like Charles Babbage, the 19th-century mathematician who, while regarded as a pioneer in computing, never managed to bring his elaborate computers to fruition.

## FURTHER READING

### In These Volumes

Related Entries in This Volume: Berners-Lee, Tim
Related Entries in the Chronology Volume: 1945: Vannevar Bush Proposes Memex, an Information Machine; 1965: Ted Nelson Coins the Terms *Hypertext* and *Hyperlink;* 1990: The World Wide Web Is Invented; 1991: The World Wide Web Is Developed at CERN
Related Entries in the Issues Volume: Copyright; Usability

### Works By Ted Nelson

"Complex Information Processing: A File Structure for the Complex, the Changing and the Indeterminate." *Proceedings of the 1965 20th National Conference of the Association for Computing Machinery,* 1965.
*The Home Computer Revolution.* South Bend, IN: T. H. Nelson, 1977.
*Computer Lib/Dream Machines.* Chicago: T. H. Nelson, 1974; rev. ed., Redmond, WA: Tempus Books of Microsoft Press, 1987.
*Literary Machines.* Swarthmore, PA: T. H. Nelson, 1981; rev. ed., Sausalito, CA: Mindful Press, 1987.
*The Future of Information.* Tokyo: ASCII, 1997.

### Articles

Ditlea, Steve. "HyperTed." *PC-Computing,* October 1, 1990.
———. "Ted Nelson's Big Step." *MIT Technology Review,* September 1, 1998.
Edwards, Owen. "Ted Nelson." *Forbes,* August 25, 1997.
Hamilton, David P. "Japanese Embrace a Man Too Eccentric for Silicon Valley." *Wall Street Journal,* April 24, 1996.

Loeb, Larry. "Bucking the System with Ted Nelson." www.ibm.com, November 1, 2001, http://www-106.ibm.com/developerworks/web/library/us-bucking/ (cited September 16, 2004).
Logan, Tracey. "Visionary Lays into the Web." *BBC News,* October 8, 2001, http://news.bbc.co.uk/1/hi/sci/tech/1581891.stm (cited on September 16, 2004).
"Orality and Hypertext: An Interview with Ted Nelson." By Jim Whitehead. *KUCI FM Program Guide* 1986, http://www.ics.uci.edu/~ejw/csr/nelson_pg.html (cited September 16, 2004).
Pollack, Andrew. "Two Men, Two Visions of One Computer World, Indivisible." *New York Times,* December 8, 1991.
Smith, Gina. "A New Xanadu." *San Francisco Examiner,* January 23, 1994.
Williams, Sam. "Welcome to the Pleasure Dome." *Upside Today,* August 25, 1999.
Wolf, Gary. "The Curse of Xanadu." *Wired* 3.06, June 1995, http://www.wired.com/wired/archive/3.06/xanadu.html (cited September 16, 2004).

### Websites

Project Xanadu. Original Xanadu site, with links to the Xanadu Archive, project description and product information, http://www.xanadu.com/ (cited September 16, 2004).
Ted Nelson and Xanadu. Entry on Nelson and Xanadu as part of The Electronic Labyrinth, a project involving hypertext and creative writing at the University of Virginia's Institute for Advanced Technology in the Humanities, http://www.iath.virginia.edu/elab/hfl0155.html (cited September 16, 2004).
Ted Nelson Home Page. Nelson's personal home page, with a few links to his writings, http://ted.hyperland.com/ (cited September 16, 2004).
Udanax.com—Enfiladic Hypertext. Home of the two open-source versions of Xanadu software: Udanax Green and Udanax Gold, http://www.udanax.com/ (cited September 16, 2004).
Xanadu Australia. The most comprehensive of the Xanadu Websites, with links to an extensive bibliography and FAQs, http://xanadu.com.au/archive/bibliography.html (cited September 16, 2004).

# Jakob Nielsen (1957–)

## USABILITY EXPERT

**Dr. Jakob Nielsen is one** of the foremost experts in the field of Web usability. A self-described "user advocate," Nielsen has been pushing for clean, simple Web design since his days at Sun Microsystems where his title was Distinguished Engineer. While at Sun, Nielsen began a regular online column, "Alertbox," to help spread the usability gospel. In 1998, he left Sun to found his own usability consulting firm with Donald Norman, a former Apple Computer executive. Over the next five years, the Nielsen Norman Group gathered in such clients as General Electric, UPS, and wsj.com, the *Wall Street Journal's* Website. Nielsen, dubbed by *Fortune* magazine "the reigning guru of Web usability" in 2001, can claim much of the credit for the firm's success. Currently, he holds more than seventy patents in the field of usability and is the author of several related books, including *Hypertext and Hypermedia* (1990) and *Usability Engineering* (1993).

## DANISH MODERN

Nielsen was born to a pair of Danish psychologists and raised in Copenhagen, Denmark. He first learned computer programming while in high school in the early 1970s, using a Danish Gier computer that ran on paper tape and filled an entire room. Though his first computer language was Algol, Nielsen quickly moved on to other languages, such as BASIC.

After graduating from high school, Nielsen attended Denmark's Aarhus University, studying in a combined bachelor's and master's degree program in computer science. During this time he discovered the writings of Ted Nelson. Nelson's books, particularly *Computer Lib* (1974) and *Literary Machines* (1981), had inspired a whole generation of computer enthusiasts. Nielsen, in particular, took to Nelson's vision of a global hypertext system. "Some of his technology ideas were a little different, but his vision—that you could reach out to everybody and bring them online—was incredibly compelling," Nielsen explained in a 2002 interview with the Web site *Chief Officer.* "I give Ted Nelson a lot of credit for just getting me fired up about what could be."

At the same time, Nielsen was extremely frustrated with the state of computing. The typical mainframe computers of the late 1970s and early 1980s were enormous, many taking up entire rooms. (The mainframe at Aarhus University was so large that it was stored in the basement.) In comparison to Nelson's visions of universal electronic libraries and personal computers, Nielsen found the reality of mainframes oppressive and boring. In many ways, he preferred the paper tape of his first Gier, with its typewriter console, to the mainframes of higher education, which he disparaged in a 2001 interview with *eContent* magazine. "They were all big mainframes, and they were used for payroll processing and things like that. So, I decided I could have a career doing payroll processing or I could have a career trying to make computers capable of doing something else."

Nielsen's drive to make computers capable of more than crunching numbers led him, after earning his joint degree in 1983, to a doctoral program in computer-human interaction at the Technical University of Denmark, one of the country's leading institutions in the field of science in engineering. There, he began to explore the little-known field of usability—the study of the user's experience when interacting with machines or systems. "I knew that it *could* feel good to use computers," he later wrote in a 2004 online column, "and I wanted to recapture that sense of empowerment and put humans back in control of the machines."

## DEFINING USABILITY

Usability engineering, as a discipline, first emerged during World War II. When the air force began losing planes not to enemy fire but to pilot failure, the government began studying the ways in which people interacted with machines, with the goal of designing a simpler, safer cockpit that could save lives and help win the war. By the 1950s, scientists had begun to employ similar usability testing on civilian products, the telephone, for example.

In the 1970s, usability research came to computer science. While IBM worked to make mainframes more user-friendly, scientists at the cutting-edge computer science lab at Xerox PARC (Palo Alto Research Center) took usability to another level, designing the Alto, the first personal computer, which was small enough to sit atop a desk, and pioneering the graphical user interface (GUI). Apple Computer began to embrace usability engineering in the early 1980s, bringing users in to test designs for its early computers, including the Apple II and the Macintosh.

Nielsen began his own forays into usability engineering in the early 1980s as well. Usability engineering brought together Nielsen's varied interests, including computers and psychology. Indeed, usability —and the related field of human-computer interaction—is interdisciplinary, drawing from computer science, cognitive science, engineering, and ergonomics.

For his doctoral dissertation, Nielsen focused on the user interface—the part of the computer system that is exposed to the user. In 1988, the year he earned his doctorate in human-computer interaction, he began teaching the university's first course on interface design and usability. The course was popular, according to Nielsen, although few students went on to pursue careers in usability. "When they graduated they all got non-usability jobs," he explained in the *Chief Officer* interview, "because there were not very many [usability] jobs in the 80s." Just as Nielsen began to tire of teaching the course, he received a phone call from Bell Communications Research (Bellcore) in the United

States, inviting him to bring his expertise in usability and user interface to the world of telecommunications research.

## COMING TO AMERICA

In 1990, Bellcore was the leading telecommunication research lab in the country. The lab, located in New Jersey, was founded in 1984, after an antitrust suit forced AT&T to split into various regional "Baby Bells." Still, Bellcore retained the formidable reputation of its predecessor, the Bell Telephone Laboratories, which had pioneered many phone technologies, such as telephone switches, since it was founded in the 1920s. By the time Nielsen arrived, Bellcore was doing world-class research in the interface field.

As a senior research scientist, Nielsen was surrounded by some of the most brilliant minds in his field. But with his long history with computers, Nielsen found Bellcore's allegiance to the "voice" aspect of telecommunications frustrating. In the 1970s, AT&T had turned its back on the new packet-switching technology and the burgeoning data networks spawned by government-funded research on the ARPANET, predecessor of the Internet. Bellcore, likewise, refrained from exploring the terrain of intranets—internal data networks—a topic that piqued Nielsen's interest. Thus, after four years at Bellcore, Nielsen was ready to leave. In 1994, when Sun Microsystems asked him to come to Silicon Valley to be one of its Distinguished Engineers for Strategic Technology at SunSoft, Sun's software division, he readily accepted. He took the job offer as a sign that leading computer companies were finally taking the field of usability seriously.

When Nielsen left Bellcore, he compiled a list of every email address and mailing list he had ever used, then sent out a mass email with his new contact information at Sun to 5,500 or more addressees. As Nielsen's various colleagues were on many of the same mailing lists, some individuals received his email dozens of times. Recalling the incident, some observers jokingly note that this email makes Nielsen the "father of spam."

> *I believe that what constitutes a good site relates to the core basis of human nature and not to the technology, not to fashion.*
>
> —*Nielsen to* Newsweek, *March 2001*

*Jakob Nielsen. (Photo courtesy of Jakob Nielsen)*

## HERE COMES THE SUN

At Sun, Nielsen was responsible for helping the company chart its technological future. Initially, he worked on making Sun's UNIX-based workstations more user-friendly. "Pretty tricky!" he remarked in a 2001 interview with *Internet Magazine*. By 1994, however, the Web browser was beginning to revolutionize the way people used the Internet. "It became clear that Sun needed to focus on the Web," Nielsen told *Internet Magazine* in 2001. "That's easy to say in retrospect, but in those days it was quite a profound thing."

Nielsen had started using the Web early on, in 1991, after witnessing a demonstration by the inventor of the World Wide Web, Tim Berners-Lee. At the time, Nielsen felt the Web was "a minor thing" and complained that the software he downloaded from CERN crashed frequently. But when the Mosaic browser emerged in 1993, Nielsen was impressed, as were scien-

tists at Sun. In a 1998 interview with *art-bin,* a Swedish online magazine, Nielsen said, "Sun was one of the companies with the most smart employees who understood the Internet revolution."

Within months of being hired, Nielsen had gone from Sun's UNIX usability expert to its Web usability expert. His first tasks were to help design the company's intranet—the SunWeb—and Sun's external Website (www.sun.com). Nielsen initially felt that much of the usability advice he offered was ignored, primarily because the culture of Sun—as well as the entire computer industry at the time—was geared toward the goal of *more* features, not necessarily *simpler* features. But the advent of the Web, and the simplicity it required, made his insight indispensable.

Design for SunWeb began in the summer of 1994 and took only weeks to accomplish. To ramp up development, the design team had used Nielsen's concept of *discount usability engineering*—a term Nielsen coined in 1993. Discount usability engineering refers to a quick-and-dirty style of usability testing that utilizes small test samples (only a few test users, most of them experts) and relies almost solely on direct observation (as opposed to statistically generated findings). Results are often incorporated on the fly, which translates into rapid prototyping during the early design phase.

In keeping with his method, Nielsen conducted four usability studies in just over a week, which included an icon recognition test and two rounds of one of his tried-and-true testing methods, the cognitive walk-through of the interface design in which users, one at a time, describe aloud their thoughts and feelings as they manipulate the interface. The testing quickly honed the design team's initial concepts, and SunWeb was soon up and running. Indeed, more than 2,500 different Sun employees had accessed Sun-Web in the month before it was officially launched.

Nielsen was also part of the team that helped Sun develop a magazine metaphor for the 1995 launch of the company's Website. (Most other Websites were menu-based.) As part of Sun's new site, Nielsen began a biweekly column on usability issues, dubbed "Alertbox." (In his first column, published on May 25, 1995, he discussed the usability of Sun's latest developments, Java and HotJava browser.) Even though Sun dropped the magazine-style design a year later, "Alertbox" has continued as an online venue for Nielsen's scathing Website critiques and regular Top 10 lists, ranging from "best intranets" to "worst Web design mistakes." Nielsen now places the readership of "Alertbox" at roughly 9 million page views per year.

Nielsen first earned a widespread reputation as a usability guru at Sun in the mid-1990s. In addition, between 1995 and 1997, he conceived and developed numerous usability methods and apparatuses, filing dozens of patents for things ranging from font sizing to eye-track-driven information retrieval. Still, for most of the early years of Web design, usability experts were, as Nielsen stated in a 2003 interview with *VNUnet* magazine, "roundly booed at Internet conferences and ignored by the prevalent 'killer design' agencies." By the

late 1990s, the unrelenting growth of the Web prompted Nielsen to strike out on his own.

## A GROUP OF HIS OWN

Just as Nielsen began entertaining thoughts of leaving Sun, Donald Norman, then the vice president of research at Apple Computer and another leading voice in the field of usability, suggested that he and Nielsen join forces to form a consulting firm focused solely on Web and software usability. "Don convinced me that usability was getting too big for any one person," Nielsen told *Chief Officer* in 2002.

> *Design is done for a reason. . . . Do it well and your business will prosper; do it badly and it will suffer.*
>
> —*Nielsen to* Sydney Morning Herald, *July 2002*

The two men founded the Nielsen Norman Group (NNG) in August 1998, with the goal of helping companies "enter the age of the consumer" by designing "human-centered products and services." NNG quickly garnered high-profile customers, including the *Wall Street Journal,* Hallmark, and General Electric.

One of the firm's primary tools has been heuristic evaluation—a key part of the discount usability engineering techniques Nielsen pioneered in the early 1990s. Heuristic evaluation involves having a small number of experts examine and evaluate a site's design based on broad guidelines or principles. Nielsen had developed a set of ten general principles for heuristic evaluation in 1990, in collaboration with another Danish usability engineer, Rolf Molich. In 1994, Nielsen, using new factors introduced by the Web, elaborated on his original heuristic evaluation tools.

The guidelines include using common, everyday language, keeping the user informed of the status of the system, making exit functions clear and easy to use, designing clear and helpful error messages, and—most important—employing simple, minimalist design. According to Nielsen, changes made after testing a site using these guidelines can improve the effectiveness of a Website anywhere from 100 to 1,000 percent.

Testing often takes place behind a one-way mirror, with Nielsen or another NNG principal watching individual users perform tasks using a given interface. The user is not prompted to answer specific questions but, instead, is asked to describe the experience, while experts behind the glass take note of how the user manipulates the site, and if and where he or she gets stuck. Nielsen has found that what designers tend to believe are simple tasks—for example, purchasing an item from an e-commerce site—can take a user ten or twenty minutes to figure out because of bad design.

Ironically, when Nielsen performed usability tests on the firm's own Website, he found that he sometimes violated his own rules. In one instance, a user pointed out that the mission statement, "The mission of the Nielsen Norman Group is simple . . ." was followed by a long paragraph of explanation. "I still remember the guy [in a testing session] who said: 'If it's simple, it shouldn't take a whole paragraph to explain it,'" Nielsen recalled in a 2000 *Business Week* interview, "He was absolutely right."

Testing also provides the basis of many of NNG's published reports and white papers, some of which boast astounding statistics. A 2002 NNG report claimed that a company with 10,000 employees could expect a gain of *$5 million a year* in employee productivity just by improving the design of a company's intranet, and that instituting top-quality intranet design globally could save the world economy *$1.3 trillion*. The numbers are not unfounded. The U.S. Defense Finance and Accounting Service (listed by "Alertbox" as one of the Top 10 Government Intranets for 2003) calculated that its usability redesign resulted in a total savings of 200 staff years.

> *I always thought technology was too complicated and too boring and should be something that was immediate and intuitive and exciting.*
>
> —*Nielsen to eContent, June, 2001*

## THE NIELSEN RATINGS

Through the 1990s, as Nielsen gathered ever-more accolades in the press; he was dubbed the "eminent Web usability guru" by CNN and the "usability Pope" by a German magazine. He also began to amass a fair number of detractors. The criticism often came from Web designers, who found Nielsen's stringent guidelines for Web design—no flash animation, no splash pages—stultifying and limiting. Many complained that if Nielsen's guidelines were actually followed, the Web would be a pretty dull place, filled with cookie-cutter, graphics-free, text-driven sites. Others claimed that Nielsen's beliefs stunted creativity. One woman at a 2001 Web conference in Washington, D.C., dubbed Nielsen the "Jerry Falwell of Web design," citing his "fundamentalist" design tastes.

Other critics believe usability is more a matter of opinion than, as Nielsen suggests, a science, and scoff at the thousands of dollars NNG and Nielsen himself charge for design suggestions that seem, to many, rather obvious. In a 2001 opinion column, Lance Concannon, a writer for *Internet Magazine,* argued: "The good Doctor [Nielsen] charges thousands of pounds to speak at conferences and his talks often consist largely of common sense."

Still others point to the startlingly spare appearance of Nielsen's Alertbox Website as proof that his tastes are out of touch with modern Web design. However, when asked about the issue by the British *Guardian* in 1999, Nielsen responded, "It's almost like a provocation that I do it the way I do."

Others, while critical of Nielsen in some areas, nevertheless respect him. In an article for *UsabilityNews.com,* George Olsen claims one of the greatest reasons behind the backlash against Nielsen's brand of usability is envy of Nielsen's enormous financial success and his marked abilities as a self-promoter. "Say what you will," Olsen writes, "[Nielsen has] definitely been giving away his thinking for years, even while others claimed they knew—but wouldn't tell—the proprietary secrets to creating successful user experience."

Indeed, much of Nielsen's wisdom can be found sprinkled among his "Alertbox" columns—which may be a testament to his unflagging belief in the gospel of usability. "Usability is not just a small issue," Nielsen explained in a 2004 interview with *eWeek* magazine. "It's one of the biggest driving factors for really getting our productivity up in the white-collar economy, the service economy, and it's really in many ways the equivalent of what was done in the old days when people were studying productivity on the assembly lines." Nielsen himself has spent more than twenty years refining his study of usability—as a consultant, columnist, speaker, author, and developer of more than seventy patents in the field. As Websites and intranets evolve, Nielsen remains at the forefront of discipline, helping users get the most out of the Web.

## FURTHER READING

### In These Volumes

Related Entries in this Volume: Berners-Lee, Tim; Nelson, Ted

Related Entries in the Chronology Volume: 1970: Xerox Palo Alto Research Center; 1984: Apple Macintosh Debuts; 1990: The World Wide Web Is Invented; 1991: The World Wide Web Is Developed at CERN

Related Entries in the Issues Volume: Usability

### Works By Jakob Nielsen

*Hypertext and Hypermedia.* Boston, MA: Academic Press, 1990.

*Usability Engineering.* Boston, MA: Academic Press, 1993.

*Multimedia and Hypertext: The Internet and Beyond.* Boston, MA: AP Professional, 1995.

*Designing Web Usability: The Practice of Simplicity.* Indianapolis, IN: New Riders, 2000.

"Alertbox: Current Issues in Web Usability," http://www.useit.com/alertbox/ (cited September 16, 2004).

with Marie Tahir. *Homepage Usability: 50 Websites Deconstructed.* Indianapolis, IN: New Riders, 2002.

### Books

Brinck, Tom, Darren Gergle, and Scott Wood. *Usability for the Web: Designing Web Sites that Work.* San Francisco: Morgan Kaufmann, 2001.

### Articles

Cirillo, Rich. "User Advocate or Enemy of Creativity?" *VARBusiness,* February 19, 2001.

Concannon, Lance. "Each to Their Own." *Internet Magazine,* December 1, 2001.

Hamilton, Joan. "Diss My Web Site, Please." *Business Week,* November 20, 2000, http://www.businessweek.com/2000/00_47/b3708076.htm (cited on September 16, 2004).

Helm, Leslie. "Nielsen's Ratings: Where Is Web Technology Taking Us?" *Los Angeles Times,* April 13, 1998.

Hill, Steve. "Use and Usability." *Internet Magazine,* February 1, 2001.

Marer, Eva. "February Q&A: Jakob Nielsen." *Chief Officer,* February 8, 2002, http://www.chiefofficer.com/particle.php?t=0 (cited September 16, 2004).

Mieszkowski, Katharine. "5 Fatal Flaws of Web Design." *Fast Company,* October 1, 1998, http://www.fastcompany.com/magazine/18/fatalflaws.html (cited September 16, 2004).

———. "Usability Makes the Web Click." *Fast Company,* October 18, 1998, http://www.fastcompany.com/magazine/18/usability.html (cited September 16, 2004).

O'Donavan, Cheryl. "Dot Ugh." *Communication World,* June 1, 2001.

Rae-Dupree, Janet. "The Web's Not So Friendly for this Design Virtuoso." *U.S. News & World Report,* December 18, 2000.

Richtel, Matt. "Making Web Sites More 'Usable' Is Former Sun Engineer's Goal." *New York Times,* July 13, 1998.

Tallmo, Karl-Erik. "The Web Is Not a Selling Medium." *The Art Bin* 16, October 26, 1998.

Walker, Leslie. "You'd Think They'd Learn: Bad Design Kills Web Sites." *Washington Post,* October 25, 2001.

Wieners, Brad. "Time for a Redesign: Dr. Jakob Nielsen." *CIO Insight,* June 1, 2004, http://www.cioinsight.com/article2/0,1397,1612183,00.asp (cited September 16, 2004).

### Websites

Nielsen Norman Group. Online home of the Nielsen Norman Group, with links to articles about NNG principals, http://www.nngroup.com/ (cited September 16, 2004).

Usability News.com. Online digest of news stories related to usability, http://www.usabilitynews.com/ (cited September 16, 2004).

useit.com: Jakob Nielsen on Usability and Web Design. Nielsen's personal Website, with extensive links to his "Alertbox" columns, news about Nielsen and usability, his books and other papers and essays, http://www.useit.com/ (cited September 16, 2004).

U.S. Department of Health and Human Services: Usability. U.S. government Website on the subject of usability, http://www.usability.gov/ (cited September 16, 2004).

# Jonathan Postel (1943–1998)

## KEEPER OF THE DOMAIN NAME SYSTEM

**For almost thirty years,** from his days as a graduate student in the late 1960s until his starring role in the Domain Name System (DNS) wars of the late 1990s, Jonathan Postel gently guided the evolution of the Internet. His accomplishments are almost too many to recount: he helped set up the first node of the ARPANET, the military-funded predecessor to the Internet; served as the editor of the Requests for Comments (RFC), the seemingly informal notes that are the basis of Internet architecture; and he also played a key role in the development of early Internet communications protocols, for example, the telnet email protocol. As the leader of the Internet Assigned Numbers Authority (IANA), Postel was, for many years, the sole keeper of the Domain Name System, the system of Web addresses—such as www.iana.org—that organizes the Internet. For much of his life Postel labored in relative obscurity; however, the struggle over the DNS in the late 1990s launched the humble "numbers czar," as Postel was sometimes called, into the spotlight. In 1998, just weeks before his plan to shift responsibility for the DNS from the U.S. government to a private organization was to be implemented, Postel died unexpectedly, at age 55.

## SCOUT'S HONOR

Postel was born in Altadena, California, just outside of Los Angeles. The one-time Eagle Scout attended Van Nuys High School in the San Fernando Valley. Postel was a class behind Vinton Cerf, a man with whom he would later help build the ARPANET.

Postel took his first computer class in junior college in the 1960s. While in a chemistry class, Postel noticed a fellow student working on something that appeared to him to be a jigsaw puzzle—it was a computer program. The following semester, Postel signed up for the class. "They had a computer club so you could submit a program that would run for about a minute at midnight on

an IBM 7094," he told *OnTheInternet* magazine in 1996. "You'd get your output the next day."

From such modest beginnings, Postel continued to pursue his interest in computers. He transferred to the University of California–Los Angeles (UCLA), where he earned a bachelor's degree in engineering in 1966. Two years later, he had completed a master's in engineering; he then enrolled in a doctoral program in the relatively new field of computer science.

Postel studied under UCLA professor Gerald Estrin and University of California–Irvine professor David Farber, and soon began working as a research assistant under professor Leonard Kleinrock—one of the leading theorists on packet switching, a then-new idea in data communications. (Packet switching involves breaking a message into small packets of information that can travel quickly through a network.) At the time, Kleinrock's lab was one of the select academic outposts to fall under the wing of the Pentagon's Advanced Research Projects Agency (ARPA), which had funded research in computer science since the mid-1960s.

Postel began work an ARPA-funded project dubbed the Snuper Computer, which involved using one computer to monitor the performance of another computer, a project headed by Estrin. By the end of 1968, however, ARPA had decided to build the ARPANET—an experimental computer communications network—and designated Kleinrock's UCLA computer lab to become the Network Measurement Center. There, Kleinrock's graduate student assistants, including Postel and other early ARPANET luminaries, including Vinton Cerf and Steve Crocker, would measure and study the performance of the ARPANET network, using some of the advances made by the Snuper Computer project.

In addition, UCLA was to become to the first node of the ARPANET. In late August 1969, the first interface message processor (IMP), a specially designed computer built to manage network traffic, was shipped from the Bolt, Beranek and Newman research facility in Massachusetts to Kleinrock's lab on the UCLA campus. Kleinrock's assistants then began to connect the IMP to the university's Sigma-7 computer. The first connection was made seamlessly—on September 2, 1969, UCLA became the first node of the ARPANET.

Over the next several months, additional nodes were added—at the Stanford Research Institute in October, at the University of California–Santa Barbara in November, and then at the University of Utah in Salt Lake City in December; the initial four-node ARPANET was complete. Graduate students and budding computer scientists began mapping out and testing the network, working out kinks in the technology and writing the basic protocols on which computer networking would be founded.

All the while, Postel was taking notes—logs of experiments gone awry, as well as technical successes and everything in between. In a 1998 memorial for Postel in *Internet Computing Online,* Kleinrock recalled, "It was Jon who took it upon himself to 'just do the right thing' and create a written log of all activities that took place as they occurred." Indeed, it was a habit Postel would keep for years to come.

## TAKING REQUESTS

Beginning in the summer of 1968, as the groundwork for the ARPANET was just being laid, scientists in ARPA-funded labs throughout the country began to come together to discuss their progress. Their meetings were informal and the discussion, initially, was highly theoretical, as work on the cornerstones of the network, the IMPs, had not yet begun. Soon, the scientists came to be known, informally, as the Network Working Group (NWG).

By the second meeting of the NWG, held in Utah in March 1969, Steve Crocker, one of Postel's colleagues at UCLA, decided he should document the process. Crocker saw the documentation as delicate: the notes would contain the ideas and critiques of mere graduate students in their 20s, commenting on the work of established researchers and Pentagon-funded scientists.

*Be liberal in what you accept, and conservative in what you send.*

—Postel in RFC 793 (1981; sometimes called "Postel's Law")

Crocker would later write, "I spent a sleepless night composing humble words for our notes. The basic ground rules were that anyone could say anything and that nothing was official. And to emphasize the point, I labeled the notes 'Request for Comments.'"

The meeting notes, which became known by their initials, and RFC 001, titled "Host Protocol," were circulated in April 1969. Soon after, Crocker handed stewardship of the RFCs over to Postel, who became the official RFC editor.

Despite inauspicious beginnings—Postel at first used a tattered notebook to keep track of the RFCs—the documents quickly became the technical notes on which the Internet is built. As editor, Postel was responsible for assuring that each RFC lived up to his standard, which, for protocols, meant clean and conservative design. He then distributed the RFCs via U.S. mail and maintained the central distribution lists. Once the ARPANET was up and running, distribution began online.

From the initial discussions on host software came some of the most important standards in computing. Postel himself penned various landmark RFCs. In April 1972, he wrote RFC 318, the first explication of the telnet protocol, a remote login program and one of the earliest communications protocols established for the ARPANET. He also wrote RFC 821 (1982), detailing the simple mail transfer protocol (SMTP; a new mail transfer protocol), and RFC 959 (1985), the finalized standard for file transfer protocol (FTP; first proposed in RFC 114 [1971]). In the early 1980s, Postel did vast amounts of work on the Internet protocol suite—RFCs 791–793. Indeed, it was Postel, along with Cerf and Danny Cohen, another scientist, who suggested the Internet protocol (IP)—the routing aspect of the existing transmission control protocol (TCP), which they isolated into a separate-but-related protocol. Many point to the development of—and, later, the shift to—TCP/IP as the true beginning of the Internet. The shift to TCP/IP was precipitated by RFC 801 (1981)—also written by Postel.

In all, Postel authored or coauthored more than 200 RFCs, more than any other individual; he edited nearly 2,500. His proficiency earned him lofty nicknames, like Jon the Protocol Czar—a name often followed by the tagline "Unfailing Arbiter of Good Taste in Protocols." In a 1996 interview with *OnTheInternet,* Postel said of all the many things he had accomplished, he was most proud of his work on the basic protocols of the Internet. "It's fairly satisfying to see that those get a widespread use," he said. "It's really very unusual in the research world that something done as a research project actually gets out there into the world to that extent."

## BECOMING IANA

After earning his doctorate in computer science in 1974, Postel left Los Angeles for Virginia to work for a time at Mitre Corporation, a Defense Department–funded research facility, and then to Stanford Research Institute (SRI), in Menlo Park, California, where he worked alongside Douglas Engelbart on an early hypertext system known as NLS (oNLineSystem). By March 1977, Postel had moved to Marina Del Rey, California, to join the computer networks division of the University of Southern California's Information Sciences Institute (ISI). There he led ARPA-funded projects on networking and provided support on ISI's other computing projects, distributed computing and multimedia teleconferencing among them.

Postel would remain at ISI until his death in 1998, rising to director of the computer networks division in January 1997. In addition to his research, he founded the Los Nettos regional computer network, which still services the Los Angeles area. As Postel explained in the 1996 *OnTheInternet* interview, "[Los Nettos] gives [ISI] a firsthand sort of truth about what's interesting and what the problems are in networking and gives our research a little more context."

While at ISI, Postel also cowrote, with Zaw-Sing Su, the first RFC to suggest structure for organizing computers on the ARPANET—RFC 819 (1982), titled "Domain Naming Convention for Internet User Applications." In it, they described a hierarchical, treelike structure that would enable addresses to be distributed across a network instead of existing on one single computer. (At the time, the ARPANET, 200 or so computers strong, was just beginning to outgrow its centrally administered list of names and addresses, HOSTS.TXT, which was established in 1973.) Postel then asked Paul

Mockapetris, an ISI researcher, to help him develop such a directory, which would link the numerical addresses (for example, 128.125.253.146) to their alphabetical names (usc.edu) and be able to scale up as traffic on the network steadily increased.

The system, which was known as the DNS, for Domain Name System, was first successfully tested at ISI on June 23, 1983. Six months later, Mockapetris and Craig Partridge, a colleague from Bolt, Beranek and Newman in Massachusetts, delivered RFC 882 (1983), which detailed the DNS system. After another year, the NWG had hashed out the details and, in October 1984, with RFC 920, Postel established several top-level domains that are familiar to anyone using a computer today—.com, .edu, .gov, .mil, .org—and opened the door for country-based domains, such as .jp for Japan, or .uk for the United Kingdom (.arpa existed as a temporary domain).

By 1985, with the official transition from NCP, the ARPANET protocol, to TCP/IP, the Internet protocol, use of the network began to explode. DARPA (ARPA was renamed the *Defense* Advanced Research Projects Agency in the 1970s) began to encourage the widespread use of the DNS. The first domain registered was symbolics.com, on March 15 of that year.

All the while, Postel was the record keeper—just as he had been at UCLA. He kept track of the domain name registrants and the corresponding Internet addresses. In the beginning, the budding Internet had 200 to 300 domain name registrants per month, but with the advent of the Web and Web browsers like Mosaic and Netscape, the volume of registrants grew exponentially—in 1998, to more than 3,000 per day. Postel began assigning blocks of IP addresses to Internet registries, allotting new country codes, maintaining America's ten root servers, where domain names and IP addresses are stored, and, in general, making sure the entire DNS system supported such growth. Postel also kept lists of the technical codes involved in Internet protocols. Those, too, began as scribbled numbers on a piece of paper. Eventually, Postel began publishing memos with the new protocol numbers, and, in the mid-1990s, began posting them on the World Wide Web.

These myriad responsibilities fell under what came to be known as the Internet Assigned Numbers Authority (IANA). The "authority" behind IANA was Postel himself, although the Internet Engineering Task Force had final say on Internet standards, and the U.S. government funded the entire effort. Indeed, IANA was an organization with no members, no bank account, no business cards and, in the beginning, no employees other than Postel. (In 1983, he hired Joyce Reynolds to assist with IANA- and RFC-related responsibilities; colleagues at ISI provided support as well.) In a 1997 *Network World* interview, Postel explained, "IANA is just a name we invented in the '80s to describe the work." (While at ISI, Postel also managed the .us domain name, under government contract, but this was separate from his IANA responsibilities.)

Because Postel so single-handedly managed these tasks, many people began to use the acronym IANA and the name Postel interchangeably. At the time, no one questioned the ability of one man to do the work of what would later take a full-fledged organization—perhaps only because that one man was Jonathan Postel.

## GOD

The DNS system worked quietly and efficiently for several years. By the 1990s, however, with the explosive growth of the Internet brought on by the Web, change had begun. In 1993, Network Solutions, Inc. (NSI), a technology consulting firm, won a five-year contract from the National Science Foundation to take over registration of new .com, .net, and .org domain names, under the name InterNIC—moving much of this responsibility away from IANA. IANA was still under government contract to manage the numerical codes and IP addresses on which the Internet ran; Postel, a.k.a. the "numbers czar," continued to allocate blocks of new IP addresses to InterNIC and number the new Internet protocols. (Under separate contracts, Postel also administered the .edu and .us domains.) Although InterNIC drew much of the media attention around domain-name issues, Postel unofficially retained ultimate control over the system.

In September 1995, InterNIC announced that it would charge for domain name registration; with businesses flocking to the Web, it soon became clear that InterNIC's parent company, NSI, had almost instantly acquired a monopoly on Internet real estate. The Internet community grumbled.

In late 1996, Postel proposed a plan for 150 new top-level domains (such as .web, .arts or .info) to be cre-

ated, with competing registries taking charge of them, and suggested that a private, global Internet management group be created in Geneva, Switzerland. That December, he helped found the International Ad-Hoc Committee—which, in 1997, became the Policy Oversight Committee—to address such issues. These changes were prompted by the impending expiration of NSI's contract with the National Science Foundation and, more important, the Clinton administration's growing desire to shift control of the Internet from the U.S. Department of Commerce to an international private organization.

Some critics of the plans for a global DNS management group claimed, in the words of Representative Charles Pickering, that the Internet was "something uniquely American" and should be kept that way, arguing that American taxpayers, companies, and government agencies had built the Internet in the first place. In the heated atmosphere preceding the transition, IANA itself drew criticism. In February 1997, Image Online Design, a San Luis Obispo, California–based company, sued Postel for reneging on a promise to award it the .web registry, claiming that IANA lacked the authority to make such decisions.

Concurrently, Eugene Kashpureff, a Seattle engineer who created his own online registry, AlterNIC, loudly proclaimed that individuals had as much right as InterNIC to register domain names, since the DNS was government-funded public property. That July, Kashpureff rerouted traffic from InterNIC to AlterNIC in an act of protest. (After fleeing to Canada, Kashpureff was later jailed for the stunt.)

After months of fact-finding meetings with engineers and researchers, including Postel, in January 1998, the Department of Commerce issued a "Green Paper," officially calling for a nonprofit organization to take over management of the DNS and general Internet architecture. As outlined by the government, the proposed organization would be responsible for the overall stability of the Internet. In many ways, the organization would be IANA's successor.

The following month, Postel—in what he claimed was a routine test of the network—rerouted information from five root servers from the NSI headquarters in Virginia to his own lab at USC, where he was able to upload the location of every single domain name in the world. The test ran from January 23 through February

4, without the knowledge of the NSI, the Department of Commerce, or any other Internet authority. Postel made headlines, with rampant speculation that he had hijacked the system or was, at the very least, flexing his muscles to show the global community who, exactly, held control of the Internet.

The test, and the ongoing struggle to find IANA's successor, put Postel center stage in the DNS wars—an uncomfortable place for someone who had functioned behind the scenes for nearly twenty years. In a 1997 *Network World* interview, Postel said, "No one wanted to talk to me yesterday. . . . Now I've got nine messages waiting from reporters."

Postel worked relentlessly throughout 1998. In June, the government published a "White Paper" that incorporated aspects of more than 600 suggestions and comments inspired by the "Green Paper." That summer, Postel began collaborating with NSI on the charter for a new nonprofit organization dubbed the Internet Corporation for Assigned Names and Numbers (ICANN). The final submission to William Daley, U.S. secretary of commerce, was made on October 2, 1998.

Five days later, Postel submitted testimony to the U.S. House of Representatives, defending ICANN's charter. He wrote:

> After several years of debate and several months of very hard work . . . we are close to accomplishing the challenge laid down in the White Paper: to create a global, consensus nonprofit corporation with an international board, transparent and fair procedures, and representation of all the various Internet constituencies, from the technical people who created and have nurtured the Internet from its earliest days, to the commercial interests who now see it as an important business tool, to individual users from around the globe.

House Committee members were still mulling over the ICANN charter on October 16, 1998, when Postel died unexpectedly in a hospital in Santa Monica, California, from complications following heart surgery.

*Being in the limelight has its minuses. I'm not pushing to have that happen.*

—Postel to OnTheInternet, September/October 1996

News of Postel's death quickly spread. But as the Internet community mourned, business proceeded. Four days after Postel's death, the Commerce Department sent a letter to ISI, where ICANN would be housed, assuring it that Postel's proposal would, with certain modifications, pass. Nine individuals were chosen to serve as the interim board, and on October 26, 1998, ICANN held its first meeting—behind closed doors.

The secrecy of that first ICANN meeting, in addition to the months of furor that preceded it and the sudden absence of Postel, the one figure who could have inspired confidence on all sides, set ICANN on a rocky course. In a 2000 interview with Australia's *Courier-Mail*, Esther Dyson, the journalist-cum-industry analyst who became ICANN's interim director, recalled, "Instead of working under the mantle of one of the Internet's 'saints,' we faced lots of criticism for being unknown and unelected." Tempers continued to flare over ICANN for weeks, months, and even years. Indeed, many believed the turmoil surrounding ICANN precipitated Postel's untimely death.

During the November 5, 1998, memorial for Postel held at USC, Postel's ARPA colleague Bob Braden claimed, "It was easy to overlook or underestimate Jon's contributions." Indeed, for most of his life, Postel toiled quietly in the background of the networking revolution, dealing with what Internet insiders call the "plumbing" of the Internet—the DNS, the root servers, the protocols. "Lots and lots of very bright people contributed ideas and words to the Internet protocol suite," Braden continued, "but it was Jon Postel who spun out the final words that define the Internet."

For those not close to Postel, the memorial and the outpouring of remembrances that followed were an opportunity to learn more about the man, in death, than they knew in life. Often, friends and colleagues recalled Postel's penchant for sandals, jeans, and a long graying beard—a look that fit nicely with the impression that Postel was god of the Internet, as suggested by *The Economist* magazine in 1997. Postel's one-time ARPA colleague Robert Metcalfe recalled a time when he and Postel were escorted out of the officer's club at an air force base for looking more like antiwar hippies than the cutting-edge scientists that they were. Ira Magaziner, President Bill Clinton's top Internet adviser, recalled when twenty minutes of cajoling were required before the Secret Service would let the bushy-bearded scientist into the White House for a meeting. Magaziner also delivered the words of the president, who wrote, "Though his life was too brief, Jon Postel made enormous contributions to the course of human progress. . . . Because of his efforts, people across America and around the world have virtually unlimited access to a universe of knowledge."

## FURTHER READING

### In These Volumes

Related Entries in this Volume: Kahn, Robert; Cerf, Vinton; Roberts, Lawrence; Taylor, Robert

Related Entries in the Chronology Volume: 1967: Plans for ARPANET Are Unveiled; 1969: The ARPANET Is Born; 1972: ARPANET's Public Debut; 1979: DARPA Establishes the Internet Configuration Control Board; 1983: Internet Is Defined Officially as Networks Using TCP/IP; 1997: Eugene Kashpureff Diverts Internet Traffic from InterNIC to AlterNIC; 1998: ICANN Is Chosen by the U.S. Commerce Department as Successor to InterNIC.

### Works By Jon Postel

RFC 318, Telnet Protocol, April 1971, ftp://ftp.rfc-editor.org/in-notes/rfc318.txt (cited September 16, 2004).

RFC 791–793, Internet Protocol Suite, September 1981, ftp://ftp.rfc-editor.org/in-notes/rfc791.txt; ftp://ftp.rfc-editor.org/in-notes/rfc792.txt; ftp://ftp.rfc-editor.org/in-notes/rfc793.txt (cited September 16, 2004).

RFC 801, NCP/TCP Transition Plan, November 1981, ftp://ftp.rfc-editor.org/in-notes/rfc801.txt (cited September 16, 2004).

RFC 819, The Domain Naming Convention for Internet User Applications, August 1982, ftp://ftp.rfc-editor.org/in-notes/rfc819.txt (cited September 16, 2004).

RFC 821, Simple Mail Transfer Protocol, August 1982, ftp://ftp.rfc-editor.org/in-notes/std/std10.txt (cited September 16, 2004).

RFC 959, File Transfer Protocol, October 1985, http://www.rfc-editor.org/cgi-bin/rfcsearch.pl (cited September 16, 2004).

"Proposal for the Internet Corporation for Assigned Names and Numbers (ICANN)". National Telecommunications & Information Administration, October 2, 1998, http://www.ntia.doc.gov/ntiahome/domainname/proposals/icann/icann.html (cited September 16, 2004).

### Books

Hafner, Katie. *Where Wizards Stay Up Late: The Origins of the Internet.* New York: Simon & Schuster, 1996.

Mueller, Milton K. *Ruling the Root.* Cambridge, MA: MIT Press, 2002.

### Articles

Bayers, Chip. "Mission Impossible." *Wired* 8.12, December 2000, http://www.wired.com/wired/archive/8.12/dyson.html (cited September 16, 2004).

Brown, Eryn. "The Net Name Game." *Fortune,* February 16, 1998.

Diamond, David. "Whose Internet Is It, Anyway?" *Wired* 6.04, April 1998, http://www.wired.com/wired/archive/6.04/kashpureff.html (cited September 16, 2004).

Hafner, Katie. "Jonathan Postel Is Dead at 55." *New York Times,* October 18, 1998.

"Interview with Jon Postel," *OnTheInternet,* September/October 1996, http://oceanpark.com/papers/postel.html (cited September 16, 2004).

Kaplan, Karen. "Pioneer's Behind-the-Scenes Toil Helped Bring Internet to Public Technology." *Los Angeles Times,* October 26, 1998.

———. "Paying a Plane Tribute to an 'Internaut.'" *Los Angeles Times,* November 9, 1998.

Kehoe, Louise. "Lord of the Net." *The Financial Post,* March 7, 1998.

Schwartz, John. "Mourning a Man Who Helped Give Life to the Net." *Washington Post,* November 2, 1998.

Wallack, Tom, and Ellen Messmer. "Industry Asks: Who Is Jon Postel," *NetworkWorld,* April 21, 1997, http://www.nwfusion.com/news/0421postel.html (cited September 16, 2004).

### Websites

Internet Assigned Numbers Authority. Online home of IANA (http://iana.org/), including links to IANA reports and tributes to Postel, http://www.iana.org/postel/postel-tribute.html (cited September 16, 2004).

Internet Society: Jon Postel. An online memorial for Postel, with links to tribute as well as Postel's achievements, http://www.isoc.org/postel/ (cited September 16, 2004).

Los Nettos. Online home of the Los Nettos regional network in Los Angeles, which Postel founded in 1988, http://www.ln.net/ (cited September 16, 2004).

Postel Center. Online home of the Institute for Information Sciences, renamed the Postel Center following Postel's death in 1998. Site includes links to many of Postel's accomplishments, his biography, and original ISI homepage, as well as RFC 2468, written by Vinton Cerf, in Postel's memory, http://www.postel.org/ (cited September 16, 2004).

RFC-Editor. Online home of the RFC series, with links to individual RFCs and a searchable database, http://www.rfc-editor.org/ (cited September 16, 2004). For the history of the RFCs, see: RFC 2555: ftp://ftp.rfc-editor.org/in-notes/rfc2555.txt (cited September 16, 2004). For RFCs in memory of Postel, see: RFC 2468: ftp://ftp.rfc-editor.org/in-notes/rfc2468.txt and RFC 2441: ftp://ftp.rfc-editor.org/in-notes/rfc2441.txt (cited September 16, 2004).

# Lawrence Roberts (1937–)

## CHIEF SCIENTIST OF THE ARPANET

**Widely regarded as one** of the founding fathers of the Internet, Lawrence Roberts was the primary architect of the ARPANET, the predecessor to the Internet. In the mid-1960s, he was lured from the Lincoln Laboratory at the Massachusetts Institute of Technology (MIT) to implement the Defense Department's Advanced Research Projects Agency's (ARPA) vision of an interconnected computer network. As one of the first engineers to have linked two computers together transcontinentally, Roberts was vital to the project. After the ARPANET was up and running, he went on to become a leading engineer of packet-switching protocols—one of the basic elements of the Internet. Through such companies as Telenet in the 1970s, NetExpress in the 1980s, ATM Systems in the 1990s, and, since 1998, Caspian Networks, Roberts has remained at the cutting edge of Internet technologies for more than four decades.

## A CONNECTICUT YANKEE IN THE PENTAGON'S COURT

The son of two chemists, Roberts was born and raised in Westport, Connecticut. Technically adept, as an eight-year-old Roberts built a Tesla coil, a high-powered electrical transformer. He later built a television set from scratch. After graduating from high school, Roberts entered MIT, where he earned three degrees in electrical engineering—a bachelor's, master's, and then, in 1963, a doctorate. While at MIT, he began to explore the inner workings of communications networks and even hacked into the New York City phone system to make free phone calls.

Upon graduating, Roberts chose to remain at MIT, working at the government-funded Lincoln Laboratory, located in a rural area just outside Cambridge, Massachusetts. Lincoln Laboratory had been set up as a lab for American air defense, in response to the growing Soviet threat after World War II. By the time Roberts arrived, however, Lincoln's priorities had come to include computing, thanks in no small part to funding from ARPA's Information Processing Techniques Office (IPTO). IPTO was founded in 1962 by J.C.R. Licklider, another of the primary visionaries of the Internet.

Roberts first encountered Licklider at a meeting in 1964. Licklider was then espousing his concept of an "intergalactic network"—a galaxy-wide network of interconnected computers. At the time, scientists at Lincoln were hard at work on developing time-sharing computers, which were then at the cutting edge of computer science. Licklider noticed how informal communities cropped up among the groups of people who worked with the same time-sharing computer, and he imagined each community connecting with others, and those connecting with still others, ad infinitum. Roberts was inspired by Licklider's vision.

At Lincoln, Roberts's primary function was to write the operating and time-sharing systems for the TX0 and TX2, two of the first transistor-based computers; these later became the prototypes for Digital Equipment Corporation's legendary PDP computers. Roberts was highly regarded at Lincoln as a brilliant and hardworking project manager. He worked alongside a friend and colleague from his days at MIT, Len Kleinrock. In the early 1960s, as part of his doctoral thesis, Kleinrock had published one of the first papers on the idea of packet switching—a method of digitizing and transmitting information in small parts through data networks. "Kleinrock was very much not understood for what his contribution was back then," Roberts told Stephan Segaller, author of *Nerds 2.0.1* (1998). Roberts kept Kleinrock's ideas in mind when he was summoned to work on a new experiment in computer networking.

## LINCOLN, WE HAVE CONTACT

Lincoln—and Roberts, specifically—received the ARPA contract to develop an experimental computer network in February 1965. In July of that year, he was joined by Thomas Marill, who was to program the network and who also believed in the promise of interactive computing. By this time, Licklider had stepped down from his post at IPTO and handed the reins to Ivan Sutherland, one of Roberts's colleagues from graduate school, and his deputy director Robert Taylor. Sutherland and Taylor continued to carry out Licklider's vision.

The experiment—simple with today's technology—involved connecting a still-in-development TX2 computer at the Lincoln Laboratory with a Q-32 mainframe computer located at the System Development Corporation in Santa Monica, California. "I was excited about trying to find out how to link computers together because Licklider had told me his vision and I was looking for a way to do that," Roberts explained in *Nerds 2.0.1.*

Roberts and Marill toiled for several months, and, by October 1965, they had succeeded in building one of the world's first digital computer communications networks. The directly linked computers did not utilize packet-switching technology. Although the connection was astonishingly slow by current standards—4.8 kilobits/second—and rather unreliable, the experiment was nevertheless considered a success—a small, important step toward Licklider's Intergalactic Network.

A year later, in October 1966, Roberts and Marill published their findings in a paper, "Towards a Cooperative Network of Time-Shared Computers," which they presented at the Fall Conference of the (now-defunct) American Federation of Information Processing Societies. The paper is widely considered to be the first published plan for the ARPANET.

## BLACKMAILED

Word of Roberts and Marill's experiment had spread throughout the computing community by early 1966. By then, Robert Taylor had taken over as director of the IPTO. In February Taylor secured $1 million in funding from ARPA's director, Charles Herzfeld, for his own experiment in networked computing, which he dubbed the ARPANET.

The ARPANET would connect several ARPA-funded universities and research centers throughout the country. Although Taylor, whose academic background was in psychology, lacked the technical skills to build such a network, he knew of someone who could: Roberts. Roberts, however, was quite happy working at Lincoln Laboratory. He famously declined Taylor's offer to become the program director for the ARPANET project.

Undeterred, Taylor repeated his offer; according to some accounts, he repeated it up to six more times. Each time, Roberts declined. In August 1966, Taylor pled his case to Herzfeld, asking him to call Roberts's lab director at Lincoln and cajole him into sending Roberts down to IPTO. For leverage he used the fact that ARPA funded more than half of Lincoln's programs. Herzfeld picked up the phone. "All of a sudden the lab director called me into his office and explained how it would be a good career move for me to go," Roberts recalled in a 1997 article for *Data Communications* magazine. Under such pressure, he did not argue and left Cambridge for Washington, D.C., two weeks later.

By the time Roberts became the chief scientist on the ARPANET project, many of the basic ideas for the network—such as the use of packet switching—were already in place. Decisions about the actual design, however, still had to be made. In April 1967, six months into his work, Roberts held a design summit during IPTO's Principal Investigator meeting in Ann Arbor, Michigan. There, he laid out many of the technical specifications—standards of identification, user authentication, transmission and retransmission procedures. The network would use dial-up modems and telephone lines, Roberts explained, and the mainframe computers at each site would be equipped to handle the networking functions.

The response to Roberts's plan ranged from decidedly tepid to nearly hostile. Several of the key investigators—particularly two artificial intelligence pioneers, Marvin Minsky and John McCarthy—were loathe to have the computing power they used for research commandeered for an experimental communications net-

> *I realized that if we got it [the ARPANET] right, it would have as great an impact as the printing press.*
>
> —Roberts to The Times (U.K.), September 1997

work. But at least one of the attendees was intrigued; Wes Clark, one of Roberts's colleagues from Lincoln, slipped Roberts a note near the end of the meeting. It read, "You've got the network inside out." Clark believed that the power of the network should not lie at its center, with the mainframe computers, but at the edges.

After the meeting, on the way to the airport, Clark explained his idea to Roberts: instead of relying on host computers at each research site, the networking functions would be handled by specialized minicomputers—interface message processors (IMPs). The IMPs would all speak the same computer language, leaving the host computer to translate data and messages only once. The host computer at the core of the ARPANET, then, would be "dumb," while the minicomputers on the outskirts would handle the technological heavy lifting—routing the messages, sending and receiving data, and checking for errors. Not only did Clark's plan simplify aspects of the network, it would go over well with ARPA-funded scientists who did not want to share their mainframes.

Roberts then revamped his project, taking heed of Clark's advice. In October 1967, he presented a paper, "Multiple Computer Networks and Intercomputer Communication," at the Association of Computing Machinery Symposium on Operating System Principles, with a newly refined vision of the ARPANET—its first public appearance. Throughout 1968, Roberts continued supervising ARPANET's development, consulting with packet-switching experts such as Paul Baran of the RAND Corporation and Roger Scantlebury of Great Britain's National Physical Laboratory. He also sought computer companies to build the IMPs. By December 1968, Bolt, Beranek and Newman (BBN) had won the contract. Roberts then wrote his official program plan, touting the benefits of networking computing and communications through the ARPANET, and delivered it to Taylor on June 3, 1969. Taylor approved it three weeks later and construction on the ARPANET officially began, with a tentative deadline set for Labor Day.

The first IMP arrived at the University of California–Los Angeles (UCLA) in August 1969. In Septem-

*Our goal [at ARPA] was always two-fold— building a better data communications network and getting the computer to share its information.*

*—Roberts to* The Times *(U.K.), September 1997*

ber, the first communication between UCLA's mainframe and the IMP took place. The following month, an IMP arrived at the Stanford Research Institute in Menlo Park, California. On October 29, 1969, the inaugural message on the two-node ARPANET was sent: "login." By December, two more IMPs had been delivered to the University of California at Santa Barbara and the University of Utah. The ARPANET was complete. Seven months later, six more nodes were added. By 1971, the ARPANET boasted twenty-three nodes worldwide.

## YOU'VE GOT MAIL

In September 1969, shortly after the first IMP arrived at UCLA and it seemed clear that the ARPANET would be a success, Taylor handed the leadership of IPTO over to Roberts. Roberts remained at the helm of the ARPANET, but also began to expand the ways in which the network could be used. From the beginning, members of the ARPANET had been able to communicate in a somewhat real-time manner. In the fall of 1971, Ray Tomlinson, a BBN engineer, sent the first email message over the ARPANET (he also introduced the "@" as part of email addresses).

Email caught on immediately. (By 1973, according to an internal ARPA study, nearly two-thirds of the ARPANET traffic was email.) Roberts, especially, seemed suited for email communications. He became so addicted that, when he traveled, he carried a forty-pound Texas Instruments computer so he could check email in airports. "We used to keep track of phone booths in airports around the world that had electrical outlets near them," he recalled in a 2001 *New York Newsday* interview. As email grew, however, it became increasingly more frustrating for users—unlike the email of today, early email messages scrolled out in one long unbroken screed, with no real differentiation between messages. It was impossible to save, delete, or reply to individual messages.

In 1971, Roberts addressed the problem by writing one of the first email applications, called RD (short for "read"). An extremely simple program, RD revolution-

*The four researchers who won the 2002 Prince of Asturias Award for Scientific and Technological Research for their groundbreaking contribution to the development of the Internet and the World Wide Web pose for photographers after their joint news conference in Oviedo, northern Spain on October 24, 2002. From left to right: U.S. citizens Vinton Cerf, Lawrence Roberts, and Robert Kahn, and British citizen Tim Berners-Lee. (Reuters/Corbis).*

ized electronic communication of the time by allowing users to file, save, delete, forward, reply to, and organize their messages. Indeed, many email managers still use some of the same basic principles.

## PACKETS AND CELLS AND SWITCHES—OH MY!

After six years, Roberts handed the reins of the ARPANET back to Licklider, who had returned to his post at IPTO. Roberts, frustrated with the shifting of ARPA priorities toward military objectives, sought to channel his ARPA experience into a commercial packet-switching network, which he believed would outperform the government-sponsored ARPANET.

In early 1973, after his proposal was turned down by AT&T, then one of the major telecommunications companies, Roberts founded Telenet, the world's first packet-data communications carrier, with funding from BBN. Over the next seven years, he championed the X.25 data protocol, developed at Telenet in 1976. X.25, which is still in use more than thirty years later, is an international network interface protocol that, in the 1970s, helped the growing number of private and commercial data communications networks communicate freely with each other. Roberts stepped down when GTE purchased Telenet in 1979. (Telenet later became the data division of Sprint.)

Roberts then worked to refine the ATM (asynchronous transfer mode) protocol—a high-speed method for transferring voice, video, and data over public networks. ATM, which permitted data to move faster, differed from the standard Internet communications protocol in its complexity and versatility. In 1983, Roberts became chairman and CEO of NetExpress, a company specializing in packetized fax and ATM equipment. In 1993, he became president of ATM Systems, a company that designed advanced ATM technologies. Roberts also championed Cells in Frames, an ATM protocol for Ethernets (local area networks).

The ATM protocol, however, proved somewhat too complex and rigid to adopt as a widespread standard; in 1998, ATM Systems's parent company shut it. Just two years later, Roberts claimed, in an interview with the telecom industry magazine *tele.com*, "It's clear that ATM isn't going to be around long enough to make it worth fixing."

The demise of ATM Systems may have also helped Roberts to make a wise choice. "After ATM, I knew being a CEO wasn't my strength," Roberts told *Wired* in 2001. "I was spending all my time building the company, and not what I do best—this facet of technology." In 1999, Roberts returned to the computer lab (along with several of ATM Systems's top engineers) to once again leverage his expertise to make the Internet run more efficiently.

With ample funding from several venture capital firms, Roberts and his cohorts founded Packetcom LLC (renamed Caspian Networks) to develop what they called a "superswitch"—a new kind of Internet router for IP networks. For most of 2000 through 2002, Caspian's activities were shrouded in mystery. Its first product, the Apeiro, debuted in April 2003, to considerable fanfare.

The Apeiro, a flow-based router, can identify the nature of a packet—be it audio, text, or video—and prioritize it accordingly. It is considered to be one of the first major advances in routing technology in nearly thirty years. In March 2004, the Apeiro passed network testing by the Defense Information Systems Agency's Joint Interoperability Test Command, which then announced that it would use Caspian's technology for the Defense Department's advanced voice-over-IP and videoconferencing needs—in many ways, bringing Roberts's work for the Department of Defense full circle.

Over four decades and in various capacities, Roberts, currently the vice chairman and chief technology officer at Caspian, has been at the forefront of designing increasingly efficient Internet architectures and protocols. It was a goal he had held for a long time. "When we looked at the issue originally in the Sixties," Roberts told *The Times* (London) in 2001, "the psychologists told us people would get annoyed if there was a wait of even half a second." Since then, Roberts has striven—from streamlining email with RD, championing data-transfer protocols such as X.25 and ATM, and developing Caspian's superswitch—to realize his earliest vision of Licklider's Intergalactic Network.

## FURTHER READING

### In These Volumes
Related Entries in this Volume: Cerf, Vinton; Licklider, J.C.R.; Postel, Jonathan; Taylor, Robert

Related Entries in the Chronology Volume: 1958: The Advanced Research Projects Agency Begins Operation; 1963: Memo to the Intergalactic Computer Network; 1967: Plans for ARPANET Are Unveiled; 1969: The ARPANET Is Born; 1973: Telenet, the First Public Packet-Switching Service; 1983: Internet is Officially Defined as Networks Using TCP/IP

### Works By Lawrence Roberts
"Connection Oriented." *Data Communications,* October 21, 1997.

### Books
Abbate, Janet. *Inventing the Internet.* Cambridge, MA: MIT Press, 1999.

Hafner, Katie. *Where Wizards Stay Up Late: The Origins of the Internet.* New York: Simon & Schuster, 1996.

Segaller, Stephen. *Nerds 2.0.1: A Brief History of the Internet.* New York: TV Books, 1999. Chapter on Taylor's work at ARPA available online at: http://www.pbs.org/opb/nerds2.0.1/networking_ner ds/bbn.html (cited September 16, 2004).

### Articles
Booth, Nicolas. "Doctor Assisted at Birth of the Net." *The Times* (U.K.), September 24, 1997.

Haring, Bruce. "Who Really Invented The 'Net?" *USA Today,* September 2, 1999.

McHugh, Josh. "The *n*-Dimensional Superswitch." *Wired* 9.05, May 2001, http://www.wired.com /wired/archive/9.05/caspian_pr.html (cited September 16, 2004).

Phan, Monty. "Email Delivers." *Newsday,* October 17, 2001.

### Websites
Larry Roberts's Home Page. Roberts's personal and professional home page, with links to ARPA and Internet histories, his published papers, and related news, http://www.packet.cc/ (cited September 16, 2004).

Caspian Networks. Online home for Caspian Networks, where Roberts spearheaded the development of the Apeiro, a next-generation IP router, http://www.caspiannetworks.com/home.asp (cited September 16, 2004).

# Aliza Sherman (1968–)

## AUTHOR; FOUNDER OF CYBERGRRL, WEBGRRLS, FEMINA.COM

**Aliza Sherman was one** of the first women to make a splash on the World Wide Web. She first logged on to the Internet in 1989 and soon turned her hobby into a career. In 1995 she pioneered the first woman-owned, full-service Internet company, CG Internet Media, Inc., also known as Cybergrrl, Inc., after Sherman's online alter ego, Cybergrrl—a comic strip superheroine who helps women use the power of the Internet. In addition to building Websites and consulting, Cybergrrl developed the first women's Internet networking group, Webgrrls International, and the first woman-oriented search engine, Femina.com. At the time, men outnumbered women online by a margin of 3-to-1 (by some accounts, even 9-to-1). Sherman helped many women overcome their fear of the Net, opening the doors for thousands of women to find community, information, resources, and education on the Web.

## THE EARLY ADVENTURES OF CYBERGRRL

Before she found the Internet, Aliza Sherman was a wayward Renaissance girl. She contemplated many paths including acting, fashion merchandising, business law, and creative writing, but after three universities, four years, and no degree, she went to work as a secretary. With dreams of becoming a published writer, she bought her first computer (an Amstrad 1640) in 1989. A friend taught her how to log on to the Internet. Soon, Sherman was spending hours in chat rooms, on bulletin boards, and in text-based interactive environments called multiuser dungeons (MUDs), learning what cyberspace had to offer. She found little in the way of content for women, aside from the occasional list of feminist resources, such as the one sponsored by the Massachusetts Institute of Technology (MIT).

After six years as a music industry publicist in Virginia, North Carolina, and New York City, Sherman switched gears and became executive director of the Domestic Abuse Awareness Project, a nonprofit organization founded by New York photojournalist Donna Ferrato. Sherman used her Internet skills to find government reports, abuse statistics, and funding information, and to network with the few women's organizations that were online in the early 1990s. She was one of 500 or so women who discovered Women's WIRE (Worldwide Information Resource and Exchange), when it began in January 1994. She logged on daily. What she could not find online, Sherman created. Noting the lack of information on domestic violence, she started SafetyNet, an online resource and forum for discussions on domestic abuse. She also started The Women's List, an email list about women's issues both on- and off-line.

Then, in November 1994, Sherman was mugged at gunpoint. She left New York to recover at her sister's home in Santa Fe, New Mexico. There, she signed up for a class in HTML (HyperText Markup Language). Soon after completing the hour-long class, Sherman began creating her first Website.

At the time, the Web was growing and businesses were getting online. However, not everyone was wired—especially women. Indeed, women comprised only 10 percent of Internet users—a fact reflected in online content. "At the time, when you typed in 'women' on Yahoo!, the only sites that popped up were triple-X-rated," Sherman told the *Palm Beach Post*. She used the hiatus from work to reassess her life and goals. By the end of 1994, she had decided she would be her own boss, and she would use her Internet skills to do it. Armed with a renewed sense of purpose, Sherman left New Mexico after two months to launch a new career in New York City.

In January 1995, Sherman founded CGI Media, Inc., in her Manhattan studio apartment. The company's first Website was her own, dubbed The Web

According to Cybergrrl, at www.cybrergrrl.com. Sherman was afraid to put her photo on the Web, so she drew a comic book rendition of herself, complete with a pink cape; thus, Cybergrrl, the alter ego, was born. From then on, Sherman marketed herself as a Web designer and Internet consultant. Few in the corporate and nonprofit sectors knew what either of those titles meant, but Sherman was determined to show her clients how the Web could transform their lives—as it had her own.

In its first month, Cybergrrl, Inc., earned only $42.50. However, the business moved quickly from consulting and online marketing to include site development and advertising. Business boomed. Sherman's first major client, Avon's Breast Cancer Awareness Campaign, for which she designed interactive forums and "live" online events, opened the door to clients like Clinique, the National Alliance of Breast Cancer Organizations, Girls Incorporated, and diet guru Dr. Robert C. Atkins. Early advertising sponsors included Chase Online and Lifetime Television. By 1996 *Newsweek* had named her one of "The Net 50: The 50 People Who Matter Most on the Internet." (Sherman was one of only three women.) The following year, Cybergrrl, Inc., had moved to offices in lower Manhattan—known as Silicon Alley—and boasted a handful of full- and part-time employees, in addition to a flood of interns. Both by accident and by design, Sherman found her company at the head of the Internet boom.

> *It was far better as a woman to start a company and be able to walk into a room as president than to try to move up the ranks.*
>
> —Aliza Sherman to the *Palm Beach Post*, January 2002

## JUST ONE OF THE GRRLS

While struggling to get Cybergrrl, Inc., on its feet, Sherman spent a lot of time seeking out other women online. She introduced herself via email and put links to their home pages on her own home page. Within months, Sherman had a small following.

One Saturday in late April 1995, Sherman arranged to meet some of these women at the @café, a cyber-cafe in Manhattan's East Village. Numbering six in total, they perused each other's home pages at the cafe's computer terminals and talked about what they did—or wanted to do—online. Two weeks later, more than a dozen women showed up, many of whom had discovered Cybergrrl while surfing the Web. By September, more than 100 had joined. Sherman dubbed them Webgrrls.

The "grr" of Webgrrls, and of Cybergrrl herself, was Sherman's way of suggesting that these women were strong, independent, capable, and irreverent. "Girl sounded too young, and women doesn't have an edge to it," explains Sherman. "When I replaced the i with an r, it seemed to have attitude." The term *grrrl*—with three Rs—had first emerged in the Riot Grrrl movement of the 1990s, which was a young, bristling feminist subset of the alternative punk rock music scene. With one less r, Webgrrls adopted a slightly gentler, more mainstream version of the word.

The initial goal of Webgrrls was simple—to meet other women who knew about the Internet—but the group became far more. The number of new women joining each month caused Sherman to move meetings from the cafe to a downtown loft donated by Jupiter Interactive, a new media company. By the seventh New York Webgrrls meeting in October 1995, a professional atmosphere was evident: representatives from *Elle* and *Self* magazines showed up, announcing immediate openings, women publicized new Web projects and offered their technical services. Women who came to learn more about the Web found peers and mentors. Those without Web skills could sign up for cheap software and programming classes not unlike the HTML course that jump-started Sherman's career. Webgrrls had tapped into a need.

As Webgrrls chapters developed across the country, Sherman founded a new company, Webgrrls International, as part of Cybergrrl, Inc. Members paid a $55 fee to gain access to exclusive job lists, online training, and industry discounts, in addition to the regular resources and information offered free at www.webgrrls.com. Webgrrls International provided server space for local chapters, which were left to design their own meetings and programs. Enterprising chapters convinced companies, such as Macromedia, to offer tutorials on the latest software, while other groups held informal networking meetings. (Chapter leaders—called Pointgrrls—kept up with larger Webgrrls issues on their own private mailing lists.) In just three years, Webgrrls International grew to more than 100 chapters, uniting women from across the United States and internationally in Europe, Japan, and Australia.

"The beauty of Webgrrls is that there is no stereo-typical profile of the women involved," Sherman told the *Chicago Sun-Times.* "Each chapter has its own flavor or slant that makes it more diverse." For example, the Webgrrls chapter in Dallas, Texas, called the Big D Webgrrls, included stay-at-home moms, PR executives, computer programmers, graphic designers, and the president of the Dallas Internet Society. Such range was found in most chapters.

The larger chapters boasted thousands of members. In Seattle, where one of the oldest satellite chapters began with a handful of women meeting at a coffee shop, the group had more than 2,700 members by the year 2000, and was adding around 50 new members each week. Chapters in Germany and the Netherlands were larger still.

Some saw the openness of the Webgrrls community as key to its success. As one of the former presidents of Webgrrls Chicago told the *Chicago Tribune,* "Webgrrls is what you want it to be. You can be on the listserv . . . or come to the meetings, or volunteer for the mentoring program. As much as you want to get involved, you can." Some women came to network, others, to learn. For the self-employed, Webgrrls offered the company of professional colleagues. For Sherman, the success of Webgrrls was evidence that she had attained a goal: creating sisterhood online.

## SISTERHOOD IN CYBERSPACE?

Sisterhood, however, was not always simple. Shortly after Sherman founded Webgrrls International, she launched Femina.com, the first searchable directory of sites "for, by and about women." It was Sherman's response to being inundated by pornography sites each time she searched for women's information on Yahoo! She was attacked by some for choosing pink—her favorite color—for the background. "I can't tell you how

*I think of myself as a writer first and technology, computers and the Internet are my tools.*

*—Aliza Sherman, in an interview with WITI (undated)*

many women emailed me, outraged that I would stereo-type women," she told the Associated Press. "Here I have created the first search engine for women's sites, and the background color has become a political issue." But such internal grumblings were among the least of Cybergrrl's problems.

Beginning in 1998, satellite chapters began to secede from Webgrrls to form their own groups dedicated to wired women. The San Francisco chapter left in August to form San Francisco Women on the Web (www.sfwow.org), followed by the Austin chapter, which formed Her Domain (www.herdomain.org). In April 1999, the Washington, D.C., chapter followed suit, forming DC Web Women (www.dcwebwomen.org). By the end of November 2000, nearly one third of Webgrrls' 30,000 members, primarily those from the United States and Canada, had defected to a single organization—DigitalEve.

The former Webgrrls levied various complaints against the organization, although, by and large, they still admired Sherman's vision. For one, Webgrrls was a dot-com—technically (though not actually) a for-profit business; many who defected believed that a woman's online community should be, on principle, nonprofit. Other groups disliked the centralized nature of Webgrrls International, with all business matters handled by a staff of twenty at Cybergrrl, Inc., in New York. However, for many, the disillusionment set in after Sherman stepped down as director of Webgrrls in 1997. Over the next three years, while Sherman focused on writing her first book for women on the Web, Webgrrls had a number of directors. In February 2000, the CEO of Cybergrrl, Inc., Kevin Kennedy, took over. Women questioned if a man should be at the helm of such a historically women-centered organization.

In a December 2000 article on *Salon.com,* Sherman lamented the split in her organization. "The thing that distresses me the most is that all of this is based on helping women, yet from the get-go with Webgrrls—and almost every women's organization I've ever belonged to—there's always this infighting." By that time, however, greater forces had begun to chip away at Webgrrls' core membership. Women, once a minority online, had quickly become 51 percent of all Internet users, ushered in by e-commerce and mainstream women-oriented sites like iVillage.com, Women.com, and Oxygen.com, all of which flourished in the late 1990s. Webgrrl.com lacked the advertising and backing to truly compete, mainly because Sherman shunned funding from what she called "corporate, often only male, investors." On the other hand, both cybergrrl.com and webgrrl.com lacked the edge of the various women- and grrl-oriented online 'zines—independent, often self-published, magazines—like www.bust.com, which played an essential role in the Third Wave feminist communities of the 1990s. Some also believed, perhaps cynically, that the need for a global, women-only technology community disappeared once women no longer felt alone on the Web.

## CYBERGRRL SEZ

When Sherman left Webgrrls in 1997, she turned her attention to one of her first loves: writing. In 1998 Ballantine Books published *Cybergrrl! A Woman's Guide to the World Wide Web.* The book was billed as a frank, jargon-free introduction to the Internet and all it offered, written for an audience of women who did not necessarily know how to use a mouse. Similar to her www.cybergrrl.com column, "Cybergrrl Sez," Sherman wrote in an easy, knowing, personal style, in a tone like that of a reassuring older sister. Much of the text came directly from Sherman's own life. "I'm living proof that you don't have to be a techie or computer whiz to benefit from the Internet," she writes in the Introduction.

Sherman's goal was to present computers, modems, the Internet, and the Web in basic, easy-to-swallow bites—and in those terms, the book was a success. (Women who were already wired—Cybergrrl's original audience—found the book too simplistic.) "The biggest barrier for women in cyberspace is fear of the unknown," was Sherman's usual response when questioned about the need for her book. She used personal anecdotes, profiles, and easy-to-understand language to dispel the five myths she believed kept women from logging on—namely, that the Internet was too hard, too expensive, too dangerous, offered no personal satisfaction, and provided no professional advantage.

Sherman followed *Cybergrrl! A Woman's Guide to the World Wide Web* with *Cybergrrl @ Work: Tips and Inspiration for the Professional You,* released in 2000, which focused on finding, getting, creating, and succeeding at jobs online. In what had become her trademark style, she wove individual success stories in with lucid, practical information and advice. In the two years since her previous book, Sherman showed that women were now using Internet technology to advance their careers and enhance their business.

While writing *Cybergrrl @ Work* Sherman had a chance to reassess her career once again. She had stepped away from leading Cybergrrl, Inc., in 2000, temporarily handing the reins of the company to Kevin Kennedy. Sherman realized that she suffered from what she called "start-up-itis"—fatigue that set in once a burgeoning start-up becomes large and unmanageable. Her role at Cybergrrl, Inc., had become largely administrative; outside the company, she found that she had become a spokesperson, not a president. Five years after founding a pioneering new media company, Sherman no longer wanted to be an entrepreneur.

## ON THE ROAD

The technology crash that began in the spring of 2000, when previously high-flying technology companies took a beating in the stock market and many information technology workers found themselves suddenly unemployed, also contributed to Sherman's change of direction. That March, Sherman, who is of Mexican descent, had announced plans to launch Eviva.net—the Web's first bilingual online network and marketplace for Hispanic women—with Web entrepreneur Amy Ormond, creator of the first online pharmacy for women in 1994. As Sherman told *Businesswire.com,* "Building community online and offline for women and helping them to integrate technology into their lives has been my mission for over seven years. But until now, I have not been able to integrate my professional passion with my personal heritage." By mid-2000, however, funding had evaporated. The project was put on hold indefinitely.

When Sherman left New York City to promote her books in September 2000, she did not look back. Kennedy would remain Cybergrrl, Inc.'s CEO. Sherman bought a 23-foot 1977 Dodge Apache motor home and spent more than a year driving across the United States for the "On the Road with Cybergrrl and Webgrrls" book tour. She also began to work on another book, *PowerTools for Women: 10 Ways to Succeed in Life and Work,* a motivational book for women in business. Ever the wired woman, in April 2001, Sherman sent the manuscript to her publishers via cellular modem from her motor home.

Disillusioned, first by the technology crash, then by the terrorist attacks of September 11, 2001, Sherman moved permanently from New York City to Cheyenne, Wyoming, in the fall of 2001. That November, she joined the Wyoming Business Council as manager of marketing and public relations, a position she left in August 2003 to work on her sixth book, about her cross-country travels. Driving solo across the United States, she told *USA Today,* "was the same feeling as when I first connected to the Internet and discovered a whole new world."

In many ways, Aliza Sherman's Internet career mirrors the boom-bust cycle of the 1990s. Sherman was young, ambitious, full of energy and ideas, and, as she would often admit, a bit of a geek. She may have lacked the credentials to succeed in the traditional corporate world, but the Internet opened doors for her and others like her to become entrepreneurs. Through Cybergrrl, Inc., Webgrrls International, Femina.com, and the two *Cybergrrl* books, Sherman tapped into the growing number of women who wanted to stake their claim in cyberspace. Once women had arrived online—and as the technology sector crashed—Sherman was one of the first to log off, in search of a quieter life, off line.

## FURTHER READING

### In These Volumes

Related Entries in this Volume: Borg, Anita
Related Entries in the Issues Volume: Digital Divide; Languages and Cultures; Online Communities

### Works by Aliza Sherman

*Cybergrrl!: A Woman's Guide to the World Wide Web.* New York: Ballantine Books, 1998.
*Cybergrrl @ Work: Tips and Inspiration for the Professional You.* Berkeley, CA: Penguin Putnam, 2001.
*PowerTools for Women in Business: 10 Ways to Succeed in Life and Work.* Irvine, CA: Entrepreneur Press, 2001.

### Books

Cherny, Lynn, and Elizabeth Reba Weise, eds. *Wired Women: Gender and New Realities in Cyberspace.* Seattle, WA: Seal Press, 1996.

### Articles

Bedell, Doug. "Big D Webgrrls Meld Technology Talent And An Attitude." *Dallas Morning News,* February 2, 1999.
Brown, Janelle. "Who Are You Calling 'Sister'?" *Salon.com,* December 21, 2000, http://dir.salon.com/tech/feature/2000/12/21/webgrrls/index.html (cited September 29, 2004).
Demos, Kristin. "Where the Grrls Are; Grass-Roots Group Attracts Women Looking to Succeed in a Technical World." *Chicago Tribune,* August 30, 2000.
Fischer, Marcelle S. "For Women and Girls, a Global Forum for Navigating the Web." *New York Times,* May 24, 1998.
Maney, Kevin. "Former Tech Pioneer Finds She's at Home on the Range." *USA Today,* January 16, 2002.
Martinez, Amy. "Pioneering CEO Serves as Exemplar for Women." *Palm Beach Post,* January 28, 2002.
McMullen, Cynthia. "Internet Is Power-Packed for Cybergrrl Sherman." *Richmond (VA) Times Dispatch,* February 7, 2001.
Stiffler, Lisa. "Webgrrls' Virtual Power Growing in Real World of Careers in Technology." *Seattle Post-Intelligencer,* January 3, 2000.

### Websites

CG Internet Media, Inc. Home of CG Internet Media, founded in 1995, http://www.cgim.com/ (cited September 29, 2004).
Cybergrrl. Formerly Aliza Sherman's personal site, Cybergrrl.com features news, information, and resources for women, http://www.cybergrrl.com (cited September 29, 2004).
Femina Web Search for Women. Home of the first women-oriented search engine, founded in 1995, http://www.femina.com (cited September 29, 2004).
Media Egg. Aliza Sherman's professional Website, http://www.mediaegg.com (cited September 29, 2004).
PowerTools for Women in Business. Site for Aliza Sherman's book, *PowerTools for Women in Business,* http://www.womenspowertools.com/ (cited September 29, 2004).
RV Girl. Aliza Sherman's personal Website, which chronicles her cross-country book tour, http://www.rvgirl.com/ (cited September 29, 2004).
Webgrrls International. Webgrrls' online headquarters, featuring job resources and information, http://www.webgrrls.com (cited September 29, 2004).
WITI Women: Aliza Sherman. Aliza Sherman's interview with *Women in Technology International,* http://dev.witi.com/wire/witiwomen/asherman/index.shtml (cited September 29, 2004).

# Richard Stallman (1953–)

## FOUNDER OF THE **GNU** PROJECT AND THE FREE SOFTWARE FOUNDATION

**Richard Stallman is the** patron saint of the free software movement. In the 1970s Stallman was one of the most prolific programmers at the Artificial Intelligence Lab at the Massachusetts Institute of Technology (MIT). The hacker ethic was born at MIT; a key tenet of that ethic is that information, including information about software, should be freely available to everyone—thus the term *free software*. In the early 1980s, Stallman founded the GNU Project and established the Free Software Foundation to create a space where the hacker ethic could still thrive. He single-handedly wrote several key pieces of free software for the GNU (pronounced "guh-noo") family of software. But as the free software movement grew in popularity, Stallman became an increasingly controversial figure. For him, free software has always been a moral crusade, and he came into increasing conflict with those who saw it as a business opportunity. By 1998, his stridency caused the movement to split in two. Although many disagree with Stallman's views, few refuse him respect. Stallman is credited with making programmers, computer hobbyists, and consumers alike rethink how software is produced, reassess the meaning of intellectual property, and reconsider what, if anything, "community" means in the world of computing.

## THE YOUNG HACKER

Richard Stallman was born and raised in New York City. When his parents divorced in 1958, he split his week between his mother's Manhattan apartment and his father's home in Queens. An exceedingly bright but often recalcitrant student—for a time, he refused to write papers—Stallman had many difficulties in New York City public schools. By junior high, he had few friends besides teachers. "I outgrew what I did with my childhood friends," Stallman explained in a 1999 interview. "I had no idea what would replace it." He found some solace

in the Columbia Science Honors Program for high school students. Each Saturday, he would study with Columbia University professors, often correcting them during lectures. During his junior year in high school, Stallman was hired by the IBM New York Scientific Center, a now-defunct research facility in Manhattan. There, he got his first real taste of programming, writing software for an IBM 7094 computer.

At age 17, Stallman graduated from high school and entered Harvard University. During his first year, he began borrowing computer manuals from Harvard's computer labs, studying the specifications to compare and contrast different software. By the end of the year, frustrated with the bureaucracy of Harvard's computer labs, he ventured down the road to neighboring MIT. He had heard that the Artificial Intelligence (AI) Lab had a powerful computer and he wanted to read the documentation. But the hacker community at the AI Lab did not document their code in manuals—they just wrote it. They did offer Stallman a job, however.

For much of Stallman's time at Harvard, he spent his weekdays in class and his weekends at the AI Lab. He was an inexhaustible programmer. In addition to finding a place to exercise his technical competence, Stallman found what he often calls his first true home. (Indeed, he often slept at the lab, particularly when he was between apartments.) The group of MIT hackers formed a built-in social network for Stallman—one with a similar worldview. "I joined this community which had a way of life which involved respecting each other's freedom," Stallman told his biographer, Sam Williams. "It didn't take me long to figure out that that was a good thing. It took me longer to come to the conclusion that this was a moral issue."

In 1974, Stallman graduated from Harvard with a bachelor's degree in physics and entered MIT as a graduate student in physics. However, programming quickly became the primary focus of his life. Though he quit graduate study after one year, Stallman stayed on at the AI Lab. In 1975, he wrote Emacs, the first extensible, real-time display text editor for the lab's mainframe computers. Emacs (short for Editing MACroS) was initially developed as a set of shortcuts for another editing program used by the lab. By 1979, it was considered a powerful stand-alone text editor in its own right.

## A WHOLE GNU WORLD

Thanks to Emacs, Stallman enjoyed a great degree of success in and recognition from the programming community. However, by the early 1980s, the hacker community at the AI Lab had begun to change. Stallman often tells a tale about a laser printer to explain the moment when he first understood that the hacker ethic he so fervently embraced was endangered. Xerox had given the AI Lab an early laser printer, which often jammed. In similar situations, Stallman or one of the other hackers would simply write a program to fix the glitch. However, when Stallman asked a Xerox employee how to gain access to the source code of the software running the printer, he was told that such information was not given out. The employee had signed a nondisclosure agreement and therefore could not let Stallman see the internal workings of the software. To Stallman, such noncooperation marked a sea change. Proprietary software, enforced legally through nondisclosure agreements, was disassembling the freedom, collaboration, and community he had always enjoyed as a hacker.

Stallman was both personally and professionally affronted. "For beings that can think and learn, sharing useful knowledge is a fundamental act of friendship," he told *InfoWorld* magazine in 2001. "This spirit of goodwill, of helping your neighbor voluntarily, is science's most important resource. It makes the difference between a livable society and a dog-eat-dog jungle."

The printer incident was not the only harbinger of things to come. By the early 1980s, more than a dozen of the AI Lab's hackers had been lured to a company called Symbolics. Symbolics—and later, its rival, Lisp

*"Free software" is a matter of liberty, not price. To understand the concept, you should think of "free speech," not "free beer."*

*— from "The Free Software Definition"*

Machines, Inc. (LMI)—offered AI Lab hackers who had been developing Lisp software for MIT commercial success if they developed proprietary versions of Lisp software for the company's use. The AI Lab and Symbolics initially had an amicable relationship—Stallman, still at the lab, was free to hack any of the company's developments and vice versa. However, as competition between Symbolics and LMI grew, that collaboration ended.

Enraged, Stallman spent the next two years single-handedly replicating the work of dozens of Symbolics and LMI programmers. From a programming perspective, his output was astonishing. From Stallman's own perspective, his work was vital to keeping the AI Lab free of proprietary software.

Stallman's anger also prompted him to post a message to various computing newsgroups, announcing a bold new idea. His email, dated September 27, 1983, began, "Starting this Thanksgiving I am going to write a complete UNIX-compatible software system called GNU (for Gnu's Not UNIX) and give it away free to everyone who can use it. Contributions of time, money, programs and equipment are greatly needed." Choosing UNIX was a good move. UNIX was a fast-growing operating system developed by AT&T in the early 1970s. It had quickly gained an ardent following among programmers. Indeed, by the time Stallman had posted his email, a variant of UNIX had been developed in a hacker-minded, communal fashion—an effort led by Bill Joy, a graduate student at the University of California–Berkeley. (Some point to Joy's development, known as BSD or Berkeley UNIX, as the actual beginning of the free software movement, even though most agree that the official movement began with Stallman a decade later.)

Few answered Stallman's call to arms. Thanksgiving came and went. By January of the following year, Stallman quit his position at MIT and committed himself fully to the GNU Project.

Stallman first developed a UNIX-compatible version of his text editor. The result, GNU Emacs, was released in 1985. Shortly thereafter, Stallman issued "The GNU Manifesto," a more fully articulated version of his

1983 Usenet posting, in which he explained the philosophy and politics behind GNU:

> I consider that the golden rule requires that if I like a program I must share it with other people who like it. Software sellers want to divide the users and conquer them, making each user agree not to share with others. I refuse to break solidarity with other users in this way. I cannot in good conscience sign a nondisclosure agreement or a software license agreement . . . So that I can continue to use computers without dishonor, I have decided to put together a sufficient body of free software so that I will be able to get along without any software that is not free.

In addition to the GNU Project, Stallman established the Free Software Foundation (FSF), the first nonprofit organization dedicated to the cause of free software. Funds flowing into the FSF support the GNU Project.

By 1989, Stallman had added to the GNU family of free software the GNU C Compiler (GCC) and the GNU Debugger (GDB). Together, GNU Emacs, GCC, and GDB provided the foundation needed for GNU software developers to create new software. Companies, including Cygnus Support, built entire businesses around GNU products, proving that "free software" did not necessarily mean "no profits." As Stallman is quick to point out, the "free" in free software does not translate as "without cost" or "gratis"; free, in Stallman's world, means freedom.

## FROM RIGHT TO LEFT

To ensure that GNU Emacs remained truly free, Stallman included a copyright notice, which he called the GNU Emacs License, when he distributed the software. The license proclaimed that users were free to change GNU Emacs as long as they published their modifications and included, with the new modified version, a similar notice, so that all derivative work would remain in the public domain. In the license's early stages, Stallman also asked for suggestions for improvements.

*Software should be free and I'm going to make it free even if I have to write it all myself.*

*—Stallman to the* Wall Street Journal, *May 1991*

In November 1986, John Gilmore, a UNIX programming whiz who later worked for the FSF and cofounded Cygnus Support, responded with a key insight. Instead of limiting the license to Emacs, Gilmore suggested that the license be reworded to encompass all the software to be created by the GNU Project. "Soon," Gilmore pointed out in an email, "Emacs will not be the biggest part of the GNU system."

Gilmore was right, and Stallman, in consultation with lawyers working with the FSF, made adjustments to his license so that it was both legally sound and in keeping with Stallman's ethos. By 1989, the preamble to the license read:

> The General Public License is designed to make sure that you have the freedom to give away or sell copies of free software, that you receive source code or can get it if you want it, that you can change the software or use pieces of it in new free programs; and that you know you can do these things.
>
> To protect your rights, we need to make restrictions that forbid anyone to deny you these rights or to ask you to surrender the rights. These restrictions translate to certain responsibilities for you if you distribute copies of the software, or if you modify it.

The General Public License (GPL) constituted a new evolutionary branch in copyright. Indeed, some call the GPL Stallman's greatest hack—a significant improvement on the original. Stallman dubbed his idea "copyleft," inspired by a wry sticker he once saw that read, "Copyleft (L) All Rights Reversed." Since 1989, all GNU programs have been protected by Stallman's copyleft principle, enforced through the GNU GPL. (Copyleft protects GNU users' right to copy and modify GNU source code however they see fit. Copyleft even protects GNU users' right to make money by selling the modified software, so long as the altered code remains in the public domain.)

In 1990, Stallman's untiring work in the name of free software received the ultimate acknowledgement—

a $240,000 MacArthur Foundation ("Genius") grant. The award, he told the *Wall Street Journal* in 1991, "caused some people to respect me more, which is very useful." It also gave him the means to expand the technical staff at the FSF and, more important, to travel, spreading the gospel of free software to the far corners of the Earth.

## A KERNEL OF TRUTH

Once Stallman received the MacArthur grant, he began spending more time evangelizing free software than writing it. One significant stop on his worldwide tour came in 1990, at the Polytechnic University in Helsinki, Finland. Among the many rapt audience members was twenty-one-year-old Linus Torvalds.

At the time, many facets of the GNU free operating system were in place, with the exception of one key element—the kernel. The kernel of an operating system is a central program that directs the flow of traffic from various applications to the microprocessor. Stallman's kernel—known as HURD—had been years in the making and was still, by the early 1990s, unfinished. (Later, Stallman would admit that HURD was "a mistake.")

In 1991, Torvalds independently developed a UNIX-compatible kernel, dubbed Linux. Initially, Torvalds did not release the Linux kernel with the GNU General Public License, but, instead, with another, more restrictive license. By 1992, however, he adopted the GPL and made the source code freely available on the Internet. With it, the complete, fully functional GNU operating system was born.

The GNU-Linux operating system was quickly embraced by UNIX hackers, programmers, and developers everywhere. Just as quickly, it became known, simply, as Linux. In time, this casual misstatement has grown to become a huge point of contention for Stallman. Indeed, he is well-known for berating well-meaning journalists or programmers for failing to use the full term "GNU-Linux" in reference to the software system built—at least in part—by the hard, often unpaid, work of GNU programmers. Since 1995, he has lobbied unsuccessfully for an official name change that would af-

*The fundamental conclusion I reached 10 years ago was that asserting copyrights or patents on proprietary software requires a betrayal of all of society.*

—*Stallman to the* San Francisco Examiner, *August 1994*

fect all GNU-Linux-related programs. (Stallman's critics point out that several aspects of Linux, such as the X-Windows window sytem, were developed independently of the GNU Project.)

## TRUE FREEDOM

The name issue reflects a deeper conflict between Stallman and a significant portion of the GNU-Linux community. In a 1998 interview with *Salon.com,* Stallman claimed that "certain people are trying to rewrite history and deny me my place in the movement." In a general sense, he meant people who believe that Linux—not the GNU Project that preceded it by almost a decade—marks the beginning of the history of free software, or that Torvalds—not Stallman—is the movement's figurehead. Many in the GNU-Linux community find Stallman's grumblings egomaniacal, strangely proprietary, and, worse, counterproductive, particularly in light of Microsoft's ongoing struggle to quash the free software market.

In time, a faction of the free software community began to distance itself from Stallman and what it deemed his "radical" politics. In 1998, this group officially splintered from Stallman and created a new movement, the Open Source movement, united under the auspices of the Open Source Initiative (OSI).

Both the OSI and FSF stand for the core principle that source code should be freely available. The open source movement, however, champions the more apolitical idea that open code equals better code on a technical level, while Stallman firmly believes that keeping source code free and nonproprietary is an ethical imperative. In the end, the open source ethos proved to be far more business-friendly, which has caused another level of problems in Stallman's eyes. As Linux products have become increasingly prevalent in the business and personal sectors, certain Linux-based software systems have started to contain or support proprietary software. Open source supporters hail this, and the continuing spread of Linux, as a victory in their battle to bring free software to the mainstream. But Stallman—ever the

*Richard Stallman. (Simon Kwong/Corbis)*

purist—rails against such systems. In a 2004 essay for *LinuxWorld,* Stallman wrote, "To free the citizens of cyberspace, we have to replace those non-free programs, not accept them. They are not contributions to our community, they are temptations to settle for continuing non-freedom."

Eric Raymond, one of the primary leaders of the Open Source movement, finds Stallman's extremism to be a detriment to both movements. "Often," Raymond stated in a 1998 *Salon.com* interview, "he renders people who *want* to be his allies unable to support him." Since the late 1990s, Stallman's rhetoric has often kept him on the sidelines of free software industry conferences and has, at times, even hindered the growth of the GNU Project. Nevertheless, he has continued to push the free software movement forward. In 2001, he founded an outpost of the Free Software Foundation in Germany and, in 2003, in India. Stallman has been especially active in India, where, he believes, free software can help build vital infrastructure

and bridge the Digital Divide, both there and in other developing countries.

## HAPPY BIRTHDAY TO GNU

The year 2004 marked the twentieth anniversary of the GNU Project. Stallman used the occasion as an opportunity to remind the computing community of his role in what has become known as the free software/open source movement. (Like GNU-Linux and Linux, the two terms have come to be used interchangeably, much to Stallman's displeasure.) In an essay for *LinuxWorld .com,* Stallman reminded readers how, in the early 1980s, computer users "lost their freedom to cooperate," and how, two decades later, without continued vigilance, such freedom was still in danger of being lost.

In the two decades since Project GNU's birth, the GNU-Linux computing community has grown to somewhere between 17 and 20 million users. Although its popularity may seem bittersweet to Stallman, the overwhelming adoption of GNU-Linux systems has made it a fierce competitor of proprietary systems, including Microsoft's Windows. Less bittersweet has been the global spread of free software ideals. In addition to advances in India, in recent years Stallman has enjoyed government support of free software in Brazil, and glimmers of support in other countries as well. Slowly, Stallman's plan to recreate the spirit of the community in which he first thrived as a programmer seems to be taking hold.

## FURTHER READING

### *In These Volumes*

Related Entries in this Volume: Joy, William; Gilmore, John; Torvalds, Linus
Related Entries in the Chronology Volume: 1976 (sidebar): The Free Software Movement; 1985: Richard Stallman Establishes the Free Software Foundation; 1991: Linus Torvalds Develops the Linux Operating System; 1998: Linux Operating System Becomes a Cause Célèbre
Related Entries in the Issues Volume: Copyright; Hackers; Open Source; Patents

### *Works By Richard Stallman*

"The GNU Operating System and the Free Software Movement." In *Open Sources: Voices from the Open Source Revolution,* Chris DiBona, ed. Sebastopol, CA: O'Reilly & Associates, 1999.

"Freedom—Or Copyright?" *MIT Technology Review,* June 2000.

## Books

Levy, Steven. *Hackers: Heroes of the Computer Revolution.* Garden City, NY: Doubleday, 1984.

Wayner, Peter. *Free For All: How Linux and the Free Software Movement Undercut the High-Tech Titans.* New York: HarperBusiness, 2000.

Williams, Sam. *Free As In Freedom: Richard Stallman's Crusade for Free Software.* Sebastopol, CA: O'Reilly & Associates, 2002.

Also available online at: http://www.oreilly.com /openbook/freedom/ (cited February 1, 2004).

## Articles

Garfinkel, Simson L. "Is Stallman Stalled?" *Wired* 1.01, March/April 1993, http://www.wired.com /archive/1.01/stallman.html (cited September 16, 2004).

Kahney, Leander. "Linux's Forgotten Man." *Wired News,* March 5, 1999, http://www.wired.com /news/print/0,1294,18291,00.html (cited September 16, 2004).

Leonard, Andrew. "The Saint of Free Software." *Salon.com,* August 31, 1998, http://archive.salon .com/21st/feature/1998/08/cov_31feature.html (cited September 16, 2004).

———. "The Richard Stallman Saga, Redux." *Salon.com,* September 11, 1998, http://archive.salon .com/21st/feature/1998/08/cov_31feature.html (cited September 16, 2004).

Schofield, Jack. "The Code of the Freedom Fighter." *The Guardian,* November 12, 1998.

Steed, Judy. "Freedom's Forgotten Prophet." *Toronto Star,* October 9, 2000.

Zachary, G. Pascal. "Free For All." *Wall Street Journal,* May 20, 1991.

## Websites

FAIFzilla.org. Free online version of Stallman's biography, *Free As in Freedom,* published under the GNU General Public License, http://www.faifzilla .org/ (cited September 16, 2004). (The book is also available, under the GNU General Public License, in the O'Reilly Online Catalog. http://www.oreilly .com/openbook/freedom/ (cited September 16, 2004).)

The GNU Project and the Free Software Foundation. Online home of both the GNU Project and the Free Software Foundation, with extensive links to articles on GNU–free software philosophy, Stallman's speaking engagements and writings, and GNU software itself, http://www.gnu.org/ (cited September 16, 2004).

Open Source Initiative. Online home of the Open Source Initiative, counterpart of the Free Software Foundation, http://www.opensource.org/ (cited September 16, 2004).

Richard Stallman's Personal Page. Richard Stallman's personal home page, offering an array of anecdotes, political articles, photos, jokes, travelogues, calls for action, and links, http://www.stallman.org/ (cited September 16, 2004).

Salon.com: The Free Software Story. The complete collection of *Salon.com* articles on open source, free software, and Linux, http://archive.salon.com/tech /special/opensource/ (cited September 16, 2004).

# Bruce Sterling (1954–)

## SCIENCE FICTION AUTHOR

**Bruce Sterling is best known** as one of the original founders of cyberpunk, an anarchic, technology-heavy subgenre of science fiction that emerged in the early 1980s. A journalist by training, Sterling spent much of the 1970s and 1980s writing futuristic stories and novels. In 1990, when the U.S. government launched an intense nationwide search-and-seizure effort against computer hackers, Sterling felt compelled to write fact instead of fiction. His 1992 nonfiction book on the subject, *Hacker Crackdown,* remains one of the landmark histories of the tensions on the wild frontier of cyberspace. Since its publication, Sterling has remained one of the most prolific and entertaining writers and speakers in both the realms of science fiction and technology.

## WHICH WAY TO TURKEY CITY?

Born in Brownsville, Texas, Sterling came from a family with roots in Texas cattle ranching, but, at the time of his birth, his mother was a nurse and his father was studying mechanical engineering at the University of Texas. His father's graduation prompted a move to Galveston, in southeastern Texas, where his father took a job with an oil company.

In middle school, Sterling discovered science fiction and it quickly transformed his life. "[Science fiction] was the greatest mental antidote to school that I ever found," he said in a 1994 interview with *Electronic Learning* magazine. "In fact, when I really discovered science fiction, my grades plummeted drastically. It . . . was a sign that my intellect was actually waking up for the first time in my life." By the age of 12, Sterling was writing his own stories.

When he was fifteen, Sterling's family moved to southern India where his father helped design a fertilizer plant. For two and a half years, Sterling studied and traveled extensively throughout the region. He

calls the time he spent in India formative, explaining, in a 1997 interview with the *Dallas Morning News,* "If you look at science-fiction writers, you'll find it's very common for them to have spent a long time in another society."

When Sterling returned to the United States, he enrolled at the University of Texas–Austin. Initially, he thought he would become a scientist, but that dream changed when he talked to astronomers about the tedious realities of their work. Instead, he decided, he would become a science writer and enrolled in the university's journalism program.

Sterling continued to read science fiction well into college and was an avid fan of the New Wave subset of British science fiction writers who pushed the boundaries of science fiction from the simply fantastical into the literary. In 1973, he joined the university science fiction club; he also joined a fledgling group of amateur sci-fi writers who called themselves the Turkey City Writer's Workshop.

At just twenty years old, Sterling was one of the youngest writers in the workshop, whose loose-knit membership came from all over the state of Texas. In 1973, he and several other workshop members traveled to College Station, Texas, to the Aggieon, a science fiction convention held each year at Texas A&M University. Sterling submitted a short story, "Living Inside," to Harlan Ellison, an acclaimed science fiction author, editor, and critic, who was leading a writing workshop at the convention. Although other panelists thought little of Sterling's writing, Ellison was extremely impressed. He bought the rights to "Living Inside" (although he never published it) and later paid for Sterling to attend another well-known science fiction writing workshop.

Sterling's first published short story, *Man-Made Self,* appeared in 1976, during his senior year in college. The following year, Sterling graduated with a degree in journalism and Ellison published Sterling's first novel, *Involution Ocean,* as the fourth installment of the Harlan Ellison Discovery series. (Sterling describes the novel as "a very distorted version of *Moby Dick,*" set on an alien planet.) Although *Involution Ocean* did not launch Sterling immediately into the pantheon of science fiction writers, it did garner him one very important fan letter from an aspiring writer, William Gibson, in Vancouver, Canada. The two men corresponded for several years before they finally met at a science fiction convention in Denver in 1982. By then, Sterling had published another novel, *The Artificial Kid,* and Gibson had written a handful of short stories for Isaac Asimov's *Science Fiction* magazine. Aside from mutual admiration, the two men also shared an interest in computers and a growing frustration with the current state of their genre. "We just felt science fiction wasn't much fun," Sterling explained to the *Dallas Morning News.* "It didn't grab you by the lapels. It wasn't fast enough, it wasn't dense enough. It seemed to talk down to its readers." Sterling and Gibson set out to turn science fiction on its head.

## THE TRUTH IS OUT THERE

In the early 1980s, Sterling began editing a science fiction newsletter, "Cheap Truth." Using the alias Vincent Omniveritas—a play on the Latin phrase *Vincit omnia veritas* (Truth conquers all)—he led a growing band of like-minded writers in challenging the status quo, calling for new writing that was not only "good" and "alive," but also "readable." The newsletter had an energetic, youthful, underground sensibility, one it shared with the punk rock music of the era.

For three years, "Cheap Truth" also functioned as the unofficial voice for a group of science fiction writers who called themselves The Movement; its members: Sterling and Gibson plus John Shirley, Rudy Rucker, Lewis Shiner, and Pat Cadigan. Soon, their ideas began to spread. In the winter of 1985, *Interzone,* a science fiction magazine, published "The New Science Fiction," an article that discussed The Movement's literary aims and principles. Although "Cheap Truth" ceased

*What looks like a prediction often is really just playing off people's ignorance.*

—Sterling to Context, February 2002

publication in 1986, The Movement lived on, and it soon earned a new, decidedly more modern, name—cyberpunk.

"Cyberpunks didn't call ourselves cyberpunks," Sterling explained in a 1997 interview for sfsite.com, a science fiction Website. "It was a name that was put on us. We were just Movement SF writers. We wanted to write commercial science fiction, or at least stuff that would pass for it, with a generational voice in it, a new sensibility to it." According to Sterling, the cyberpunk sensibility emerged from the realm where the computer hacker and the punk rocker intermingled. By the mid-1980s, with the PC revolution well under way, computers were becoming a fact of everyday life. Cyberpunk authors tapped into the creeping mix of hope and anxiety that widespread computing technologies stirred up in the culture at large. In his article, "Cyberpunk in the Nineties," Sterling explained, "The techno-social changes loose in contemporary society were bound to affect its counterculture. Cyberpunk was the literary incarnation of this phenomenon." At the same time, cyberpunk authors, much like punk rock musicians, flouted the established rules of their genre with an aggressive attitude that invigorated many of their fans.

In 1986, two years after the release of Gibson's wildly popular cyberpunk novel, *Neuromancer,* and a year after Sterling published yet another novel, *Schismatrix,* Sterling edited *Mirrorshades,* a collection of short stories by various authors that is widely regarded as the seminal cyberpunk anthology. Indeed, Sterling's prologue, in which he defines the subgenre's ethos, has been described as the de facto cyberpunk manifesto. He depicted cyberpunk as "an unholy alliance of the technical world and the world of organized dissent—the underground world of pop, visionary fluidity, and street-level anarchy." Among the nearly two dozen full-length books Sterling has either written or edited, *Mirrorshades* remains his most popular work.

By the late 1980s, the unbridled spirit of the young cyberpunk movement had settled into a rather established—though still dystopian—set of literary conventions. Indeed, Sterling's next novel, *Islands in the Net*

(1989), struck many fans as a move toward more run-of-the-mill science fiction. Sterling and Gibson, however, were hard at work, collaborating intensely on a new novel. For more than two years, they sent notes, floppy disks, and manuscripts between Texas and Vancouver via Federal Express. The plot of this new collaborative work was set not in the future but in 19th-century England, and, to make clear the departure from their earlier writing style, they called their work a "steampunk" novel. *The Difference Engine* (1990) is set during the Industrial Revolution, and it imagines what might have happened if early computer scientist Charles Babbage had been able to construct his machine, the Analytical Engine (in reality, it remained only a theory on paper until many years after Babbage's death). The novel garnered both authors a great deal of acclaim.

## WHEN BIG BROTHER COMES KNOCKING

The same year *The Difference Engine* was published, Sterling became engrossed in a real-life drama in his hometown of Austin, Texas. In March 1990, U.S. Secret Service agents, collaborating with local authorities, raided the offices of Steve Jackson Games, a small-time publisher of computer games and science fiction–fantasy books. Authorities confiscated computers, hard drives, and floppy disks under a sealed search warrant. No charges were ever filed. Several months passed before any equipment was returned and, in the meantime, Jackson nearly lost his company.

Sterling was dumbfounded. Why, he wondered, would the Secret Service target a man who created science fiction role-playing games? The facts soon emerged: the Secret Service was in search of a classified telephone company document that held vital information about the emergency 911 system for the state of Illinois. The E911 document, as it was called, had been, according to the authorities, stolen by computer hackers and disseminated via the Web. The government feared hackers

*The job of [science fiction] is not to reveal your destiny. It's to expand the spectrum of possibility and refresh your thinking.*

—Sterling on Slashdot.org, October 1999

would use the document to infiltrate the telephone system and wreak havoc nationwide. The link to Steve Jackson was indirect—one of his employees was believed to have maintained an electronic bulletin board system on which the E911 document allegedly appeared.

As Sterling quickly discovered, the raid in Austin was not unique. Beginning in January 1990, more than two dozen similar raids had been made in fourteen American cities. In all, nearly 25,000 floppy disks and more than forty entire computer systems were confiscated. Each raid was part of a quietly aggressive government plan to crack down on computer hackers—Operation Sun Devil.

Once Sterling put the initial pieces of the story together, he was outraged. "It was a wake-up call," he told the *Dallas Morning News*. "I decided the time had come for me to begin to act as a journalist." Over the next year, he tracked the events that led up to Operation Sun Devil—beginning the day AT&T's long-distance telephone switching system mysteriously crashed—and its fallout, including the 1990 formation of the Electronic Frontier Foundation, a nonprofit organization dedicated to establishing and protecting civil rights in cyberspace. *Hacker Crackdown,* the resulting book of investigative journalism (Sterling's first piece of professional nonfiction writing), was published in 1992.

Sterling has since joked that the notion of computer "break-ins" was so new in the early 1990s that it took a science fiction writer to adequately capture the zeitgeist of that era. In many ways, Sterling succeeded. In addition to chronicling the events in and around May 1990, *Hacker Crackdown* illustrates the particular anxiety about computers that existed at that time. "A computer hacker puts the face on the menace that is represented by computers," he said in a 1991 interview with *Compute* magazine, adding that he, himself, understands and even shares fear of the computer. "Computers are a challenge and a threat, and they're changing our society in ways that we can't control and don't understand."

That *cyberpunk* became shorthand for a computer hacker of the late 1980s and early 1990s spoke to the same anxiety. Although Sterling has allowed that cyberpunk, as a literary genre, often deals extensively with computer crime and features (anti) heroes with highly antisocial values, he believes that the evolution of the term *cyberpunk* to mean "computer criminal" was grounded in ignorance and fear on the part of the authorities. "If [Steve Jackson] hadn't been making a game called Cyberpunk," Sterling explained to *Compute*, "this innocent guy would not have been struck down on his own property by agents of the federal government, not to mention the state of Illinois and Bell [Telephone Company]."

*Hacker Crackdown* went through several printings and launched Sterling's subsequent careers as a nonfiction writer and public speaker on the issue of civil liberties in cyberspace. (An active proponent of electronic distribution, Sterling later released the entire text of *Hacker Crackdown* online, as noncommercial "literary freeware"; many of his speeches are posted free online as well.) In the process of writing the book, Sterling became an active member of the Austin outpost of the Electronic Frontier Foundation.

In January 1993, Sterling appeared on the cover of *Wired* magazine's premiere issue, which bore the headline, "Bruce Sterling Has Seen the Future of War." (Sterling has been a contributing writer ever since and maintains a blog, or online diary, on *Wired*'s Website.) In April of that year, Sterling appeared before the House Subcommittee on Telecommunications and Finance to testify about the long-term effects of the National Research and Education Network, a proposed academic computer network. In his testimony, Sterling posed as network administrator "Bob Smith," from the year 2015, but when he finished, he took questions as acclaimed science fiction novelist Bruce Sterling, circa 1993—by his own admission, an equally unlikely expert.

> *I came to terms about being a "cyberpunk writer" a long time ago. It's like asking William Burroughs, "When they bury you, you're going to be a beatnik, aren't you?"*
>
> —Sterling to
> www.sfsite.com,
> August 1997

## IT'S NOT EASY BEING GREEN

Sterling's success with nonfiction did not lure him from science fiction. Instead, it freed him to write *Heavy Weather*, a dark novel about global warming and environmental chaos, published in 1994. His vision was prescient; indeed, in a 2000 interview with the *Austin American-Statesman*, Sterling admitted, "It scares me about how accurate that book looks now."

Two years prior to that interview, in the summer of 1998, Sterling was horrified by the smell of raging forest fires in Mexico wafting across the porch of his home in Austin. The rising temperatures and destructive weather patterns that had ravaged parts of North America had radicalized Sterling's beliefs about the environment. "This is freakish," he told the *Austin American-Statesman* in 1998. "There's no place to hide from CO2. Everybody's under the same sky. This is a stupid problem to have, like marinating in your own trash."

By the fall of 1998, Sterling had begun expressing his outrage at rampant global warming during conference speeches. He also offered a plan. Sterling called for the beginning of a radical new ecological movement—one driven less by politicking than by art and design. (Just as *Heavy Weather* marked the beginning of Sterling's interest in the environment, his book *Holy Fire*, published in 1996, was prompted by his growing interest in industrial design.) He called members of the movement "Viridian Greens"—viridian being a harsh, high-tech, chemical-hued shade of green, far different from the earth-toned greens of more middle-of-the-road eco-groups.

Sterling's movement began with a Website and a mailing list known as "Viridian Notes." On January 3, 2000, the Viridian Design movement released its manifesto. (At the time, "Viridian Notes" had about 1,200 subscribers worldwide.) With a built-in expiration date of 2012, Sterling called for the development of high-tech, industrial design products that address the failing environment. "I want to start a groundswell that says it's sexy to have a solar-powered Website, that an electric car is cooler than the kind of car where you turn the key and it kills you if you stand in the garage with it," Sterling explained in a 1999 *Los Angeles Times* interview. "If there's consumer demand for usable, Viridian goods, then everything will follow."

While the Viridian Movement continued to thrive in the 21st century, with Sterling involved in the Viridian blog (www.viridiandesign.org) and working with pro-Viridian product designers, another of his online movements, the Dead Media Project, has languished. A preservation project of another kind, the Dead Media Project aimed to amass a catalog of all types of media—from ancient to modern times—that have perished in the forward march of technology. Much like the Viridian Movement, the project was conceived, back in August 1995, over the course of several of Sterling's speeches and was sustained through an online mailing list and Website. In a speech, "The life and death of media," given in September 1995, Sterling proclaimed, "We're using the Internet to bring people together to catalog and study extinct forms of human communication. We're in the media autopsy business. We're into media forensics."

In just over three years, Sterling, project cofounder and fellow science fiction writer Richard Kadrey, and a loose group of fellow self-named *necronauts,* catalogued hundreds of different kinds of dead media, including Morse code, carrier pigeons, smoke signals, old video games, the Xerox Alto (the first personal computer), and, Sterling's particular favorite, the Incan *quipu,* an ancient communication device made of cotton and wool, the workings of which, to this day, are not understood. By 1998, however, the project was stalling somewhat, and plans for a book of dead media were shelved. "I think we have a pretty good grip on the dead media of previous centuries," he told the *Irish Times* in 2001. "Still," he added, referring to the never-ending wave of computer technology that becomes obsolete each year, "it's a bit like shoveling the sidewalk before it stops snowing."

## TOMORROW AND TOMORROW AND TOMORROW

Science fiction authors have long predicted, in their fantasy writing, objects and technologies that have appeared in reality years later. For instance, Jules Verne, the seminal 19th-century science fiction writer, famously predicted the submarine and space travel, and Arthur C. Clarke foresaw the coming of the satellite age. In a similar vein, Sterling has made a small place for himself as futurist, by way of being a novelist. Although he has scoffed at the idea of selling predictions in his work, he admitted, in a 1999 online chat for *Slashdot,* "I don't have to read the future in tea leaves. I just have to stand a little closer to the trends than most of my readers do."

After publishing the novel *Zeitgeist* in 2000, Sterling began work on a nonfiction book of predictions. The finished product, *Tomorrow Now: Envisioning the Next 50 Years,* released in 2002, emerged as a complex, book-length essay on how the 21st century might look and feel. Some critics claimed *Tomorrow Now* was merely an inventory of the themes brought up in his recent books—environmental problems, from *Heavy Weather;* life extension, from *Holy Fire;* political and social disintegration, from *Distraction.* Sterling, for his part, likened writing the book to "going on a spiritual retreat." In a 2002 interview with the *Austin American-Statesman,* he explained, "I'm tackling new themes, my writing seems to be warmer."

He has since continued to expand his horizons. In 2003, Sterling was appointed a professor at the European Graduate School in Switzerland. In 2004, he released a new novel based on the terrorist attacks of September 11, 2001, titled *The Zenith Angle.* Not quite science fiction, the book, set in 2001 and 2002, is part of a new genre Sterling has dubbed *nowpunk*—contemporary historical fiction, written while the history is still being lived.

Although Sterling's writing style continues to evolve, he remains, for many of his colleagues and fans, the embodiment of the 1980s cyberpunk. More than twenty years ago, he and his fellow Movement writers managed to capture the zeitgeist of the networked age. Since then, Sterling has consistently been at the forefront of new Internet-era movements, tackling the essential questions facing modern culture.

## FURTHER READING

### In These Volumes

Related Entries in this Volume: Gibson, William

Related Entries in the Chronology Volume: 1984: William Gibson Introduces the Term *Cyberspace;* 1990: Operation Sun Devil
Related Entries in the Issues Volume: Hackers

## *Works By Bruce Sterling*

### Novels

*Involution Ocean.* New York: Ace Books, 1977.
*The Artificial Kid.* New York: Harper & Row, 1980.
*Schismatrix.* New York: Arbor House, 1985.
*Islands in the Net.* New York: Arbor House, 1988.
*The Difference Engine.* New York: Bantam Books, 1991.
*Heavy Weather.* New York: Bantam Books, 1994.
*Holy Fire.* New York: Bantam Books, 1996.
*Schismatrix Plus.* New York: Ace Books, 1996.
*Distraction.* New York: Bantam Books, 1998.
*Zeitgeist.* New York: Bantam Books, 2000.
*The Zenith Angle.* New York: Del Rey, 2004.

### Short Story Collections

*Mirrorshades: A Cyberpunk Anthology.* New York: Arbor House, 1986.
*Crystal Express.* Sauk City, WI: Arkham House Publishers, 1989.
*Globalhead.* Shingletown, CA: Mark V. Ziesing, 1992.
*The Hacker Crackdown: Law and Disorder on the Electronic Frontier.* New York: Bantam Books, 1992.
*Tomorrow Now: Envisioning the Next Fifty Years.* New York: Random House, 2002.

## *Articles*

Barry, Courtney. "Bruce Sterling Is Helping Dream the Future into Existence." *Science Fiction Weekly,* March 12, 2001, http://www.scifi.com/sfw /issue203/interview.html (cited September 16, 2004).
Beach, Patrick. "A Futurist at 44." *Austin American-Statesman,* December 15, 1998.
Brown, Dwight, Lawrence Person, and Michael Sumbera. "Under Heavy Weather: An Interview with Bruce Sterling." *NovaExpress #13* (1997), http://www.sflit.com/novaexpress/13 /bsi-1.html (cited September 16, 2004).
Chapman, Gary. "Local Warming, Global Warming— It's the Same Thing." *Austin American-Statesman,* September 7, 2000.
Fisher, Marc. "Bones of Invention." *Washington Post,* July 17, 1998.
Godwin, Mike. "Cybergreen." *Reason,* January 2004, http://reason.com/0401/fe.mg.cybergreen.shtml (cited September 16, 2004).

Krantz, Michael. "Cyberpunk Spinmeister with Distraction." *Time,* December 21, 1998.
Marvel, Bill. "Sam Houston Meets the Cyberpunks' Cyberpunk." *Dallas Morning News,* January 30, 1992.
———. "Charting Cyberpace With a Wink and a Wit." *Dallas Morning News,* November 9, 1997.
Myer, Thomas. "Chatting with Bruce Sterling at LoneStarCon 2." *SF Site Convention Report,* August 29, 1997, http://www.sfsite.com/09a/bru16.htm (cited September 16, 2004).
Salamon, Jeff. "Bruce Sterling and the Magnificent Compost Heap." *Austin American-Statesman,* November 16, 2000.
Walsh, Dave. "Doing Sterling Work." *Irish Times,* August 27, 2001.
Weise, Elizabeth. "Writer Rants for Green Scene to Sprout From E-Roots." *USA Today,* June 16, 1999.

## *Websites*

Bruce Sterling: Science Fiction Writer and Futurist Thinker of Cyberspace. Sterling's faculty Web page at the European Graduate School, http://www.egs.edu/faculty/sterling.html (cited September 16, 2004).
The Bruce Sterling Online Index. Comprehensive fan site for Sterling created by Austin-based fan Chris Waltrip. Extensive links to Sterling's columns, blogs, interviews, Websites, and miscellany, http://www .chriswaltrip.com/sterling/bsinfo.html (cited September 16, 2004).
The Dead Media Project. Online home of the Dead Media Project, http://www.deadmedia.org/ (cited September 16, 2004).
History of Cyberpunk. Comprehensive site about the history of cyberpunk, created in 1996 by the two men behind The Cyberpunk Project, http://project .cyberpunk.ru/idb/history.html (cited September 16, 2004).
The Infinite Matrix | Bruce Sterling | Schism Matrix. Sterling's former blog on *Infinite Matrix,* a science fiction online magazine, http://www.infinitematrix .net/columns/sterling/sterlingi.html (cited September 16, 2004).
Mirrorshades Postmodern Archive. Digest of Sterling's work and related science fictions items, hosted on The Well. Extensive links, http://www.well.com /conf/mirrorshades/ (cited September 16, 2004).

The Turkey City Writer's Workshop Homepage. Online home of the Turkey City Writer's Workshop, http://home.austin.rr.com/lperson/TC.html (cited September 16, 2004). Additional information on Turkey City can be found on the Science Fiction and Fantasy Writers of America site: http://www.sfwa.org/writing/turkeycity.html (cited September 16, 2004).

The Viridian Design Movement. Online home of Sterling's Viridian Design Movement, including "Viridian Notes" and transcripts of Sterling's original speeches on the Viridian Greens, http://www.viridiandesign.org/ (cited September 16, 2004).

*Wired* News Blog: Beyond the Beyond. Sterling's current blog for *Wired*, started in September 2003, http://blog.wired.com/sterling/ (cited September 16, 2004).

# Robert Taylor (1932–)

## FOUNDER OF ARPANET AND THE XEROX PARC COMPUTER SCIENCE LAB

**Although many contributed to** the birth of the ARPANET, the government-funded predecessor to the Internet, Robert Taylor was the one who actually put the project in motion. As director of the Information Processing Techniques Office (IPTO) at the Pentagon's Advanced Research Projects Agency (ARPA) in the mid-1960s, Taylor was in a position to put funding and researchers behind the idea of the distributed computing network he dubbed the ARPANET. By the time the finishing touches were being made, Taylor had left ARPA and founded Xerox's Palo Alto Research Center (PARC), where some of the most important advances in computing—including the first personal computer—were developed during his tenure (1970 to 1983). After thirteen years of leading Xerox PARC and its band of computer science luminaries, Taylor founded Digital Equipment Corporation's System Research Center, another forward-thinking computer research lab. Taylor, a psychologist by training, was a leader and visionary; his uncanny ability to draw the best in the field to his vision of computing earned him the sobriquet, "the impresario of computer science."

## WALKING IN LICKLIDER'S FOOTSTEPS

Taylor was born in Texas and adopted as a newborn by a Methodist minister and his wife. For much of his childhood, the family moved from parish to parish. At age 16, Taylor entered Southern Methodist University in Dallas, Texas, intending to become a minister like his father. When he returned from a tour of duty with the navy during the Korean War, he entered the University of Texas at Austin to finish his degree under the GI Bill. Taylor tried many different majors and minors—according to him, he was a "professional student"—but eventually graduated with a bachelor's degree in experimental psychology, with minors in math, philosophy, English, and religion. While studying for his master's degree in psychoacoustics, a discipline within psychology that focuses on how the brain turns vibrations in the air into intelligible sound, Taylor encountered the work of J.C.R. Licklider, a professor of psychoacoustics then at the Massachusetts Institute of Technology (MIT). In the coming years, the two men would become two of the most important figures in the history of the Internet.

After a short time teaching math and coaching basketball at a Florida prep school, Taylor moved into the field of engineering. In 1960, he accepted a position as a systems design engineer at the Martin Company in Florida. The following year, he joined ACF Electronics in Maryland, but left shortly thereafter to accept a position at NASA's Office of Research and Technology. NASA offered Taylor, not yet 30 years old, a post after reviewing a proposal he had written for a flight-control-simulation display.

At NASA, Taylor was responsible for exploring new areas of technology, including some work in computer science. Indeed, this was Taylor's first real brush with the world of early computing. He channeled key funding to Douglas Engelbart's research in computer display technology at the Stanford Research Institute (SRI)—work that eventually led to the development of the first mouse. In a 2003 interview, Engelbart explained, "What saved my program from extinction then was the arrival of an out-of-the-blue support offer from Bob Taylor. . . . I had visited him months before . . . and I had been unaware that meanwhile he had been seeking funds and a contracting channel to provide some support."

Taylor also met Licklider, who was doing similar, though more broad-based, work at ARPA as founder of the Information Processing Techniques Office (IPTO). Though Taylor was already familiar with Licklider's work in psychoacoustics from the 1940s and 1950s, what truly inspired him was Licklider's seminal March 1960 paper, "Man-Computer Symbiosis." Licklider offered a vision of real-time, interactive computing, where computers augmented human capabilities instead of simply crunching numbers. Taylor, like so many young scientists of his generation, was enthralled. In no small way, Licklider's paper set the course of the rest of Taylor's career.

## ONE NETWORK TO RULE THEM ALL

In 1965, Taylor was lured from NASA to become deputy director of IPTO under Ivan Sutherland, Licklider's successor (Licklider had stepped down in 1964). When Sutherland left, less than a year later, Taylor was poised to take over. At the time, IPTO was funding several large projects, including Project MAC at MIT—an experiment in computer time-sharing that was in keeping with Licklider's man-computer symbiotic vision—as well as various smaller scale projects in artificial intelligence and computer graphics. As director, Taylor increased IPTO's budget by 100 percent, even as ARPA's budget was shrinking. Taylor stated, "Our rule of thumb was to fund people who had a good chance of advancing the state of information processing by an order of magnitude." Indeed, what followed has been called the golden age of computer research.

IPTO entered its golden age in February 1966. At the time, Taylor was able to connect with different time-sharing computers located across the country at MIT, the University of California–Berkeley, and the Systems Development Corporation in Santa Monica, California. However, each time-shared computer required its own terminal and had its own set of user commands. For Taylor, communicating with these different projects often became a game of musical chairs—he had to move from computer to computer, emailing to MIT and then to Berkeley, neither of which could communicate directly with the other. Although working with the cutting-edge of computing technology, Taylor was frustrated, and he dreamed up the idea of one network that could connect all the different terminals: the ARPANET.

*By 1975, most of the technology was available; by 1985 it was affordable; but it didn't take off until 1995. . . . My timing was awful.*

—*Taylor to* The Almanac, *October 2000*

Taylor, who was a psychologist and not a programmer, brought his idea to Charles Herzfeld, then the director of ARPA, explaining the problem and suggesting his somewhat nebulous solution. In less than twenty minutes, Taylor's pitch had worked—Herzfeld gave him a budget of $1 million.

Taylor immediately got to work on the official proposal, titled "Cooperative Network of Time-Sharing Computers," and sought out Larry Roberts, an MIT scientist with experience in computer networking, to carry out his plan. Roberts, who had recently built a transcontinental network between two computers, initially declined Taylor's offer. Undeterred, Taylor lobbied Roberts's superiors at MIT until Roberts finally relented and joined IPTO, where, soon after, he took over the development of Taylor's ARPANET.

While Roberts worked on the proposal, Taylor joined Licklider in writing a seminal article in the history of the Internet, "The Computer as Communications Device." At the time, many in the world of computing condemned the idea of the ARPANET as idiotic. Indeed, industry leaders like AT&T and IBM did not believe it could work. Still, Taylor and Licklider defended Internet-style computing as the wave of the future. Their article outlined the social impact of computer networking—the formation of virtual communities, drawn together not by geographic proximity but by interests.

In the three years it took to get the ARPANET up and running, Taylor continued to seek out and fund computing advances, including much of Engelbart's work at SRI. Taylor helped put the funding in place for Engelbart's demonstration of interactive computing at the 1968 Joint Fall Computer Conference in San Francisco, which opened the eyes of countless computer scientists still accustomed to computers that ran on paper tape and punch cards. Though he did not support the Vietnam War, Taylor also traveled to Vietnam, at the request of the federal government, to help set up a military computer center near Saigon. All the while, the U.S. government increased pressure on ARPA to move away from generalized computing goals toward more specifically military ones. Taylor felt stymied. "It crippled a lot of research," he told *The Almanac,* a Bay Area community paper, in 2000. Convinced that the ARPANET was well on its way to success, in 1969, Taylor handed IPTO's reins to Larry Roberts and headed west.

## THE OFFICE OF THE FUTURE

In 1969, Xerox Corporation announced a bold new initiative—to create "the architecture of information" for the future. (According to legend, Xerox's CEO then turned to a vice president of research and development and said, "All right, go start a lab that will find out what I just meant.") Taylor, who had joined the University of Utah after leaving ARPA, had already earned a reputation as a computing visionary through his work with Engelbart and the ARPANET. Although his vision of computing was vital to the job, Xerox was more focused on Taylor's ability to recruit the necessary talent to create the lab and architecture promised. In 1970, Taylor was brought in as the founding manager of Xerox's new computer science research lab, one of three labs in Xerox's new Palo Alto Research Center (PARC). Over the next thirteen years, Taylor would make Xerox's vision a reality.

In the beginning, Taylor and his two dozen early recruits—many of them graduate students Taylor had met while at NASA and ARPA—toiled in a temporary office near Stanford University filled with rented furniture, one telephone, and no computer. (Xerox PARC later moved into the Stanford Industrial Park, alongside companies like Hewlett-Packard.) From such humble beginnings, Taylor steadily built the lab into the jewel of Xerox PARC, bringing in the top computer scientists of the day.

By 1971, PARC had already become a hotbed of new developments in computing. Under Taylor's direction, computer scientists—most notably Alan Kay, Butler Lampson, and Charles Thacker—began work on the first personal workstation, the Alto, which they delivered in 1973. This groundbreaking computer utilized a WYSISYG (what-you-see-is-what-you-get) dis-

> *The Internet is not about technology; it's about communication.*
>
> —*Taylor to* The Almanac, *October 2000*

play, point-and-click commands and cut-and-paste word processing technology. The Alto worked with a laser printer and could also transmit documents and messages over an Ethernet, a local area network developed by legendary PARC scientist Robert Metcalfe. The laser printer and Ethernet were also developed at PARC.

Taylor set the tone that allowed the creative burst of those first two years at PARC. First and foremost, PARC was one of the rare labs where scientists worked collaboratively on a single project—at most research and academic labs individuals worked on single aspects of a larger project. To bring PARC's growing number of scientists together, Taylor held weekly "dealer" meetings. The Dealer's Room was outfitted with beanbag chairs, thick carpet, and a floor-to-ceiling dry erase board—a place where PARC scientists came together to speak their minds about the project at hand and to challenge their colleagues. (The original beanbag chairs were eventually donated to the Smithsonian.) Dialogue in the room could sometimes be brutal, but it was always constructive. As Alan Kay told National Public Radio in 2004, "Bob deserves the credit for putting together the environment that made us all smarter."

By the late 1970s, Taylor's goals for PARC were clearly beginning to diverge from those of Xerox. For several years, the copier business had become increasingly competitive, and Xerox's budget for PARC's open-ended computing research shrank in response. Taylor also felt that Xerox had begun to ignore PARC's work. Taylor finally stepped down as PARC's founding manager in 1983, leaving a formidable legacy. Although Xerox never exploited many of the inventions developed at PARC under Taylor's leadership, PARC research produced products that are the basis of some of today's leading technology companies, including Adobe, Novell, 3Com, and Apple Computer. Steve Jobs and a group of Apple engineers took generous inspiration for the Macintosh computer from the Alto, despite the wishes of many PARC staffers to keep their ideas to themselves. As Taylor explained to National Public Radio in 2004, "I happened to be out of town the day that he visited, and I didn't know about any of its arrangements or anything, and it's probably fortuitous, because if I had been in town, I would have tried to throw him out of the lab."

## THE NEXT STAGE

In 1983, when Taylor left to found the Systems Research Center (SRC), Digital Equipment Corporation's new computer research facility, he brought with him a dozen PARC scientists. Again, he amassed the talent necessary to tackle the next stage of personal computing. (SRC became part of Compaq and is currently part of Hewlett-Packard.)

During Taylor's twelve years at the helm of SRC, the lab became an industry leader, developing several key advances in computing, including the first fault-tolerant LAN (local area network), the first electronic book, and an object-oriented programming language—Modula 3—that was a predecessor to Java. Toward the end of Taylor's tenure, in the spring of 1995, SRC developed Alta Vista, the first searchable, full-text index of the World Wide Web. By the time of his retirement, in 1996, Taylor's visions for networked computing, nearly 30 years old, had finally become a mainstream reality.

Since leaving SRC, Taylor has kept in the public eye as the go-to man for writers covering the history of computing and the Internet. "Several books and stories are quite off the mark," he quipped to *The Almanac* in 2000. In retirement, he has received some of the highest accolades in his field; in 1999, he was awarded the National Medal of Technology, the nation's highest honor for technological innovators. Five years later, he and three other scientists from Xerox PARC were awarded the Charles Stark Draper Prize by the National Academy of Engineering for their work on the Alto. Indeed, since developing the original ARPANET, Taylor has continued to be one of the most important figures in the history of the Internet and modern computing.

## FURTHER READING

### In These Volumes
Related Entries in this Volume: Kahn, Robert; Licklider, J.C.R.; Metcalfe, Robert

Related Entries in the Chronology Volume: 1958: The Advanced Research Projects Agency Begins Operation; 1963: Memo to the Intergalactic Computer Network; 1966: Bob Taylor's Computers Inspire Plans for the ARPANET; 1967: Plans for ARPANET Are Unveiled; 1969: The ARPANET Is Born; 1970: Xerox Palo Alto Research Center; 1990: ARPANET Is Decommissioned

### Works By Robert Taylor
with J.C.R. Licklider. "The Computer as Communications Device." *Science and Technology,* April 1968.

### Books
Hafner, Katie. *Where Wizards Stay Up Late: The Origins of the Internet.* New York: Simon & Schuster, 1996.

Hiltzik, Michael A. *Dealers of Lightning: Xerox PARC and the Dawn of the Computer Age.* New York: HarperBusiness, 1999.

Rheingold, Howard. *Tools for Thought.* New York: Simon & Schuster, 1985.

Also available online at: http://www.rheingold.com /texts/tft/10.html (cited September 16, 2004).

Waldrop, M. Mitchell. *The Dream Machine: J.C.R. Licklider and the Revolution That Made Computing Personal.* New York: Viking, 2001.

### Articles
Markoff, John. "Robert W. Taylor: An Internet Pioneer Ponders the Next Revolution." *New York Times,* December 20, 1999.

Pitta, Julie. "Who and What Made PARC an Industry Legend." *Los Angeles Times,* September 13, 1995.

Plotnikoff, Dave. "A Father of the Net Looks Back and Asks, 'What Took So Long?'" *San Jose Mercury News,* March 12, 2000.

Softky, Martin. "Building the Internet." *The Almanac,* October 11, 2000, http://www.almanacnews.com /morgue/2000/2000_10_11.taylor.html (cited September 16, 2004).

### Websites
KurzweilAI.net: Big Thinkers—Robert Taylor. Raymond Kurzweil's profile of Taylor, with link to his 1968 paper coauthored by J.C.R. Licklider, "The Computer As Communications Device," http://www.kurzweilai.net/bios/frame.html?main= /bios/bio0146.html (cited September 16, 2004).

National Academy of Engineering 2004 Charles Stark Draper Prize: Robert Taylor. Transcript of Taylor's acceptance speech for the 2004 Draper Prize, http://www.nae.edu/NAE/awardscom.nsf/weblinks /LRAO-5X4TSP?OpenDocument (cited September 16, 2004).

PBS.org: Nerds 2.0.1. "Twenty-Minute Pitch." Segment of Robert Cringley's "Nerds 2.0.1" focused on Taylor and the ARPANET, http://www .pbs.org/opb/nerds2.0.1/networking_nerds/taylor .html (cited September 16, 2004).

The Tech Museum of Innovation | Robert Taylor. Profile of Taylor at the time of his winning the National Medal of Technology in 1999, http://www .thetech.org/nmot/detail.cfm?ID=99&STORY= 3&st=awardDate&qt=1999 (cited September 16, 2004).

"Xerox PARC." Segment of Robert Cringley's "Nerds 2.0.1" focused on Taylor at PARC, http://www.pbs .org/opb/nerds2.0.1/serving_suits/parc.html (cited September 16, 2004).

# Linus Torvalds (1969–)

## CREATOR OF LINUX

**As a twenty-one-year-old** graduate student in Finland in 1991, Linus Torvalds created the central part of a new, UNIX-like operating system (OS) and, with it, a vital new movement in computing. He posted his software code free on the Internet, and soon, programmers and hackers from around the world began to respond with suggestions and new pieces of code. Torvalds gathered the pieces into an entire operating system known as Linux (pronounced "linn-ucks"). Linux was not the first software to be developed collaboratively, but it is, by far, the most successful. Indeed, some have claimed it is the single largest collaborative project in history. Since the mid-1990s, Linux has grown into a robust and reliable operating system, favored by large corporations, governments, and hackers alike. Indeed, Linux has come to be seen as a legitimate competitor to the dominant Microsoft Windows OS.

## THE MIGHTY FINN

Torvalds was born and raised near Helsinki, Finland. At age 11, he was given his first computer by his grandfather, a professor of statistics at the nearby University of Helsinki. It was a Commodore Vic–20—a relatively cheap predecessor to the popular Commodore 64. Using the computer language BASIC, Torvalds programmed the computer to write, "Sara [his younger sister] is the best." Within months, he was writing his own simple computer games.

Torvalds continued to program throughout high school and into college. In 1988, he entered the University of Helsinki as a computer science major. In the fall of his sophomore year, he took a course in UNIX, a popular OS. UNIX had been developed in the United States twenty years earlier—the very year Torvalds was born. Because it was both powerful and flexible, UNIX had become a favorite of scientists, academics, and computer science students, Torvalds included. He wanted to

run UNIX on his computer at home, but at the time UNIX software did not run on PCs, and, at $10,000 each, UNIX workstations were prohibitively expensive.

Torvalds began experimenting with Minix, a scaled-back UNIX-clone that had been developed as a teaching tool by Andrew Tanenbaum, a computer science professor at Vrije Universiteit in Amsterdam. When Minix proved inadequate for his computing needs, Torvalds used its architecture to build his own programs from scratch. By spring 1991, Torvalds realized that although he had not intended to write an entire UNIX-like kernel, he was, indeed, close to having done so. A kernel is the central program of an operating system; it acts like a traffic cop, controlling which programs have access to the central processing unit (CPU). Torvalds decided that, for fun, he would build the rest of an operating system.

It was arduous work. Torvalds spent endless hours writing code in his room with the curtains drawn. "Forget about dating! Forget about hobbies! Forget about life!" Torvalds told *Time* in 1998. "We are talking about a guy who sat, ate and slept in front of the computer." By August, Torvalds had a working draft, and he announced his project in an email posted to the Minix newsgroup on Usenet. "Hello everybody out there using Minix," he wrote. "I'm doing a (free) operating system (just a hobby, won't be big and professional like gnu) . . . I'd like any feedback on this people like/dislike in Minix, as my OS resembles it somewhat." Torvalds closed by writing that suggestions were welcome, but that he could not promise to use them.

The "gnu" Torvalds referenced was the GNU family of free, UNIX-based software being developed by Richard Stallman and others at the Free Software Foundation in Cambridge, Massachusetts. Since the mid-1980s, Stallman had dedicated himself to creating a nonproprietary software system that computer users could freely run, copy, distribute, study, change, and improve. He called it GNU—a recursive acronym for "GNU's Not UNIX." Although many types of GNU programs were in circulation, Stallman had not yet finished the kernel for a full GNU operating system.

To ensure that his programs would remain free, Stallman developed his own copyright license—the GNU General Public License (GPL). The GPL allowed programmers to use whatever GNU code they desired, as long as any changes and improvements were returned to the public domain for others to use and adapt. Though Torvalds was aware of Stallman's work, he decided to release his code with his own, more restrictive license, which forbade anyone from profiting financially from it. (Under the GPL, users were free to profit from GNU software.)

By September 1991, a rough version of the kernel was ready. A colleague of Torvalds volunteered to post the kernel and its source code for download from a Helsinki University of Technology Website. Although Torvalds had privately called the system Linux—Linus, plus Minix—he thought the name too egotistical for publication, so he renamed the kernel Freax—free, plus freak (another name for hacker), plus Minix. His colleague disliked the name Freax and posted the kernel with the name that has, by now, become legendary—Linux.

*When I released Linux, I thought maybe one other person would be interested in it.*

*—Torvalds to the* New York Times Magazine, *February 1999*

The first fully functional (though still faulty) version of Linux appeared that October, and new releases followed almost monthly, each with enhanced capabilities. In these early days, the Linux community numbered fewer than a dozen. Torvalds only occasionally received feedback and suggestions. One early request was to write a compression program that would allow Linux to run on computers with limited memory—a project Torvalds worked on through Christmas day 1991.

By January 1992, Linux had fewer than 100 users, but the community was growing. The compression program, for example, drew dozens of new users. With the release of version 0.12, Torvalds switched Linux's copyright license to the GNU GPL. With that, the GNU operating system was complete, and a whole new community of Linux users was born.

As word of Linux began to spread, suggestions—and code—began to flow into Torvalds' email inbox. By April 1992, the first Linux newsgroup was founded on Usenet. By 1993, Torvalds began to receive truly substantial contributions of code from the community that had sprung up around his creation. By March 1994, with the official release of Linux 1.0, Torvalds' lines of code comprised a mere fraction of the entire Linux operating system. In the end, Torvalds admits, his contribution was not so much the code, but the idea. "I got it started," he told the *New York Times Magazine* in 1991. "Then people had something to concentrate on."

## STONE SOUP

More than thirty months passed between the release of the first version of Linux and the release of Linux 1.0. In that time, the number of Linux enthusiasts grew exponentially, from the ten or so people who responded to the initial Usenet posting to an estimated 100,000 by 1994. (Because Linux is freely distributable and can be downloaded free from the Internet, the exact number of Linux users is difficult to determine.) Similarly, the operating system had itself steadily grown from the initial 10,000 lines of code to more than 170,000, including code for a graphical user interface and networking capabilities.

The new code came directly from the Linux community; each member of the community was free to send in software patches and other code that could then be incorporated into the greater system. Much of Linux was developed by an informal group of about 100 programmers, who Torvalds called the core development team. The result of this volunteer, collaborative effort often nearly matched—and sometimes surpassed—the commercial versions of, for instance, UNIX. The development of Linux followed the model of the Stone Soup parable: when individuals contribute what they can, no matter how meager, a greater good can be achieved. Ambitious Linux programmers eventually began to port the operating system to platforms other than the PC and create their own Linux variants. As

*Linus Torvalds. (Jim Sugar/Corbis)*

Linux continually expanded, changed, and evolved, the glue that held it together was the GPL and Torvalds.

By placing all source code in the public domain, the GPL ensured that, even if a Linux variant branched off, it would still be able to rejoin the primary branch—which was maintained by Torvalds. "My branch is to some degree . . . the trunk of the tree," he explained to the *San Jose Mercury News* in 2003. "People try to join back into my tree."

Torvalds managed his tree in his signature unassuming style. He would spend much of his time—an estimated 95 percent—slogging through hundreds of daily emails, arbitrating disputes, and giving the final say on which code made it into his tree and which did not. (To help him manage the rather unwieldy process, Torvalds amassed an informal group of deputies—dubbed "maintainers"—who specialized in different aspects of the operating system.) Through a mix of self-deprecating humor, diplomacy, and a dedication to the maxim, "the best code always wins," Torvalds provided just the

right mix of guidance and freedom to allow Linux to evolve in its chaotic, but surprisingly efficient, manner. Indeed, in the words of one early Linux contributor, the key to Linux's success was not the collaborative development style, the GPL licensing strategy, or the merit of the code itself. The key, he told *Wired Magazine* in 1997, " . . . is spelled L-I-N-U-S."

By the mid-1990s, the Linux community began hitting its stride. Linux was running on an estimated 3,000,000 machines, on various platforms developed by Intel, Digital Equipment Corp., and Sun Microsystems. After more than seven years working on Linux as a graduate student, Torvalds had earned his master's degree from the University of Helsinki in 1996. (Higher education is free in Finland, so Torvalds had little incentive to finish.) In 1997, he left Finland for California's Silicon Valley, for a job at the then-unknown computer company, Transmeta. Many in the Linux community expressed concern that Torvalds would be too distracted to continue leading Linux—just as they had done when he married and with the birth of each of his children. Their concerns have always proved to be unfounded. Torvalds had arranged to work part-time on Linux while at Transmeta. He remained, as always, at the helm, with Linux's most vital years yet to come.

## HAPPY HALLOWEEN

Torvalds's work in his early years at Transmeta was kept highly secret. (In the beginning, some in the community even speculated the he had gone to the "dark side"—the side of proprietary software—particularly because Paul Allen, one of the founders of Microsoft, was one of Transmeta's primary investors.) While Transmeta kept a very low profile throughout much of the late 1990s, Linux was bursting into the mainstream, as was the collaborative computing movement founded around the OS, known as "open source." By the summer of 1998, Torvalds and Linux had been featured on National Public Radio, in *Wired, The Economist,* and on the cover of *Forbes.* That September, Intel and Netscape, two industry leaders, made significant investments in Red Hat, a commercial Linux enterprise. Linux was fast becoming a known quantity.

That fall, Linux received press from an unlikely source. Just after Halloween, Eric Raymond, a leader of the open-source movement, posted an internal Microsoft memo on the Internet. The memo, dated August 11, 1998, was written by Vinod Valloppillil, a Microsoft engineer who analyzed industry trends. In it, Valloppillil outlined how Linux and other open-source projects, such as the Apache Web server, constituted a threat to the dominant Windows operating system—Microsoft's bread and butter. "OSS [open-source software] poses a direct, short-term revenue and platform threat to Microsoft, particularly in server space," he wrote. Valloppillil went on to suggest that Microsoft itself could benefit by adopting a Linux-like development process, which, he wrote, "promotes rapid creation and development of incremental features and bug fixes in an existing code/knowledge base." A former Microsoft employee leaked a second "Halloween" memo to Raymond, this one more specifically focused on Linux, just a few days later. Also dated August 11,

1998, the "Halloween II" memo, as it has come to be known, called Linux the "best-of-breed UNIX" and outlined the various ways in which Linux outperforms Microsoft's NT technology.

The Halloween memos made big news in industry and mainstream media. Microsoft was forced to respond publicly about the memos' authenticity to *Wired,* the *New York Times,* and the *Wall Street Journal* among others. Microsoft downplayed the memos as much as possible, save the occasional spin that the growing success of Linux proved that Microsoft had legitimate competitors and was therefore not a monopoly. Intended to bolster Microsoft, which was in the midst of heated antitrust litigation, the spin was a great boon to Linux.

Torvalds kept quiet on the matter. True to form, he remained calm when conversation turned to Microsoft, even as Linux continued to gain market share throughout the latter part of the 1990s. By 1999, with the introduction of desktop software packages by two competing Linux development groups, GNOME and KDE, Linux was making small but significant inroads not only into the server market but into the desktop PC market as well. With Red Hat's initial public stock offering (IPO) a success, Torvalds, formerly just a well-paid programmer, joined the league of Silicon Valley millionaires. (Red Hat and VA Linux, another Linux company, had both given Torvalds gifts of stock; other than these shares, Torvalds had never profited commercially from his creation.) After Transmeta finally unveiled its mysterious product—the Crusoe microprocessor, a low power "smart chip" intended for mobile devices—in January 2000, Torvalds was free to turn most of his attention back to his true calling—the next version of Linux, Linux 2.4.

*My reasons for putting Linux out there were pretty selfish . . . I wanted help.*

—Torvalds to Wired, November 2003

## PROPRIETARY RIGHTS

Torvalds released the long-awaited Linux 2.4 in January 2001. The following month, IBM announced an investment of $300 million into the development of Linux services, in addition to the almost $1 billion it and Japanese companies, such as Hitachi, had already committed to open-source software products. Linux purists feared that IBM's investment, despite its no-strings-attached agreement, would lead to corporate attempts to control Linux's destiny. Others suspected that such vested corporate interests would topple Torvalds' de facto reign over Linux.

What IBM's investment did accomplish—akin to the way Intel and Netscape lent legitimacy to open-source software with their investments in Red Hat in the 1990s—was the spread of open-source gospel across the globe. By 2002, more than two dozen countries, including Germany and China, were in the process of adopting or had already adopted Linux as the operating system of choice for their governments. The global embrace of open-source software placed Linux a close second to Microsoft in the server market, with no sign of slowing down.

In March 2003, the SCO Group, a Utah-based company that owned the rights to the UNIX operating system, filed a lawsuit against IBM, alleging that IBM had placed thousands of lines of proprietary UNIX code into Linux—a violation of intellectual property law. The charges called into question the very legality of Linux itself, as well as the open-source model of software development.

Torvalds's initial response to the case was measured. He called it little more than contract dispute between IBM and SCO. Over time, however, Torvalds became increasingly impassioned. As part of its case, SCO claimed that UNIX source code appears in the Linux kernel—a charge that Torvalds vehemently disputes, arguing that the open-source development model speaks for itself. "With Linux code, you can see how it's been developed," he told the *San Jose Mercury News.* "You can see who applied patches. You can see when they got applied. It's all in the open." When SCO later amended its case, the company called Torvalds himself into question. Citing an email from August 2001—in which Torvalds writes, "I do not look up any patents on principle because (a) it's a horrible waste of time and (b) I don't want to know"—SCO claimed that Torvalds encouraged the Linux community to defy existing copyright law. Torvalds has derided the suit as "complete crap," and branded SCO as "extortionists."

Microsoft joined the fray in May 2003 by purchasing a license for UNIX from SCO, a move that chan-

neled money into SCO's legal fund. That June, Torvalds left Transmeta to join the Open Source Development Lab (OSDL), a consortium founded by industry leaders, including Intel, IBM, and Hewlett-Packard, to foster the development of Linux for the corporate environment. Over the next six months, the case mushroomed—Red Hat sued SCO, IBM countersued SCO, and allegations about whose copyrights preceded whose flew back and forth between SCO, IBM, Novell (which had sold UNIX to SCO), and even AT&T, the original creator of UNIX. In a bold move, the SCO legal team levied claims against the GNU GPL—the glue of the entire open-source movement—claiming that it is "preempted by copyright law." By January 2004, with no clear end of the SCO saga in sight, the OSDL announced the establishment of a $10-million legal fund for Linux customers and, if sued, Torvalds himself.

Ultimately, Torvalds owns little more than the rights to the name "Linux." The system itself belongs to the loose-knit development community, numbered in the tens of thousands, who have contributed lines of code. (The Linux 2.6 kernel, released in December 2003, clocks in at more than 2,000,000 lines of code, which still pales in comparison to the hundreds of millions lines of code for Microsoft Windows.)

> ***Really, I'm not out to destroy Microsoft. That will just be a completely unintentional side effect.***
>
> —*Torvalds to the* New York Times Magazine, *February 1999*

Through the efforts of countless anonymous individuals, what began in 1991 as one programmer's hobby has evolved into a world-class operating system. Today, Linux is the force behind the search engine Google, the animation of *Shrek,* and the car designs of Daimler-Chrysler. Linux was even used in the Rover mission to Mars. Despite Linux's high-profile success, Torvalds has remained a humble leader (and an ongoing foil to industry titans like Bill Gates). His affable nature, as well as his general concern for code over cash, has allowed Torvalds to retain the trust and loyalty of the Linux community for more than a decade.

## FURTHER READING

### In These Volumes

Related Entries in this Volume: Gates, Bill; Joy, William; Stallman, Richard

Related Entries in the Chronology Volume: 1976 (sidebar): The Free Software Movement; 1985: Richard Stallman Establishes the Free Software Foundation; 1991: Linus Torvalds Develops the Linux Operating System; 1998: Linux Operating System Becomes a Cause Célèbre

Related Entries in the Issues Volume: Copyright; Hackers; Open Source

## Works By Linus Torvalds

*Just for Fun: The Story of an Accidental Revolutionary.* New York: HarperBusiness, 2001.

## Books

Moody, Glyn. *The Rebel Code: The Inside Story of Linux and the Open Source Revolution.* Cambridge, MA: Perseus, 2001.

Raymond, Eric S. *The Cathedral & The Bazaar: Musings on Linux and Open Source by an Accidental Revolutionary.* Cambridge, MA: O'Reilly, 1999.

## Articles

Diamond, David. "The Peacemaker." *Wired* 11.07, July 2003, http://www.wired.com/wired/archive /11.07/40torvalds.html (cited September 16, 2004).

Harmon, Amy. "The Rebel Code." *New York Times Magazine,* February 21, 1999.

Learmonth, Michael. "Giving It All Away." *Metro,* May 8, 1997, http://www.metroactive.com/papers /metro/05.08.97/cover/linus–9719.html (cited September 16, 2004).

McHugh, Josh. "For the Love of Hacking." *Forbes,* August 10, 1998, http://www.forbes.com/forbes /1998/0810/6203094a.html (cited September 16, 2004).

Moody, Glyn. "The Greatest OS That (N)ever Was." *Wired* 5.08, August 1997, http://www.wired.com /wired/archive/5.08/linux.html (cited September 16, 2004).

Rivlin, Gary. "Leader of the Free World." *Wired* 11.11, November 2003, http://www.wired.com/wired /archive/11.11/linus (cited September 16, 2004).

Scoville, Thomas. "Martin Luther, Meet Linus Torvalds." *Salon.com,* November 12, 1998, http://archive.salon.com/21st/feature/1998/11/12fe ature.html (cited September 16, 2004).

Takahashi, Dean. "Linux Creator an Open Source." *San Jose Mercury News,* July 4, 2004, http://www .mercurynews.com/mld/mercurynews/6238207.htm (cited April 11, 2004).

## Websites

Linus Torvalds. Linus Torvalds's bare-bones home pages, http://www.cs.helsinki.fi/u/torvalds/ (cited September 16, 2004) and http://www.linustorvalds .net/ (cited September 16, 2004).

*Linux Journal*—The Premier Magazine of the Linux Community. Online home of the *Linux Journal,* a monthly magazine for the Linux community, founded in March 1994, http://www.linuxjournal .com/ (cited September 16, 2004).

LWN: Linux Weekly News. Online, weekly digest of news in the Linux community, http://lwn.net/ (cited September 16, 2004).

The Open Source Development Lab. Online home of the OSDL, where Torvalds is a fellow. The OSDL "strives to be the center of gravity for Linux," http://www.osdl.org (cited September 16, 2004).

OSI: The Halloween Documents. Eric Raymond's digest of the "Halloween memos," with running commentary, http://www.opensource.org /halloween/ (cited September 16, 2004).

# Ann Winblad (1950–)

## VENTURE CAPITALIST

**In 1989, Ann Winblad** cofounded Hummer Winblad Venture Partners, the first venture capital firm dedicated solely to software businesses. In the ensuing decade, she became one of the most important venture capitalists in the history of the Internet. Indeed, Winblad has been counted among *Time Digital*'s top fifty Cyber Elite, *Business Week*'s Elite 25 Silicon Valley power brokers, and *Upside Magazine*'s 100 Most Influential People of the Digital Age. Winblad has spent more than twenty-five years in the software industry and nearly fifteen in the realm of venture capital, and is considered one of the top players in each of these primarily male-dominated realms.

## ALL-AMERICAN GIRL

Winblad grew up in Farmington, Minnesota, a small, semirural town twenty-five miles south of the Twin Cities. The eldest daughter of a high school football and basketball coach, she often wore a T-shirt that read "little coach" and hung out in the men's locker room as a child. In high school, she led the cheerleading squad, lettered in track and field, and was valedictorian at her graduation. To many, she was the embodiment of the All-American girl.

After graduation, Winblad attended the College of St. Catherine, a women's college in nearby St. Paul, where she earned degrees in both math and business administration and learned basic computer programming. In her sophomore year she proved to be so adept at FORTRAN programming that she was asked to teach the course the following year, when the professor fell ill. Winblad worked her way through college with a variety of part-time jobs—waiting tables at a bar, keeping books for a hardware store, and

> *Nerds are finally cool... and I'm proud to be one of them.*
>
> —*Winblad to Soft-Letter, May 1994*

fielding customer service requests for the local phone company. She then enrolled at the University of St. Thomas, a college in St. Paul mostly attended by men, where she earned master's degrees in both international economics and education. Often, Winblad was the only woman in her graduate math, computer science, and business classes. Holding her own in the company of men would set Winblad apart for the rest of her life.

Upon graduation, Winblad was courted by the FBI to become a gun-toting tax agent, but she chose to begin her career as a systems programmer for the Federal Reserve Bank in Minneapolis. After little more than a year, however, she grew tired of her job and contemplated starting her own company—a software firm specializing in accounting programs. It was risky: Winblad was young and had little actual sales or management experience. The software industry had yet to burst into the mainstream. Her parents were skeptical, telling her to "get a real job."

In 1976, however, Winblad borrowed $500 from her younger brother and, with colleagues from the Federal Reserve, founded Open Systems, Inc. in her apartment. Open Systems soon opened its first office, located above the Moon Sound music studio in downtown Minneapolis, where the then-seventeen-year-old pop star known as Prince recorded his first demo album. After a rough start, Winblad helped make Open Systems into a top-selling accounting software firm during the next six years. In 1983, when she sold Open Systems to the Dallas-based UCCEL Corporation, her initial $500 investment was worth $15 million.

Winblad took her profits and headed west to San Francisco, where the software industry had begun to take off. While writing a technical book on object-oriented programming, Winblad began consulting for top-level companies, including Price-Waterhouse, IBM, Apple, Microsoft, and various other start-ups. The *San Francisco Examiner* named her as one of California's up-and-coming entrepreneurs. It was during this time that she began dating Bill Gates, whom she had met at the PC Forum, a computer industry conference, in early 1984.

## A Humdinger of a Firm

During a consulting job in the late 1980s, Winblad met John Hummer, a former center for the Seattle Supersonics. Hummer had earned his MBA after retiring from the NBA in the 1970s and was an experienced venture capitalist. He took to Winblad almost immediately and suggested that they start a firm together. Winblad, however, had little interest in venture capital, jokingly telling Hummer that she did not even like venture capitalists. Eventually, though, Hummer's persistence and monthly phone calls won her over. In 1987, Winblad joined Hummer to develop the first venture capital firm focused solely on the software industry.

As Hummer and Winblad worked to launch their firm, they quickly earned a reputation as the odd couple of the venture capital world. At 6'10," Hummer stood more than a foot and a half above the 5'3" Winblad. Some ridiculed them as "Goldilocks" and "Lurch." Even Winblad jokes that, when she and Hummer would arrive for meetings with entrepreneurs, it might seem as though the circus had come to town. Their firm, initially located in Emeryville, on the unfashionable side of San Francisco Bay, sat far from the primary locus of Silicon Valley venture capital firms, which had sprung up along Sand Hill Road in Menlo Park. Most important, however, industry leaders believed focusing solely on software start-ups was foolhardy. In a 1998 interview with *Fortune,* Winblad recalled, "Everyone basically told us, 'Software is a stupid thing to invest in because the assets walk out the door at night.'"

> In math class, it was me and the guys. In business class, it was me and the guys. In the computer center, it was me and the guys. . . . It prepared me for an industry that started out as me and the guys.
>
> —*Winblad to* USA Today, *June 1996*

Eighteen months, 110 investors and $35 million later—in early 1988—Hummer Winblad Venture Partners had its first venture capital fund. By the late 1990s, the ten-year fund had returned more than $350 million to its major investors, and the firm's early detractors were well aware of what Hummer Winblad believed all along—software was the economy's next big thing.

Winblad's first investment had been in T/Maker, a software company founded by Heidi Roizen, whom Winblad befriended when she first moved to California. Roizen, still in her 20s, was one of the only other highly successful women in the world of computing. (She later became a top-level executive at Apple Computer and is now a successful venture capitalist herself.) Winblad's early $1 million investment, made in 1989, paid off handsomely. Roizen sold T/Maker for more than $20 million in 1994.

T/Maker was just the first of Winblad's triumphs. Indeed, she displayed an uncanny gift for spotting potential in struggling start-ups. In 1990, she invested in Wind River Systems, a fledgling software company she called "the Rodney Dangerfield" in her portfolio of companies, because it garnered so little respect from venture capitalists. Soon, however, Wind River Systems became the industry leader in the embedded systems market. Similarly, Winblad's guidance helped PowerSoft Corporation, a database developer, grow from nothing into a $20-million company in less than a year. (PowerSoft was later acquired by Sybase, a leading information management corporation.) Berkeley Systems, Inc., another early Hummer Winblad success, did not initially experience such spectacular growth, but when Winblad joined the board of directors, she helped the small screen-saver software company evolve into a full-fledged Internet entertainment company. Indeed, Berkeley Systems executives claim that Winblad's patience and vision was vital to the company's growth. In addition, Winblad proved to be an expert marketer, particularly when the first round of companies in Hummer Winblad's portfolio held initial public stock offerings (IPOs)

*Ann Winblad. (Roger Ressmeyer/Corbis)*

in the mid-1990s. She is credited with the wildly successful Arbor Software IPO in 1995.

Amid such early success, failure wasn't unknown. In 1990, Winblad poured $400,000 into Slate, a pen-based computing company. (Kleiner Perkins, the other top Silicon Valley venture capital firm, also invested.) Though PalmPilots employ similar pen-based technology today, in the early 1990s, the idea failed to catch fire. By the time Slate was acquired by Compaq, Hummer Winblad had lost more than $1.2 million.

The firm regrouped, and, by fall of 1996, Hummer Winblad had yielded a 50 percent annual return for its investors. Competing firms averaged returns of around 20 percent. Riding high from its early accomplishments, Hummer Winblad opened a second fund, focusing on Internet companies, worth $65 million in 1994. A third fund of around $100 million, begun in mid-

1997, would earn Hummer Winblad's limited partners twice their original investment in just two years. Their fourth fund of $320 million, begun in 1999, coincided with the peak—and the ensuing crash—of the Internet boom.

## THE DOT-BOMB YEARS

Between 1997, when Hummer Winblad began to invest seriously in Internet companies, and early 2000, before the Internet market crashed, venture capital firms investing in Silicon Valley earned, on average, annual returns of 35 and 45 percent—nearly double normal returns from just two years earlier. In 1999, some firms reported triple-digit returns.

The heady economic atmosphere resulting from such high returns led many to push companies on the market prematurely. Whereas in the early 1990s, start-ups would mature for five or six years before making an IPO, in the late 1990s, companies often went public just six months after getting off the ground.

Though Hummer Winblad was no newcomer to the volatile world of Internet investing, it too fell prey to dot-com fever. One of the firm's highest profile consumer-oriented Internet companies, Pets.com, which had been added to its portfolio in March 1999, was a disaster. The Pets.com IPO was held in February 2000, even though Pets.com had lost more than $60 million over the previous year and the online pet supply retail market was rife with competition. Winblad remained a vocal champion of Pets.com's potential until, nine months after the IPO, the company announced that it was auctioning off its assets and closing its Website for good.

Other Hummer Winblad Internet investments quickly followed suit. By 2001, Gazoontite Inc., an online allergy-product retailer in which Hummer Winblad had invested $26.5 million in late 1999, had filed for bankruptcy, as did Homes.com, a real estate site, and eHow.com, a how-to site that was one of Winblad's early Internet stars. Site Technologies, a Website management firm, and Rival Networks, an online sports-related company, also folded. On the seven companies Winblad had helped go public since 1997, the return to investors was—an 89% loss. In a letter to investors, Winblad wrote, "These were the biggest losses we have ever taken. . . . It is an understatement to say how bad we feel about this."

While some investors were inclined to chalk up the losses to a boom gone bust, others found fault in the firm itself—particularly with respect to Winblad. Her critics claimed that her outside commitments—she sat on the boards of more than seven Hummer Winblad firms, kept at least two standing public speaking engagements each month, and was a regular at weekly industry events—distracted her from basic obligations to the firm. Those jumping to her defense claimed her widespread networking had always been a key to Hummer Winblad's success.

## REVENGE OF THE KILLER APP

Although in retrospect the Internet bubble seems to have burst almost overnight, in the early months of 2000 many believed recovery was always just around the corner. So, in May 2000, when Hummer Winblad made a bold, $15-million investment in Napster, Inc., the controversial music file-sharing company, the move was deemed risky not because of the market situation but because of the investment itself. Since December 1999, Napster had been embroiled in lawsuits brought by the largest record labels in America.

Still, Winblad saw potential in Napster. In August 2000, on the CNNfN news show *Digital Jam,* she told interviewers that, despite the "precarious and overexciting" nature of the market, her firm was, "viewing [Napster] as a killer application." The firm already claimed one online music distribution company, Liquid Audio, founded in 1996. Unlike Napster, which circumvented the record industry by allowing peer-to-peer file sharing, Liquid Audio worked with the record labels to find secure means to distribute music digitally. By the late 1990s, Liquid Audio, while moderately successful, had been quickly overshadowed by the popularity of MP3 file sharing. Some saw Napster as Hummer Winblad's attempt to recoup its losses, while others saw the Napster investment as yet another incidence of dot-com fever. Hilary Rosen, head of the Recording Industry Association of America, told the *New York Post* in 2000 that Hummer Winblad was just "looking for a

quick buck," adding that the company also bought itself some trouble.

Indeed, Hummer Winblad had. As part of the deal, Hummer Winblad placed one of its partners, Hank Barry, as Napster's interim CEO, and John Hummer joined the board of directors, making it difficult to disentangle the venture capital firm from the actions of Napster, Inc. As lawsuits against Napster mounted, various plaintiffs began to see Hummer Winblad—with its multimillion-dollar funds—as a deep-pocketed potential defendant.

Napster's downfall came as somewhat of a surprise to Winblad and the firm's other partners. In interviews, Hummer Winblad partner Dan Beldy claimed from the beginning that the firm had done a five-month investigation into Napster's case and believed the law to be on its side. The courts, however, forced Napster to shut down in July 2001; various injunctions kept the site off-line. On April 21, 2003, the Universal Music Group and other record companies brought civil charges against Hummer Winblad Venture Partners, Hank Barry, and John Hummer himself, alleging that each party contributed to copyright infringement by supporting Napster financially. By March 2004, the number of plaintiffs had risen to sixteen.

The case against Hummer Winblad and its partners has sent chills through the venture capital world. If successful—many legal experts believe that it will not be—the suit would mark the first time investors and shareholders have been held accountable for the actions of a company. Since the lawsuit, Winblad has stepped out of the spotlight, usually declining to comment on the case.

> *Shame on us—the established venture capitalists who already had the discipline and the experience—for performing like amateurs.*
>
> —Winblad to Fast Company, *March 2004*

## ON THE REBOUND

By early 2001, as the dot-com frenzy died down, Hummer Winblad began to turn back to its roots—software companies. (Winblad remained on the board of The-Knot.com and Deananddeluca.com, the firm's primary Internet survivors.) The firm began to show signs of slow recovery. In January 2002, the firm invested in Jareva Technologies, followed by a modest investment in

Yosemite Technologies, Inc., in April and, in August, made a $7.5-million investment in Knowmadic, Inc. In July 2003, Winblad herself led the funding for Voltage Security Inc., a Palo Alto company specializing in encryption software. Industry magazines hailed the Voltage deal, only the fourth investment by Hummer Winblad in an eighteen-month period, as the firm's comeback.

If so, Winblad would have come full circle. She began when few believed software was a worthwhile investment, won big with early firms like Wind River and Berkeley Systems, then lost even bigger with a slew of ill-conceived mass-market dot-coms. In many ways, she is the poster child for the Internet boom and bust.

## FURTHER READING

### In These Volumes

In this Volume: Fanning, Shawn; Gates, Bill; Lessig, Lawrence

In the Chronology Volume: 1999: Napster Roils the Music Industry; 2000: The Dot-Com Crash; 2000 (sidebar): Pets.com Is Put to Sleep; 2003: Music Sharers Sued by Recording Industry

In the Issues Volume: Copyright; E-business; E-commerce

### Works By Ann Winblad

*Object-Oriented Software.* Boston: Addison-Wesley, 1990.

*Implementing Object Technology.* Boston: Addison-Wesley, 1996.

"Foreword." In *Zero Gravity 2.0: Launching Technology Companies in a Tougher Venture Capital World,* by Steve Harmon. Princeton, NJ: Bloomberg Press, 2001.

### Books

Gupta, Udayan. *Done Deals: Venture Capitalists Tell Their Stories.* Boston: Harvard Business School Press, 2000. Excerpts available at: http://www.derbymanagement.com/knowledge/pages/knowing/perspective.html (cited September 16, 2004).

### Articles

Anders, George. "Finding the Needles." *Wall Street Journal,* November 22, 1999.

Diamond, David. "Adventure Capitalist," *Wired,* September 1996, http://www.wired.com/wired/archive/4.09/winblad.html (cited September 16, 2004).

Hennes, Doug. "Always a Coach." *St. Thomas Magazine,* Winter 2001, http://www.stthomas.edu/magazine/showarticle.cfm?ArticleID=1007063115 (cited September 16, 2004).

Judge, Paul. "Newsmakers Q&A: Ann Winblad on Women in the New Marketplace." *Business Week,* February 17, 2000, http://www.businessweek.com/bwdaily/dnflash/feb2000/nf00217c.htm (cited September 16, 2004).

Tate, Ryan. "Hummer Winblad Could Answer for Napster's Sins." *UPSIDE Today,* August 4, 2000.

Tully, Shawn. "Big Man Against Big Music." *Fortune,* August 14, 2000.

Wingfield, Nick. "When Venture Capitalists' Investments End Up as Dot-Com Debris." *Wall Street Journal,* April 16, 2001.

Wylie, Margaret, and Rose Aguilar. "Venturesome Capitalist." *CNet News.com,* September 11, 1996.

### Websites

Hummer Winblad Venture Partners. Online home of Winblad's venture capital firm, with links to many of the companies in its portfolios, http://www.humwin.com/ (cited September 16, 2004).

Tech Museum of Innovation: The Revolutionaries: Ann Winblad. Online transcript of Winblad interview for the Tech Museum of Innovation, http://www.thetech.org/revolutionaries/winblad/ (cited September 16, 2004).

Venture Capital Firms: Silicon Valley Daily. A digest of information about Silicon Valley venture capital firms, with news updates and other resources, http://www.svdaily.com/capital.html (cited September 16, 2004).

# Jerry Yang (1968–) and David Filo (1966–)

## FOUNDERS OF YAHOO!

**In 1993, Jerry Yang and** David Filo, two electrical engineering doctoral students at Stanford University, began putting together a list of their favorite sites on the burgeoning World Wide Web. In just over two years, what began as a hobby quickly turned into one of the most successful companies in Internet history—Yahoo! Although Yahoo! initially made its name as one of the first navigational guides to the Web, as the company added online services, like free Web-based email, and expanded internationally, it grew into one of the Web's largest portals. For more than a decade, Yang and Filo, known as the "chief yahoos" of the company, have continued to expand yahoo.com, making it one of the most visited sites on the Web.

## YAHOO!'S YIN & YANG

Jerry Yang was born in Taipei, Taiwan, where his parents had fled from mainland China in the late 1960s. After the death of his father in 1970, his mother supported the family by teaching English and drama at a local university. In 1978, fearing her two sons would be drafted into Taiwan's military, she brought the family to the United States; the Yangs were part of the first wave of Taiwanese immigration after the United States and China reestablished diplomatic relations. The family settled near San Jose, California.

Although Jerry arrived knowing only one word of English—"shoe"—he quickly acclimated to his new surroundings and was soon earning straight As in school, rising from remedial English to Advanced Placement English in a matter of a few years. He was bright, industrious, and outgoing, earning extra money with a paper route and, by high school, involved in extracurricular activities like student government and the tennis team. One year, he worked in a summer program at NASA. Yang graduated valedictorian of his class.

Having taken enough Advanced Placement courses in high school, Yang entered Stanford University as a sophomore and was able to finish both a bachelor's and a master's degree in electrical engineering in four years.

Filo was born in Wisconsin, but spent most of his youth in a cooperative living community in Moss Bluff, Louisiana, with his father, an architect, and mother, an accountant. A shy but academically gifted student, Filo earned a full scholarship to study computer engineering at Tulane University, where he graduated summa cum laude and was named Tulane's top graduate of 1988. Filo then headed to Stanford for graduate study.

Yang claims that he and Filo first met in 1989, when Filo, two years Yang's senior, was a teaching assistant in one of Yang's undergraduate computer courses. Although Filo does not recall meeting Yang this early on, Yang remembers it vividly—he received his lowest grade at Stanford in that class and often pleaded with the various teaching assistants to change his grades. (None, including Filo, did.) A couple years later, Yang joined Filo in the electrical engineering doctoral program, and Yang courted Filo as a lab partner. "You always knew that if you were ever going to do something, you'd want to do it with David because he's one of the brightest guys you'll ever meet," Yang said in a 1997 interview with *CNET News.com.* Soon, both students were working under the same adviser, researching computer-aided design (CAD) software for semiconductors—a growth area of technology in the late 1980s.

In 1992, Yang and Filo went to Kyoto, Japan, on a six-month academic exchange program. There, Yang and Filo truly bonded, talking endlessly about sports and technology; Yang met his future wife, Akiko Yamazaki; and the two men also struck up a friendship with fellow Stanford student Srinija Srinivasan, who would later play a vital role in Yahoo!'s evolution.

When Yang and Filo returned to Stanford in 1993, much had changed. The CAD industry had consolidated into just a handful of companies by the early 1990s. Their shared dream of working at a small CAD start-up once they graduated had vanished. At the same time, the World Wide Web was catching on. Both Yang and Filo had been using the Internet since the late 1980s, but with the advent of the Web browser Mosaic, a whole new world emerged online. As Yang explained in a 2000 *Fortune* interview, "In 1993 and 1994, even though we were trying to get through the last of our Ph.D.s in [our] narrow field, we were seriously asking each other what else is cool out there."

## NOT YOUR EVERYDAY TRAILER TRASH

Back at Stanford, Yang and Filo were relegated to ramshackle offices in a temporary building—nothing more than a double-wide trailer (notably located in the shadow of what is now the Bill Gates building). They toiled on their computer workstations amidst a tangle of golf clubs, sleeping bags, and discarded take-out containers. One friend described the scene as "a cockroach's picture of Christmas."

Exploring the World Wide Web became the duo's favorite method of avoiding their dissertations. (They had already exhausted other time-consuming hobbies, including golf.) They looked up basketball statistics and sought out news on their favorite sumo wrestlers (they had named their workstations after legendary sumo wrestlers—Yang's was called Akebono, while Filo's was Konishiki). Although the Web was still so new that only a day or so was needed to surf almost all existing Websites, such searching took skill and a lot of patience, with losing your way a real possibility.

Soon, Filo started keeping a list of his favorite URLs, as did Yang. "I bugged Dave into telling me which sites he found that were any good," Yang told the *Daily Telegraph,* an Australian newspaper, in 2001. "And his sites were always better." The Web grew by the day, and so did the lists. In January 1994, they combined their lists

and posted it to the Web as "Jerry and David's Guide to the World Wide Web."

Yang directed friends to the list, and friends at Stanford and across the country at the Massachusetts Institute of Technology told their friends, and soon strangers were sending Yang and Filo emails, asking for *their* favorite sites to be added to the list. Building the list was a rather casual endeavor—as Yang recalled in a 1995 CNN interview, "They said, 'Well, you know, I have this sailing page. Can you add it?' And we said, 'Well, sure.'"

Within a month, the "Jerry and David's Guide to the World Wide Web" boasted more than 100 sites. Once the list had grown to more than 200 entries, it became unwieldy—and still there were more sites to be added. Since the early version of the Mosaic browser could not sort or arrange bookmarks, Yang and Filo began to develop software that would let them categorize their links. (Previously, a user had to scroll through the list sequentially.)

As word-of-mouth spread, requests poured in. The initial list was broken into subject categories, then, when those became too full, Filo and Yang created new subcategories. In a matter of months, what began as a new way to avoid writing their dissertations turned into a full-time, twenty-hour-per-day job—one Yang and Filo worked without pay. "We had close to five hundred to a thousand requests a day to please be added," Filo told CNN in 1995. "Between that and everything else, there was no way the two of us could keep up." Indeed, the Jerry and David site had exploded—by May 1994, it had been accessed by more than 100,000 users from more than forty countries.

> *There may be a textbook on how to found a company, but we didn't read it.*
>
> —*Yang to* Fortune, *March 2000*

## RIGHT UNDER THEIR NOSES

In the rare moments when they were not cataloguing the Web, Yang and Filo were busy brainstorming ideas for a start-up company. Each week, they would get together with friends to discuss potential ways to exploit the nascent Web for profit—inspired, in part, by the new company Netscape, which had secured funding to build a commercial version of the Mosaic browser. Yang and Filo drafted various business plans for online shopping malls and even online bookselling software not unlike the as-yet-to-be-founded Amazon.com. But nothing stuck. "Jerry and David's Guide," however, grew longer and more complex by the day.

Because they did the list for fun—a time-consuming but rewarding hobby—neither had thought that the list could be their start-up. Plus, they had no idea how to make money with it. "There was no real business model that fit it," Yang explained to *Fortune* in 2000. "We knew we had to keep it free because everything on the Internet was free. So there was no way to charge people for using it or even to charge people for their sites to be listed on it. And besides, we were just two guys slaving away on the technology. What did we know?"

One thing they did know by September 1994 was that their site had become overwhelmingly popular. That month, the site achieved its first million-hit day, claiming almost 100,000 unique visitors. The swelling Web traffic had begun to strain the university's computer network, and Stanford officials firmly suggested that Filo and Yang remove the site from school's server. Filo and Yang began to consider partnering with a company that could host their site.

By then, Filo had already voiced concern about using their names for the site, which, to him, seemed egotistical. In keeping with a long-standing tradition among computer geeks and programmers, he wanted to rename it using a witty acronym, one that began with YA, which stands for "Yet Another . . ." (The computer program YACC, for instance, is an acronym for Yet Another Compiler Compiler.) Late one night, Yang and Filo broke open a dictionary and started looking up words that started with YA. When they read the definition of *yahoo*—"rude, unsophisticated, uncouth"—they stopped. "Yahoo seemed about right, since that's what we considered ourselves, a couple yahoos," Yang said in a 1996 *Dallas Morning News* interview.

Using the letters Y-A-H-O-O as a guide, they came up with a fitting root for the acronym: Yet Another Hi-erarchical Officious Oracle. The term itself was irreverent, and easy to remember. Later, they added the exclamation point—Yahoo!—to emphasize the thrill of discovering a great new Website.

## THE SUPPORTING CAST

By early 1995, investors were paying attention to the Internet and Web search engines such as Lycos, Infoseek, and Excite had already received funding. Yahoo! was a searchable Web directory, but not, strictly speaking, a search engine. Search engine companies like Lycos and Excite employed technology that searched the contents of Web pages to match a keyword query; Yahoo! started out as a directory of prescreened Web pages chosen, so to speak, by hand. (Yahoo! added a conventional Web search function in June 1996, when the company partnered with the search engine Alta Vista.)

Kleiner Perkins, an esteemed venture capital firm in Silicon Valley, expressed interest in Yahoo!, but wanted to merge it with a search engine developed by Stanford graduates. America Online attempted a buyout as well. "I said to David, if other people are willing to fund us as a business we should start to convince ourselves of it," Yang said in a 2002 interview with the *Australian*. "That was when we realized it was serious."

That spring, Tim Brady, a friend from Stanford and Yahoo! employee number three, helped Yang and Filo put together their first business plan, and the company incorporated in March 1995, moving into a 1,700-square-foot office near the railroad tracks in Mountain View, California. After speaking with half a dozen venture capital firms, Yahoo! went with young venture capitalist Mike Moritz at Sequoia Capital. Sequoia could count among its successes Apple Computer, Atari, Oracle, and Cisco Systems, and Moritz, more than any other venture capitalist they spoke to, seemed to understand where Yang and Filo were coming from. Indeed, when Yang asked Moritz whether they should change the name Yahoo! to something less irreverent,

> *There's this huge, fast-moving train called the Internet. And we're just half a mile ahead laying the tracks to make sure it doesn't go off the cliff. It's felt like that since the very beginning.*
>
> —*Yang to CNET News.com, November 1997*

*David Filo (left) and Jerry Yang. (Ed Kashi/Corbis)*

Moritz said that if they did, he would take back his $1-million investment. On April 10, 1995, Yang and Filo, just six months shy of attaining their doctorates, announced that they were taking a leave of absence from Stanford and beginning full-time work on Yahoo! Inc.

Long before Sequoia's $1-million investment, the company had grown too large for Yang and Filo, neither of whom had any real business experience. Yahoo! was, after all, the first real job for both Yang and Filo. Thus, the first order of business was to build the Yahoo! team.

To help organize the ever-growing "hierarchical oracle," Yang and Filo hired Srinija Srinivasan, the fellow Stanford graduate they had met while in Kyoto. Srinivasan, who had studied artificial intelligence and information organization, was appointed Ontological Yahoo (that title was printed on her business cards); she developed Yahoo!'s underlying organizational structure based on the ad hoc categories she inherited from the "Jerry and David's Guide" days. She set up various other organizational guidelines as well, such as cross-linking sites by topic as well as region, so that a Los Angeles dry cleaner is listed under "Los Angeles," with a cross-link from "dry-cleaners." Her goal, simply put, is to "know the Web."

For "adult supervision," Yang and Filo recruited Tim Koogle, a Stanford engineering department alumnus and former Motorola executive who was then running the $400-million tech company Intermec. Koogle was 43 years old—sixteen years Yang's senior. Koogle recalled, in an interview for the Stanford Engineering Department 1995–96 annual report, "When I first met Jerry and David, what struck me immediately was that they had filled a fundamental need and they had done it intuitively. That's what you look for in starting a business." Koogle became president and chief executive officer in June 1995. The next key hire was Jeff Mallett as chief operating officer.

Yang and Filo took positions as Chief Yahoos—a title that upheld their positions as founders, while allowing them to function within the company in their own ways. Filo became, in essence, Yahoo!'s chief technologist, responsible for the Website's infrastructure and overall functionality. (Filo later renamed himself Cheap Yahoo, a nickname he earned for his preference for no-frills PCs.) Yang (nickname: Grumpy) became the public face of Yahoo!—both on magazine covers and in business meetings. Although Koogle and Mallet made many of the critical business decisions, Yang still played a vital role. "People always ask me why I took myself out of the day-to-day operating responsibility," Yang told *Fortune* in 2000. "I knew so little about business that I didn't want to slow things down when the company began to scale up. And anyway, to me, the broader the role, the more exciting it is. Yahoo is in everything from pets to old people to finance to communications to e-commerce and more, and I really thrive on that."

## THE SPECIAL SAUCE

The Yahoo! team quickly went to work on the company's business strategy—subsidizing the site's free content and online services with money from online advertising. Yahoo! began serving ads on its site in August 1995, forever altering the way Internet sites do business. Like more traditional media companies—such as network television and radio stations—most of Yahoo!'s revenue came from advertising. Yahoo! was one of the first to develop and implement technology that tracked a user's Web surfing patterns to help target advertising—so that, for example, a user searching for "sneakers" would see a banner ad for Reebok. Yahoo! also pioneered a fee schedule in which advertisers paid them based on the number of times a user visited a site, not just how many times an ad is viewed, and developed technology to help show advertisers which ads drew the most clicks. Such techniques are common now, but in the early days of the Web, they were practically revolutionary.

Still, Yahoo! faced stiff competition. Already, a handful of other Websites offered similar services—namely, how to find something on the Web. Search engine companies, such as Excite, Infoseek, and Lycos, had superior search technology, including advanced "spider" software and complex mathematical algorithms that could find

and deliver hundreds if not thousands more Websites per query than Yahoo! But, as Randall Stross wrote in his 1998 *Fortune* magazine article, "Yahoo! had the best name, the worst technology, and a quaint belief that while other companies' machines surveyed Website addresses by the thousands every second, the human touch could somehow win out."

Indeed, even with its less powerful technology, Yahoo! boasted at least twice as many visitors as its rivals. Part of this success was the human touch—on the ever-growing Web, delivering *more* results did not always translate into delivering *better* results. For example, although Yahoo! delivered fewer results for a search on "Madonna," Yahoo!'s suggested sites would take into account whether the user was looking for the spiritual figure or the rock star. Software-based searches, by contrast, tended to treat all Madonnas equally. At Yahoo!, the humans doing the indexing made searches simpler for the human end users.

The other part of the site's success is attributable to the Yahoo! brand. In 1996, an aggressive advertising campaign based on the phrase, "Do You Yahoo?" established Yahoo! as one of the most hip companies online. Later on, publicity stunts, like Yahoo!'s annual yodeling contest, kept its image light and irreverent, particularly with mainstream audiences. On the other end of the spectrum, serious technophiles appreciated Yahoo!'s being built on open-source software, like the Apache Web server, the FreeBSD operating system, and the Perl scripting language.

## MONEY MATTERS

In April 1996, Yahoo! held its initial public stock offering, hot on the heels of Lycos and Excite, which each held their stock offerings just days before. Yahoo!'s price jumped from $13 to $33 per share on the first day of trading, but when prices fell the following day, some analysts balked—one even dismissed Yahoo! as "Yet Another Highly Overhyped Offering." But Yang and Filo persevered, pushing the Yahoo! brand and adding new online services every month. Soon, the site offered Yahoo! chats, Yahoo! maps, Yahoo! mail, online stock trading, and e-shopping. In two short years, Yahoo! had international sister sites in Japan, Germany, the United Kingdom, Singapore, Korea, and Australia, and, by November 1997, Denmark, Norway, and Sweden. By

December of 1977, the company had 1,700 paying advertisers, with revenues topping $67 million. Sequoia Capital's initial $1 million investment had grown 560 percent. In January 2000, Media Metrix ranked Yahoo! the number one site on the Web.

Yahoo!'s stock hit an all-time high of $250.06 per share on January 7, 2000, then fell (along with the rest of the industry) when the stock market took a nosedive later that year. Yahoo! weathered the economic downturn somewhat better than many of its rivals, although in 2001 the stock dipped when news broke about Koogle's departure as CEO and the company announced that earnings would be $50 million less than expected. Still, that year, Yahoo! could claim 123.8 million unique global visitors each month, compared with MSN's 113.1 million and AOL's 99.3 million. (Domestically, Yahoo! trailed AOL.) By 2002, Yahoo! was back on the road to posting regular quarterly profits.

Much of Yahoo!'s success can be attributed to the flexibility of the site, which, Yang states, was something Filo built in from the beginning. Users can land on www.yahoo.com in any number of ways—from email to a Web search. "[David] figured that it didn't really matter how you got to Yahoo as long as you got there," Yang told *Fortune* in 2000. "That flexibility in our product—that it does lots of things, that it's there and quickly responsive when you need it, and that it lets the user choose without censorship or being told what he ought to look at—all of that started as early as 1994. It wasn't a business decision we made later."

In 2004, the tenth anniversary of its founding, Yahoo! was still actively expanding, with an average of 2.4 billion page views per day and more than 274 million unique users worldwide. With the heyday of the Web portal giving way to the era of the search engine—exemplified by the success of Google—Yahoo! announced, in February 2004, its own search technology. (Previously, Yahoo! had used Google's technology.) With the new technology, some of the human touch that made Yahoo! a user favorite in its early days has given way to mathematical algorithms—but Yahoo!'s original human-fashioned Web directory remains, as does its ever-popular brand image, which includes the yin and yang personalities of its founders, Filo and Yang.

## FURTHER READING

### In These Volumes

Related Entries in this Volume: Brin, Sergey and Larry Page

Related Entries in the Chronology Volume: 1990: World Wide Web Is Invented; 1994: Stanford Graduate Students Found Yahoo! Search Engine

### Books

Angel, Karen. *Inside Yahoo! Reinvention and the Road Ahead.* New York: John Wiley & Sons, 2002.

### Articles

Beach, Michael. "Yahoo! Gets Serious" *Daily Telegraph,* March 17, 2001.

"Found You on Yahoo!" *Red Herring,* October 1, 1995.

Galarza, Paul. "The Search Engine That Could." *Money,* April 1, 1999.

Hansell, Saul. "Is Yahoo Flying High with a Bull's-Eye on Its Back?" *New York Times,* February 1, 1998.

———. "Obsessively Independent, Yahoo Is the Web's Switzerland." *New York Times,* August 23, 1999.

Harmon, Amy. "The Oracle of Yahoo Has Internet Surfers Going Gaga." *Los Angeles Times,* April 10, 1995.

Krantz, Michael, Patrick E. Cole, Wendy Cole, William Dowell, Aixa M. Pascual, and David S. Jackson. "Click 'Till You Drop." *Time,* July 20, 1998.

Leonard, Andrew. "Yay for Yahoo!" *Salon.com,* February 22, 1999, http://archive.salon.com/21st/feature /1999/02/cov_22featurea.html (cited September 28, 2004).

Levy, Steven. "Surfers, Step Right Up!" *Newsweek,* May 25, 1998.

Mangalindan, Mylene, and Suein L. Hwang, "Gang of Six." *Wall Street Journal,* March 9, 2001.

Nelson, Brian. "Yahoo! Founders Have Something to Whoop About." *CNN,* September 16, 1995, http://www.cnn.com/TECH/9509/yahoo/ (cited September 28, 2004).

Olsen, Stefanie. "Yahoo Co-founder Filo Muses on the Early Days." *CNET News.com,* December 3, 2000, http://news.com/2100–1038_3–5113121.html (cited September 28, 2004).

Piller, Charles. "Sitting on Top of the World." *Los Angeles Times,* October 12, 1998.

Schibsted, Evantheia. "The Legends." *Business 2.0,* June 1, 1999.

Schlender, Brent. "The Customer Is the Decision-Maker." *Fortune,* March 6, 2000.

———. "How a Virtuoso Plays the Web Eclectic," *Fortune,* March 6, 2000.

Steinberg, Steve G. "Seek and Ye Shall Find (Maybe)." *Wired 4.05,* May 1996, http://www.wired.com /wired/archive/4.05/indexweb.html (cited September 28, 2004).

Stross, Randall E. "How Yahoo! Won the Search Wars." *Fortune,* March 2, 1998.

Swisher, Kara. "Cyberspace Success Ignites Yahoo! Shares." *Wall Street Journal,* April 14, 1998.

———. "A Glimpse Behind the Portal at Yahoo!" *Wall Street Journal,* April 19, 2000.

———"Can This Rare Bird Fly in a Harsh Climate?" *Wall Street Journal,* October 16, 2000.

Weiss, Aaron. "Boo for Yahoo!" *Salon.com,* February 22, 1999, http://archive.salon.com/21st/feature/1999 /02/cov_22featureb.html (cited September 28, 2004).

Wylie, Margie. "Barefoot Millionaire Boys." *CNET News.com,* November 10, 1997, http://news.com .com/2009–1082_3–233605.html (cited September 28, 2004).

## Websites

Yahoo! Main Yahoo! Website, http://www.yahoo.com (cited September 28, 2004).

Yahoo! Inc. Yahoo! Inc. company information site, including biographies and Yahoo! timeline, http:// docs.yahoo.com/info/ (cited September 28, 2004).

# Acronyms

| | | | |
|---|---|---|---|
| ANSI | American National Standards Institute | DNS | Domain Name System |
| AOL | America Online | DSL | Digital Subscriber Line |
| ARPA | Advanced Research Projects Agency | | |
| ASCII | American Standard Code for Information Interchange | EDVAC | Electronic Discrete Variable Automatic Computer |
| AUP | Acceptable Use Policy | EFF | Electronic Frontier Foundation |
| | | EGP | External Gateway Protocol |
| BASIC | Beginner's All-Purpose Symbolic Instruction Code | ENIAC | Electronic Numerical Integrator and Computer |
| BBN | Bolt, Beranek and Newman | | |
| BBS | Bulletin Board System | FCC | Federal Communications Commission |
| BITnet | Because It's Time/There Network | FSF | Free Software Foundation |
| | | FTC | Federal Trade Commission |
| CDA | Communications Decency Act | FTP | File Transfer Protocol |
| CDC | Control Data Corporation | | |
| CEO | Chief Executive Officer | GUI | Graphical User Interface |
| CERN | Centre Européen pour la Recherche Nucléaire (European Laboratory for Particle Physics) | | |
| | | HES | Hypertext Editing System |
| CIX | Commercial Internet eXchange Association | HTML | Hypertext Markup Language |
| | | HTTP | Hypertext Transfer Protocol |
| CORE | Council of Registrars | | |
| CPU | Central Processing Unit | I2 | Internet2 |
| CREN | Corporation for Research and Educational Networking | IAB | Internet Activities Board |
| | | IBM | International Business Machines |
| CSS | Content Scrambling System | ICANN | Internet Corporation for Assigned Names and Numbers |
| DARPA | Defense Advanced Research Projects Agency | ICCC | International Conference on Computer Communications |
| DCA | Defense Communications Agency | IEEE | Institute of Electrical and Electronics Engineers |
| DeCSS | Decryption of Content Scrambling System | IETF | Internet Engineering Task Force |
| DMCA | Digital Millennium Copyright Act | IM | Instant Messenger |

| IMP | Interface Message Processor |
| INWG | International Network Working Group |
| IP | Internet Protocol |
| IPTO | Information Processing Techniques Office |
| ISN | Internet Shopping Network |
| ISOC | Internet Society |
| ISP | Internet Service Provider |
| LAN | Local Area Network |
| MBONE | Multicast Backbone |
| MIT | Massachusetts Institute of Technology |
| MITS | Model Instrumentation Telemetry Systems |
| MMOG | Massively Multiplayer Online Games |
| MP3 | Moving Picture Experts Group Layer-3 Audio |
| MS-DOS | Microsoft Disk Operating System |
| MUD | Multiuser Dungeon, Domain, or Dimension |
| NC | Network Computer |
| NCSA | National Center for Supercomputing Applications |
| NPTN | National Public Telecomputing Network |
| NSF | National Science Foundation |
| NSI | Network Solutions, Inc. |
| OEM | Original Equipment Manufacturer |
| OOP | Object-Oriented Programming |
| OS | Operating System |
| P2P | Peer-to-Peer |
| PARC | Palo Alto Research Center |
| PC | Personal Computer |

| PDA | Personal Digital Assistant |
| PDP | Programmed Data Processor |
| PGP | Pretty Good Privacy |
| RFC | Request for Comments |
| RIAA | Recording Industry Association of America |
| RSA | Rivest, Shamir, and Adleman |
| SABRE | Semi-Automatic Business Research Environment |
| SAFE | Safety and Freedom Through Encryption |
| SAGE | Semi-Automatic Ground Environment |
| SF | Science Fiction |
| SGML | Standard Generalized Markup Language |
| SoPAC | Society for Public Access to Computing |
| TCP | Transmission Control Protocol |
| TMRC | Tech Model Railroad Club |
| UNIVAC | UNIversal Automatic Computer |
| URL | Universal Resource Locator |
| UUCP | Unix to Unix Copy Protocol |
| VoIP | Voice over Internet Protocol |
| W3C | World Wide Web Consortium |
| WELL | Whole Earth 'Lectronic Link |
| WLAN | Wireless Local Area Networks |
| WO/RE | Write Once/Run Everywhere |
| WWW | World Wide Web |
| XML | Extensible Markup Language |
| XOC | Xanadu Operating Company |

# Glossary

**Artificial intelligence** The simulation of human intelligence by computers.

**Bandwidth** The width of the channel that transmits data over a network. The higher the bandwidth, the more quickly data can be transmitted.

**Binary** Numerical system using base 2 (as opposed to the decimal system, which uses base 10). The binary system has only two digits, 0 and 1 (the decimal system has digits 0 through 9). Because computers are digital (dual state), the binary system is the natural way to represent both their data and their instructions.

**Bit** Short for binary digit. A bit can only be a 0 or a 1. Eight bits equal one byte.

**Blog** See Weblog.

**Bot** A slang term for robot that has been extended to describe computer programs that perform specific tasks. For example, "cancelbots" delete messages on the Usenet system, and "shopbots" search the Web to find the best price for a product.

**Browser** A piece of software that enables the user to view and navigate the World Wide Web.

**Bulletin board system** Usually referred to by its acronym, BBS. A computer system where users log on and view messages left by other users and post their own messages.

**Byte** A measurement of computer memory space. One byte equals 8 bits.

**Chat room** An online environment where users can interact with each other in "real time" by typing messages and transmitting them over a network.

**Compiler** A program that translates a high-level computer programming language into a device-specific machine language.

**Cookie** A small file left on computers by Web pages. Typically, cookies can identify users and, sometimes, gather information about their surfing habits.

**Cracker** A slang term for a person who breaks into computer networks with malicious intent. Compare to "hacker."

**Cryptography** The practice and study of encryption and decryption (encoding data so that it can only be decoded by the recipient).

**Cyberpunk** Term applied to a group of forward-thinking science fiction writers, whose work combines the sensibilities of computer hackers and punk rockers.

**Cyberspace** A term coined by author William Gibson in *Neuromancer* to describe the theoretical space created by a network.

**Decryption** The process of translating a piece of encrypted code into meaningful information. See Encryption, Cryptography.

**Digital** A description of data stored in discrete states; in a binary digital system the states are two, e.g., modern electronic computers.

**Domain name** The name assigned to a unique Internet protocol (IP) address; for example, the IP address 209.239.59.120 is assigned the domain name History OfTheInternet.com. A domain name server is used to translate, or resolve, the domain name to the corresponding IP address.

**Dot-com** Term used to describe any business venture whose products or services relate to or are sold on the Internet.

**E-book** Electronic book; a book that has been made available in an electronic format and can be viewed through a variety of devices such as PCs, personal digital assistants, or e-book readers.

**Encryption**    The process of converting information into secret code; only those with the right key can translate it.

**Ethernet**    A computer network communications protocol invented by Robert Metcalfe.

**Gopher**    A menu-style program for accessing data on the Internet.

**Graphical user interface**    A system that allows users to communicate instructions to their computers through windows, icons, and mouse clicks—rather than the text-only system used by early computers. Also known as "GUI," pronounced "gooey."

**Hacker**    Slang for a person with a high degree of computer skill who writes programs and/or breaks into networks. In the media, the term is often synonymous with "criminal," but among computer users, "hacker" is a complimentary description.

**Hardware**    The material parts of a computer or other system. Computer equipment. Compare to "software."

**Host computer**    A computer that provides services such as computation and database access to other computers connected to it.

**Hypertext**    A format for documents in which users can follow information nonsequentially through links.

**Integrated circuit**    A silicon chip comprised of numerous minitransistors.

**Interpreter**    A computer program that not only translates a high-level language but also executes the resulting program. Typical interpreted computer languages include Basic and Java.

**IP number**    The numerical address of a particular Website or computer. Frequently associated with a domain name. For example, instructing a browser to seek out the domain name HistoryOfTheInternet.com will direct the browser to 209.239.59.120, which is the IP number of the host for that site.

**Java**    An architecture-neutral programming language for Internet software. (See "Write Once/Run Everywhere.")

**Killer app**    Short for "killer application," a particular application or functionality that makes the entire system worthwhile. For example, the spreadsheet is some-times considered the first killer app of the personal computer.

**Local area network**    A computer network confined to a limited number of computers, such as those at a particular place of business.

**Mainframe**    Typically thought of as a large-capacity, powerful computer designed to accommodate many users and tasks at one time. Although early mainframes took up whole floors of buildings, their smaller size and greater power today reflect the state-of-the-art in hardware and technology.

**Market share**    The proportion of industry sales of a good or service that is controlled by a particular company.

**Meme**    The term used by zoologist Richard Dawkins in his 1976 book *The Selfish Gene* to describe a discrete unit of culture, for example, religious beliefs, words and phrases, building techniques, or fashion trends.

**Microprocessor**    The circuit in a personal or microcomputer that contains the CPU, or central processing unit, where instructions are decoded.

**Modem**    Short for MOdulator/DEModulator. A device that translates the computer's digital signals into an audio signal so they can be transmitted over analog media such as telephone lines.

**Name server**    A computer whose function is to translate domain names into their numeric counterparts; when surfing the Internet, a personal computer would query a name server to locate a particular site.

**Netizen**    Slang for citizen of the Internet.

**Newsgroup**    A bulletin board focused on a particular topic. Usenet is a system hosting thousands of different newsgroups.

**Nodes**    A specific computer, router, or location along a network. The original ARPANET structure, for example, had four nodes.

**Object-oriented programming**    A programming methodology incorporating the use of collections of programs and data that are encapsulated (or represented) as objects. For example, a programmer developing a financial application might use a "spreadsheet object" as part of the new application.

**Online communities**    *See* **Virtual communities.**

**Open source**    A method and philosophy for software licensing and distribution that guarantees anybody rights to freely use, modify, and redistribute the code.

**Operating system**    The program that controls the central functions of a computer and allows other applications to run. Examples include UNIX and Windows.

**Packet switching**    A networking methodology under which data travels in small, independent pieces (known as "packets") through the network. A packet-switched network is more reliable since packets can travel many different routes to reach their destination.

**Parallel processing**    A computational method that utilizes more than one processor to run one or more programs. Not to be confused with multitasking, which is the ability of a system to time-slice its processor usage among competing tasks.

**Phreaking**    Hacking as applied to telephone systems.

**Ping**    Packet INformation Groper; a protocol that sends a message to another computer and waits for acknowledgment, often used to check if another computer on a network is reachable.

**Protocol**    A standardized method of information transmission, such as the Internet protocol.

**"Push" technology**    A technology that delivers individualized information packets (gathered from the Internet) straight to a computer user's desktop.

**Real time**    In general terms, "real time" refers to computer/user interactions that happen instantaneously: instant messaging over the Internet, for example, occurs in real time. In more technical terms, a real-time system is a computer system that processes information so quickly that it seems to be instantaneous, although it often is not.

**Router**    A device that acts as a traffic coordinator among networks or devices. Frequently a router will link and direct traffic to disparate networks.

**Semiconductor**    A type of solid-state device that exhibits electrical conductivity and resistivity within a well-defined range. Examples of semiconducting materials include silicon and germanium.

**Social software**    Broadly, any computer program that helps people socialize. Social software is designed to work in conjunction with real-world interactions (unlike virtual communities, which can exist independently of any real-world interaction).

**Software**    Computer programs. See "Hardware."

**Spam**    Unsolicited e-mail or posts sent indiscriminately to multiple mailing lists, individuals, or newsgroups, usually appearing in great, unwanted quantities.

**Supercomputer**    A term used to describe the fastest, most powerful computer of a particular era. Typically, supercomputers are used for scientific applications that manipulate hundreds or thousands of variables at one time. A typical example of supercomputer application is weather modeling.

**Sysop**    Short for system operator, the person who runs a computer system or network.

**Time-sharing system**    A method of computing that enables users simultaneous access to a central computer through remote terminals. From the days before personal computers.

**UNIX**    An operating system developed in the 1970s that was immensely popular among programmers because of its flexibility, but difficult for the uninitiated to use.

**Venture capital**    Money provided by outside investors to assist small companies in growing into bigger ones.

**Virtual communities**    Various groups of people who interact via Internet Websites, chat rooms, newsgroups, email, discussion boards, or forums; also called online communities. *See* **Social software.**

**Virus**    A piece of computer code that is designed to infect a computer and then to replicate itself through other programs or data, either for no meaningful reason or with malicious intent.

**Warez**    Copyrighted digital material—whether software, music, movies, or e-books—that has been "cracked," or illegally stripped of its copy-protection software. Warez traders usually swap these files over the Internet.

**Weblog**   Probably coined by programmer Jorn Barger in 1997 to describe his personal Website, a collection of links, photos, and his own brief comments. The term caught on as a way to describe similar sites.

**Wide area network**   Collections of computers or local area networks that are connected over large physical distances.

**Wi-fi**   Short for "wireless fidelity." An international standard for wireless networking. Wi-Fi is effectively a wireless version of Ethernet, the most popular method of networking PCs with cables.

**Write Once/Run Everywhere**   The holy grail of computer programmers: a Write Once/Run Everywhere (WO/RE) program will work on any type of computer system.

# Index

Note: italic page numbers indicate pictures.

# About the Author

Laura Lambert is a writer living in New York City. Drawing from a rich liberal arts background, she has written on a variety of topics from incidents of domestic terrorism, in the award-winning *Encyclopedia of Terrorism* (2003), to profiles of groundbreaking American women, in *A History of Women in the United States: A State by State Reference* (2004). Laura earned her master's degree in journalism from Columbia University.